AF820349

Methods and Protocols in Food Science

Series Editor
Anderson S. Sant'Ana
University of Campinas
Campinas, Brazil

For further volumes:
http://www.springer.com/series/16556

Methods and Protocols in Food Science series is devoted to the publication of research protocols and methodologies in all fields of food science.

Volumes and chapters will be organized by field and presented in such way that the readers will be able to reproduce the experiments in a step-by-step style. Each protocol will be characterized by a brief introductory section, followed by a short aims section, in which the precise purpose of the protocol will be clarified.

Fungi

Edited by

Pablo Navarro Simarro

Botanical Institute, University of Castilla-La Mancha, Albacete, Spain

Humana Press

Editor
Pablo Navarro Simarro
Botanical Institute
University of Castilla-La Mancha
Albacete, Spain

ISSN 2662-950X ISSN 2662-9518 (electronic)
Methods and Protocols in Food Science
ISBN 978-1-0716-4988-6 ISBN 978-1-0716-4989-3 (eBook)
https://doi.org/10.1007/978-1-0716-4989-3

© The Editor(s) (if applicable) and The Author(s), under exclusive license to Springer Science+Business Media, LLC, part of Springer Nature 2026

This work is subject to copyright. All rights are solely and exclusively licensed by the Publisher, whether the whole or part of the material is concerned, specifically the rights of translation, reprinting, reuse of illustrations, recitation, broadcasting, reproduction on microfilms or in any other physical way, and transmission or information storage and retrieval, electronic adaptation, computer software, or by similar or dissimilar methodology now known or hereafter developed.
The use of general descriptive names, registered names, trademarks, service marks, etc. in this publication does not imply, even in the absence of a specific statement, that such names are exempt from the relevant protective laws and regulations and therefore free for general use.
The publisher, the authors and the editors are safe to assume that the advice and information in this book are believed to be true and accurate at the date of publication. Neither the publisher nor the authors or the editors give a warranty, expressed or implied, with respect to the material contained herein or for any errors or omissions that may have been made. The publisher remains neutral with regard to jurisdictional claims in published maps and institutional affiliations.

This Humana imprint is published by the registered company Springer Science+Business Media, LLC, part of Springer Nature.
The registered company address is: 1 New York Plaza, New York, NY 10004, U.S.A.

If disposing of this product, please recycle the paper.

Preface to the Series

The Methods and Protocols in Food Science series is devoted to the publication of research protocols and methodologies in all fields of food science. The series is unique as it includes protocols developed, validated, and used by food and related scientists as well as provides theoretical basis for each protocol. Aspects related to improvements in the protocols and adaptations and further developments in the protocols may also be approached.

The Methods and Protocols in Food Science series aims to bring the most recent developments in research protocols in the field as well as very well-established methods. As such the series targets undergraduates, graduates, and researchers in the field of food science and correlated areas. The protocols documented in the series will be highly useful for scientific inquiries in the field of food sciences and presented in such a way that the readers will be able to reproduce the experiments in a step-by-step style.

Each protocol will be characterized by a brief introductory section, followed by a short aims section, in which the precise purpose of the protocol is clarified. Then, an in-depth list of materials and reagents required for employing the protocol is presented, followed by comprehensive and step-by-step procedures on how to perform that experiment. The next section brings the dos and don'ts when carrying out the protocol, followed by the main pitfalls faced and how to troubleshoot them. Finally, template results will be presented and their meaning/conclusions addressed.

The Methods and Protocols in Food Science series will fill an important gap, addressing a common complain of food scientists, regarding the difficulties in repeating experiments detailed in scientific papers. With this, the series has a potential to become a reference material in food science laboratories of research centers and universities throughout the world.

Preface

Fungi play a crucial role in nutrient recycling within terrestrial ecosystems and, for centuries, have helped us upcycle agricultural by-products. Thanks to their powerful enzymatic machinery, they can break down substrates such as straw or compost, giving rise to fruiting bodies, mycelium, and even spores that can be both nutritious and beneficial to health. Their immense morphological and chemical diversity is particularly noteworthy, though they also share common traits, such as a high content of carbohydrates like β-glucans and chitin. A key challenge in the current landscape is to enhance extraction methods for these bioactive compounds and to better characterize the chemical profiles of commercial, wild, and filamentous fungi used in the food industry.

The metabolic versatility of fungi is reflected in a wide range of health effects. Particularly notable are the prebiotic effects of certain polysaccharides and the anti-inflammatory, antioxidant, and antimicrobial potential of some metabolites. In addition, metabolites from certain so-called medicinal fungi have shown remarkable neurological effects, opening the door to the screening of promising fungal species for applications in the food or pharmaceutical industries.

Given their nutritional composition, many researchers and companies are working to develop functional foods based on mushrooms, or even to use filamentous fungi to valorize other food matrices and enhance their properties. In the first case, we highlight the use of mushroom flours to fortify bakery products, their application as flavor enhancers, and their use as a basis for biodegradable food packaging films. In the second case, techniques such as liquid and solid state fermentation stand out as some of the most promising methods for producing pigments, microbial protein, or meat analogues. These techniques are inspired by traditional fermentation processes used to prepare foods such as tempeh or oncom.

This book is intended as a comprehensive guide for researchers at all stages who are drawn to the fascinating intersection of fungi, biotechnology, and food science. Across its 21 chapters, it explores three fundamental areas: the extraction, quantification, and identification of fungal bioactive compounds; the evaluation of their effects on health; and the development of novel functional foods. As the scientific community continues to unveil the extraordinary capabilities of fungi, it becomes increasingly clear that these organisms hold the keys to solving some of the most pressing challenges in food, health, and sustainability. Only through collective effort and interdisciplinary collaboration can we fully unlock their vast, untapped potential.

Albacete, Spain | *Pablo Navarro Simarro*

Contents

Contributors

MANUEL ÁLVAREZ-ORTÍ • *Higher Technical School of Agricultural and Forestry Engineering and Biotechnology, University of Castilla-La Mancha, Albacete, Spain*

ANA ANDRÉS • *University Institute of Food Engineering - FoodUPV, Universitat Politècnica de València, Valencia, Spain*

HELENA ARAÚJO-RODRIGUES • *Universidade Católica Portuguesa, CBQF - Centro de Biotecnologia e Química Fina –Laboratório Associado, Escola Superior de Biotecnologia, Rua de Diogo Botelho, Porto, Portugal; Cell Biology, i3S-Instituto de Investigação e Inovação em Saúde, Universidade do Porto, Porto, Portugal*

JAQUELINE AUER • *Department of Molecular Sciences BioCenter, Swedish University of Agricultural Sciences, Uppsala, Sweden*

JAVIER AVALOS • *Facultad de Biología, Departamento de Genética, Universidad de Sevilla, Sevilla, Spain*

OM BAGADE • *Department of Pharmacy, School of Pharmacy, Vishwakarma University, Kondhwa, Laxminagar, Pune, Maharashtra, India*

ISHIKA BERA • *Center of Excellence in Fungal Research, Mae Fah Luang University, Chiang Rai, Thailand*

PATRICIA BERMÚDEZ-GÓMEZ • *Mushroom Technological Research Center of La Rioja (CTICH), Autol, Spain; IPOA Research Group, Institute for Agri-Food and Agri-Environmental Research and Innovation, Miguel Hernández University (CIAGRO-UMH), Orihuela, Spain*

DIEGO BOGANTES-LEDEZMA • *Centro para Investigaciones en Granos y Semillas (CIGRAS), Universidad de Costa Rica, San José, Costa Rica*

ISMAEL BORREGO-SERRANO • *Facultad de Biología, Departamento de Genética, Universidad de Sevilla, Sevilla, Spain*

LAURA ALEJANDRA FERNANDEZ CASTANEDA • *Department of Molecular Sciences BioCenter, Swedish University of Agricultural Sciences, Uppsala, Sweden*

FATIH ÇAYAN • *Department of Chemistry and Chemical Processing Technologies, Muğla Vocational School, Muğla Sıtkı Koçman University, Muğla, Türkiye*

ANTONIO CILLA • *Bionutest Research Group—Nutrition and Food Science Area, Faculty of Pharmacy and Food Sciences, University of Valencia, Valencia, Spain*

EBRU DEVECI • *Chemistry and Chemical Processing Technology Department, Technical Sciences Vocational School, Konya Technical University, Konya, Türkiye*

CRISTIAN MARTÍNEZ FAJARDO • *Botanical Institute, University of Castilla-La Mancha, Albacete, Spain*

SAIMA FARIHA • *Department of Applied Nutrition and Food Technology, Islamic University, Kushtia, Bangladesh*

ANDREIA SILVA FERNANDES • *Centro de Biotecnologia e Química Fina Laboratório Associado, Escola Superior de Biotecnologia, Porto, Portugal*

JUANA FERNÁNDEZ-LÓPEZ • *IPOA Research Group, Institute for Agri-Food and Agri-Environmental Research and Innovation, Miguel Hernández University (CIAGRO-UMH), Orihuela, Spain*

MIGUEL ÁNGEL GALLARDO • *Higher Technical School of Agricultural and Forestry Engineering and Biotechnology, University of Castilla-La Mancha, Albacete, Spain*

AMPARO GAMERO • *Bionutest Research Group—Food Technology Area, Faculty of Pharmacy and Food Sciences, University of Valencia, Valencia, Spain*

ANTONI GANDIA • *Institute for Plant Molecular and Cell Biology, Polytechnic University of Valencia, Valencia, Spain*

MÓNICA GANDÍA • *Bionutest Research Group—Food Technology Area, Faculty of Pharmacy and Food Sciences, University of Valencia, Valencia, Spain*

JORGE GARCÍA-HERNÁNDEZ • *Advanced Center of Applied Microbiology (CAMA)—UPV, Universitat Politècnica de València, Valencia, Spain*

DEECKSHA GOMDOLA • *Center of Excellence in Fungal Research, Mae Fah Luang University, Chiang Rai, Thailand; School of Science, Mae Fah Luang University, Chiang Rai, Thailand*

DIDSANUTDA GONKHOM • *Center of Excellence in Fungal Research, Mae Fah Luang University, Chiang Rai, Thailand; School of Science, Mae Fah Luang University, Chiang Rai, Thailand*

SUMAIYA HASSAN • *Department of Applied Nutrition and Food Technology, Islamic University, Kushtia, Bangladesh*

ANA HEREDIA • *University Institute of Food Engineering - FoodUPV, Universitat Politècnica de València, Valencia, Spain*

DÁMASO HORNERO-MÉNDEZ • *Department of Food Phytochemistry, Instituto de la Grasa, CSIC, Campus Universidad Pablo de Olavide, Sevilla, Spain*

ANDREA IRÍAS-MATA • *Centro para Investigaciones en Granos y Semillas (CIGRAS), Universidad de Costa Rica, San José, Costa Rica*

AMPARO JIMÉNEZ-QUERO • *Division of Industrial Biotechnology, Department of Life Sciences, Chalmers University of Technology, Gothenburg, Sweden*

SABIN KHYAJU • *Center of Excellence in Fungal Research, Mae Fah Luang University, Chiang Rai, Thailand; School of Science, Mae Fah Luang University, Chiang Rai, Thailand; Mushroom Research Foundation, Chiang Rai, Thailand*

ENIKO KOVACS • *INCDO INOE 2000, Research Institute for Analytical Instrumentation, Donath 67, Cluj-Napoca, Romania*

MAUD LANGTON • *Department of Molecular Sciences BioCenter, Swedish University of Agricultural Sciences, Uppsala, Sweden*

M. CARMEN LIMÓN • *Facultad de Biología, Departamento de Genética, Universidad de Sevilla, Sevilla, Spain*

ALBERTO JOSÉ LÓPEZ-JIMÉNEZ • *Botanical Institute, University of Castilla-La Mancha, Albacete, Spain; Department of Agroforestry Science, Technology and Genetics, Higher Technical School of Agricultural, Forestry and Biotechnology Engineering, University of Castilla-La Mancha, Albacete, Spain*

SUSANA LÓPEZ-LÓPEZ • *Faculty of Medicine, Department of Inorganic, Organic and Biochemical Chemistry, University of Castilla-La Mancha, University Campus s/n, Albacete, Spain*

MUSSA MAKRAN • *Bionutest Research Group—Food Technology Area, Faculty of Pharmacy and Food Sciences, University of Valencia, Valencia, Spain; Bionutest Research Group—Nutrition and Food Science Area, Faculty of Pharmacy and Food Sciences, University of Valencia, Valencia, Spain*

ASADULLAH MARRI • *Institute of Food Sciences and Technology, Sindh Agriculture University, Tandojam, Pakistan*

M. ESTHER MARTÍNEZ-NAVARRO • *Higher Technical School of Agricultural and Forestry Engineering and Biotechnology, University of Castilla-La Mancha, Albacete, Spain*

HELENA MARTIN-GÓMEZ • *CETT Barcelona School of Tourism, Hospitality and Gastronomy, University of Barcelona, Barcelona, Spain*

ERIC H. C. MCKENZIE • *Manaaki Whenua-Landcare Research, Auckland, New Zealand*

CAMILA MENDOZA • *Bionutest Research Group—Food Technology Area, Faculty of Pharmacy and Food Sciences, University of Valencia, Valencia, Spain; Bionutest Research Group—Nutrition and Food Science Area, Faculty of Pharmacy and Food Sciences, University of Valencia, Valencia, Spain*

A. T. M. MIJANUR RAHMAN • *Department of Applied Nutrition and Food Technology, Islamic University, Kushtia, Bangladesh*

SARA MUÑOZ-PINA • *University Institute of Food Engineering - FoodUPV, Universitat Politècnica de València, Valencia, Spain*

PABLO NAVARRO SIMARRO • *Botanical Institute, University of Castilla-La Mancha, Albacete, Spain*

ÒSCAR BORONAT NIELSEN • *CETT Barcelona School of Tourism, Hospitality and Gastronomy, University of Barcelona, Barcelona, Spain*

NATALIE NUSSBAUM • *Eidgenössische Technische Hochschule Zürich, Institute of Food, Nutrition, and Health, Zürich, Switzerland*

JOSÉ EMILIO PARDO • *Higher Technical School of Agricultural and Forestry Engineering and Biotechnology, University of Castilla-La Mancha, Albacete, Spain*

PAMELA FREIRE DE MOURA PEREIRA • *Division of Industrial Biotechnology, Department of Life Sciences, Chalmers University of Technology, Gothenburg, Sweden*

MARGARITA PÉREZ-CLAVIJO • *Mushroom Technological Research Center of La Rioja (CTICH), Autol, Spain*

MONTHIEN PHONEMANY • *Center of Excellence in Fungal Research, Mae Fah Luang University, Chiang Rai, Thailand; School of Science, Mae Fah Luang University, Chiang Rai, Thailand*

MANUELA PINTADO • *Universidade Católica Portuguesa, CBQF - Centro de Biotecnologia e Química Fina –Laboratório Associado, Escola Superior de Biotecnologia, Rua de Diogo Botelho, Porto, Portugal*

ZEINAB QAZANFARZADEH • *Division of Industrial Biotechnology, Department of Life Sciences, Chalmers University of Technology, Gothenburg, Sweden*

JOÃO B. RELVAS • *Cell Biology, i3S-Instituto de Investigação e Inovação em Saúde, Universidade do Porto, Porto, Portugal; IBMC-Instituto de Biologia Molecular e Celular, Universidade do Porto, Porto, Portugal*

SURAIYA HASSAN RICHI • *Biotechnology & Genetic Engineering Discipline, Khulna University, Khulna, Bangladesh*

MONTSERRAT RIU-AUMATELL • *Department of Nutrition, Food Science and Gastronomy, Faculty of Pharmacy and Food Sciences, University of Barcelona, Santa Coloma de Gramenet, Spain; Institute of Nutrition and Food Safety (INSA-UB), University of Barcelona, Santa Coloma de Gramenet, Spain*

PATRICK A. RÜHS • *Eidgenössische Technische Hochschule Zürich, Institute of Food, Nutrition, and Health, Zürich, Switzerland*

THAVIPHONE SALICHANH • *Center of Excellence in Fungal Research, Mae Fah Luang University, Chiang Rai, Thailand; School of Science, Mae Fah Luang University, Chiang Rai, Thailand; Mushroom Research Foundation, Chiang Rai, Thailand*

ANA SOFIA SALSINHA • *Universidade Católica Portuguesa, CBQF - Centro de Biotecnologia e Química Fina –Laboratório Associado, Escola Superior de Biotecnologia, Rua de Diogo Botelho, Porto, Portugal*

PARAMJEET SAROJ • *Department of Forest Biomaterials and Technology, Swedish University of Agricultural Sciences, Umeå, Sweden; HAMK Bio, Häme University of Applied Sciences (HAMK), Hämeenlinna, Finland*
MARIN SENILA • *INCDO INOE 2000, Research Institute for Analytical Instrumentation, Donath 67, Cluj-Napoca, Romania*
NIDA SHAIKH • *Institute of Food Sciences and Technology, Sindh Agriculture University, Tandojam, Pakistan*
AIJAZ HUSSAIN SOOMRO • *Institute of Food Sciences and Technology, Sindh Agriculture University, Tandojam, Pakistan*
SABEENA HUSSAIN SYED • *Department of Pharmacy, School of Pharmacy, Vishwakarma University, Kondhwa, Laxminagar, Pune, Maharashtra, India*
POONAM TARU • *Department of Pharmacy, School of Pharmacy, Vishwakarma University, Kondhwa, Laxminagar, Pune, Maharashtra, India*
FRENI K. TAVARIA • *Centro de Biotecnologia e Química Fina Laboratório Associado, Escola Superior de Biotecnologia, Porto, Portugal; Universidade Católica Portuguesa, CBQF - Centro de Biotecnologia e Química Fina –Laboratôrio Associado, Escola Superior de Biotecnologia, Rua de Diogo Botelho, Porto, Portugal*
GÜLSEN TEL-ÇAYAN • *Department of Chemistry and Chemical Processing Technologies, Muğla Vocational School, Muğla Sıtkı Koçman University, Muğla, Türkiye*
SUPRIYA UNAVANE • *Department of Pharmacy, School of Pharmacy, Vishwakarma University, Kondhwa, Laxminagar, Pune, Maharashtra, India*
MARIA VINAS • *Centro para Investigaciones en Granos y Semillas (CIGRAS), Universidad de Costa Rica, San José, Costa Rica*
MATTEO VITALI • *Bionutest Research Group—Food Technology Area, Faculty of Pharmacy and Food Sciences, University of Valencia, Valencia, Spain; Bionutest Research Group—Nutrition and Food Science Area, Faculty of Pharmacy and Food Sciences, University of Valencia, Valencia, Spain*
MANUEL VIUDA-MARTOS • *IPOA Research Group, Institute for Agri-Food and Agri-Environmental Research and Innovation, Miguel Hernández University (CIAGRO-UMH), Orihuela, Spain*
ARTTAPON WALKER • *Center of Excellence in Fungal Research, Mae Fah Luang University, Chiang Rai, Thailand*
SHAOJUN XIONG • *Department of Forest Biomaterials and Technology, Swedish University of Agricultural Sciences, Umeå, Sweden*
GALIA ZAMARATSKAIA • *Department of Molecular Sciences BioCenter, Swedish University of Agricultural Sciences, Uppsala, Sweden*
CAROLE ZERMATTEN • *Eidgenössische Technische Hochschule Zürich, Institute of Food, Nutrition, and Health, Zürich, Switzerland*

Part I

Taxonomy of Fungi and Methods for Extracting, Identifying, and Isolating Bioactive Compounds

Chapter 1

Exploring Mushroom Diversity: Systematics and Chemical Insights

Deecksha Gomdola, Eric H. C. McKenzie, Ishika Bera, Arttapon Walker, Sabin Khyaju, Didsanutda Gonkhom, Monthien Phonemany, and Thaviphone Salichanh

Abstract

This chapter provides a comprehensive exploration of the diversity, systematics, and chemical insights of seven key genera of macrofungi; *Antrodia*, *Auricularia*, *Hericium*, *Lentinus*, *Morchella*, *Pleurotus*, and *Schizophyllum* from *Ascomycota* and *Basidiomycota*. These fungi play vital roles in ecosystem processes, such as nutrient cycling, and symbiotic interactions with plants. The chapter reviews their distribution, classification, morphological characteristics, and ecological significance, emphasizing the taxonomic challenges and ongoing discoveries in fungal diversity. Additionally, the bioactive compounds found in these fungi, including terpenoids, polyphenols, and polysaccharides, are documented for their therapeutic properties, demonstrating their importance in pharmaceuticals, nutraceuticals, and functional foods. The study highlights the vast diversity and importance of macrofungi with respect to their applications in sustainable agriculture.

Key words Bioactive compounds, Biodiversity, Macrofungi

1 Introduction

Fungi are a highly diverse group of heterotrophic organisms that play essential ecological and economic roles in various ecosystems [1–3]. The global diversity of fungi has been a subject of considerable studies [4–8], with current estimates suggesting there may be between 1.5 and 3.8 million species [9, 10]. The most widely recognized estimate is approximately 2.5 million species [9]. However, only about 140,000 species, encompassing both microfungi and macrofungi, have been formally described [11]. Macrofungi, which are characterized by visible fruiting bodies, are mainly found within phylum *Basidiomycota*, with a few species belonging to *Ascomycota* [12]. Estimates of macrofungal diversity vary depending on classification criteria, with some researchers suggesting that up to 25,000 species exist within *Basidiomycota* [13], while others

Pablo Navarro Simarro (ed.), *Fungi*, Methods and Protocols in Food Science, https://doi.org/10.1007/978-1-0716-4989-3_1,
© The Author(s), under exclusive license to Springer Science+Business Media, LLC, part of Springer Nature 2026

estimate a broader range of 53,000 to 110,000 species, including those within *Ascomycota* [14]. Ainsworth and Bisby's Dictionary of the Fungi listed 31,515 species [15], and in recent classifications, over 36,000 species have been identified [16]. Subsequently, this number was further revised to 41,270 in the most recent classification of *Basidiomycota* [17, 18]. Further research predicted that by 2030, the number of *Basidiomycota* species alone could exceed 54,000, bringing the total fungal species estimate to as high as 4.2 million [19]. These figures underscore the need for continued taxonomic research to fully document fungal diversity.

Macrofungi contribute significantly to ecosystem functions such as nutrient cycling, wood decay, and symbiotic relationships with plants [2, 3]. Notable edible macrofungi include species from genera like *Agaricus*, *Flammulina*, *Lentinula*, and *Pleurotus* [20]. These fungi also serve as valuable sources of bioactive natural compounds, offering substantial benefits in the pharmaceutical, nutraceutical, and functional food industries [21]. Their secondary metabolites, such as terpenoids, polyphenols, and polysaccharides, have been shown to combat oxidative stress, reduce inflammation, inhibit tumor growth, and suppress microbial and viral activities [22]. Some bioactive compounds derived from macrofungi also possess immune-boosting properties [23]. For example, *Auricularia* species are known for their antioxidant properties, which are attributed to phenolic acids, flavonoids, and ascorbic acid that neutralize reactive oxygen species [24]. Similarly, *Boletus* and *Pleurotus* produce polysaccharides and polyphenols with notable antioxidant activities [25].

The distribution of these macrofungi is shaped by environmental factors, habitat preferences, and biogeographical patterns, leading to significant diversity across different continents. Asia, particularly Southeast Asia, is one of the richest biodiversity hotspots, where tropical forests, temperate zones, and high-altitude ecosystems contribute to high species richness [3, 19]. Numerous macrofungal species from this region are also used for medicinal purposes [26, 27]. Eurasia showcases an overlap in macrofungal diversity, particularly in temperate and boreal forests [28]. While Africa remains relatively underexplored, its tropical rainforests harbor unique fungal diversity [12]. In North and South America, fungal diversity is particularly high, with research concentrated on tropical rainforests and temperate forests, where ectomycorrhizal fungi play crucial ecological roles [14]. Australia's distinctive macrofungal diversity is also notable [29], and even Antarctica is home to psychrophilic fungi, which have adapted to the harsh, cold environment [30].

The taxonomy of macrofungi continues to evolve with advancements in genome-based techniques, but phenotype-based methods, physiological data, and DNA sequence analysis remain significant [31]. Despite the progress in fungal taxonomy, many taxa remain undiscovered, particularly in tropical and subtropical regions, offering opportunities for further taxonomic, ecological, and biochemical discoveries. While genera like *Agaricus* [32], *Pleurotus* [33], *Ganoderma* [34], and *Lentinula* [35] have been extensively studied due to their economic importance, other genera, such as *Russula* [36] and *Favolus* [37], remain underexplored due to their morphological variability and cryptic diversity, presenting taxonomic challenges.

This chapter focuses on seven genera of macrofungi, representing both *Ascomycota* and *Basidiomycota*, with a particular emphasis on their ecological importance, systematics, and chemical properties. The genera selected for each case study are *Antrodia*, *Auricularia*, *Hericium*, *Lentinus*, *Morchella*, *Pleurotus*, and *Schizophyllum*. These fungi play vital roles in maintaining ecosystem balance. Their diverse evolutionary relationships, morphological characteristics, and molecular classifications offer valuable insights into fungal taxonomy and biodiversity. Furthermore, the bioactive compounds produced by these fungi hold great potential in various applications in medicine, industry, and agriculture. By combining taxonomic, ecological, and chemical perspectives, this chapter aims to provide a comprehensive overview of these genera and their significance in mycology and applied sciences.

Taxonomic Classification of Fungi

According to the most recent outline by Hyde et al. [11], the kingdom fungi includes 19 phyla, 83 classes, 1220 families, 10,685 genera, and approximately 140,000 recognized species. Focusing on *Basidiomycota*, He et al. [18] outlined its classification as comprising four subphyla, 20 classes, 77 orders, 297 families, and about 2130 genera. Compared to He et al. [17], this represents an increase of two classes, nine orders, 56 families, and 206 genera. A simplified overview of fungal classification is presented in Fig. 1, adapted from Hyde et al. [11]. Fig. 1 highlights only selected examples from each rank, primarily those discussed in this book.

2 Case Studies

2.1 Antrodia *P. Karst., Meddn Soc. Fauna Flora fenn. 5: 40 (1879)*

2.1.1 Distribution and Taxonomy

The polypore genus *Antrodia* (*Fomitopsidaceae*, *Polyporales*, *Incertae sedis*, *Agaricomycetes*, *Agaricomycotina*, *Basidiomycota*) was established by Karsten in 1879 [38, 39]. However, classification of this genus remained unclear until Donk [40, 41] reinstated the genus and designated *Daedalea serpens* (=*Antrodia albida*) as the type species. As *Antrodia* gained recognition, various brown rot polypores were included in the genus without a well-defined classification, leading to taxonomic uncertainties [38, 39]. The polypores with either effused or effuse-reflexed basidiomata and dimitic

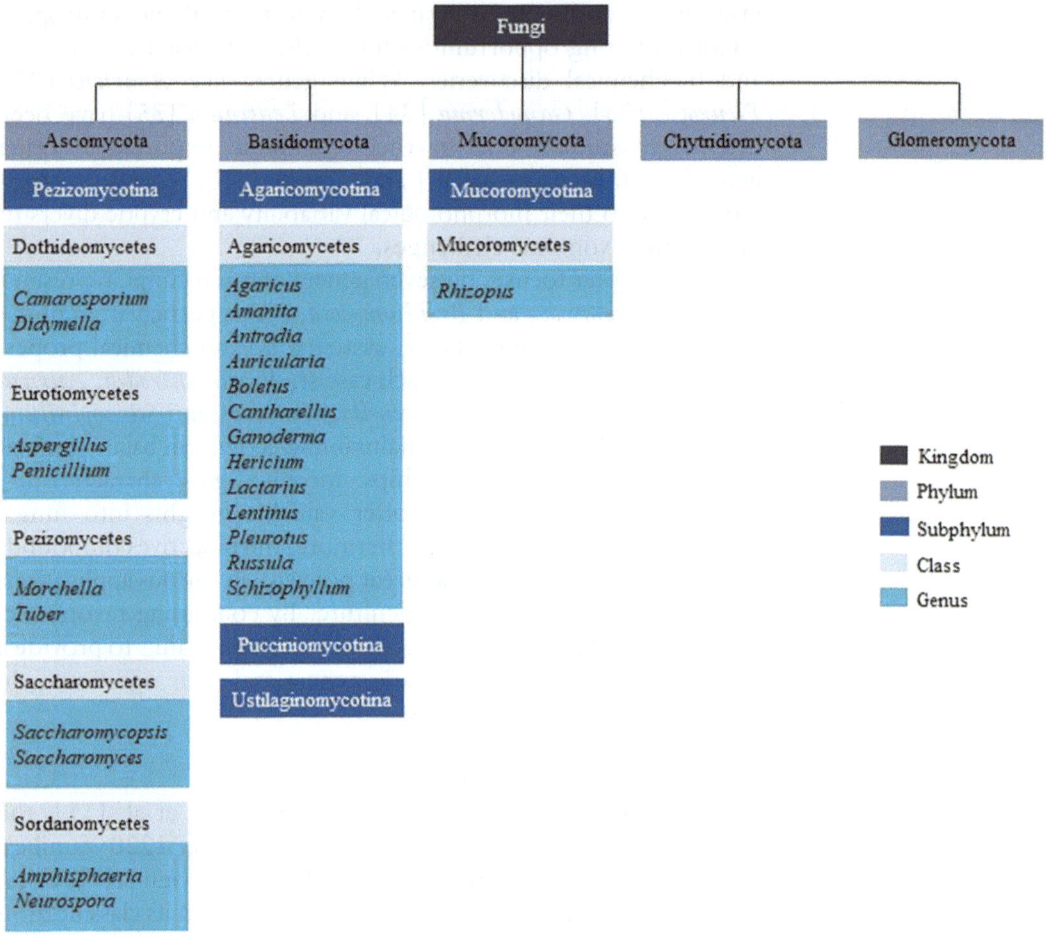

Fig. 1 A simplified taxonomic classification of fungi

hyphal system were considered species of *Antrodia* solely based on morphology [38]. The initial molecular study by Hibbett and Donoghue [42] identified the *Antrodia* clade as a subgroup in the polypore clade that comprises 11 brown rot genera other than *Antrodia* [43]. *Antrodia* was recognized as a polyphyletic genus where species in *Antrodia* sensu lato were characterized by the effuse-reflexed to resupinate basidiomata, clamped generative hyphae and hyaline skeletal hyphae in dimitic hyphal system, and inamyloid basidiospores [39]. Spirin et al. [38] clarified the type species and defined a core group, *Antrodia* sensu stricto around it. Their study revealed that *A. albida* forms a distinct clade of brown rot polypores within a larger group of dimitic brown rot polypores in *Polyporales* [38, 39, 43].

Most *Antrodia* species are closely related to other genera, such as *Amyloporia* (currently *Antrodia*), *Daedalea*, and *Fomitopsis*, than to *Antrodia* sensu stricto [38]. Many species of *Antrodia s. str.* are so closely related that it is quite challenging to distinguish them morphologically and phylogenetically [38]. On the other hand, Ortiz-Santana et al. [43] identified several groups in the *Antrodia* clade that may guide generic reclassification of the brown rot fungi. Recent molecular studies have tried to resolve certain morphospecies within *Antrodia*, such as the *Antrodia crassa* group and the *Antrodia serialis* group [39, 44, 45]. These led to the documentation of several novel taxa within these groups. Runnel et al. [46] used three genetic markers [loci of the internal transcribed spacer (ITS) and large subunit (28S), and the gene for the RNA polymerase II largest subunit (*RPB1*)] to revise *Antrodia s. str.* A corticioid species with a smooth hymenial surface, *Phlebia griseoflavescens* (=*Antrodia griseoflavescens*), clustered with the core polypore group of *Antrodia s. str* [38, 46]. This is the only corticoid species reported in this polypore genus. Previous studies suggested the affinity of corticoids with the *Antrodia* clade. However, documented taxa always formed distinct lineages [43, 47]. This highlights the need for molecular data on taxa with diverse hymenial structures, rather than limiting assumptions solely to polypores. Currently, *Antrodia* is classified under Fomitopsidaceae, accounting for 80 species primarily found on deciduous hosts in North America, Europe and East Asia [11, 38].

2.1.2 Importance and Ecological Roles

This interesting wood-inhabiting genus is ecologically significant in balancing the forest ecosystem. Members of this genus break down lignin and cellulose from fallen dead wood, facilitating the return of nutrients to the soil. By decomposing wood, they also regulate carbon flow in forest systems. Generally, species of *Antrodia* are quite common in forest ecosystems and play important functional roles [39]. However, some species are rare and host-specific. For example, *A. cinnamomea* only inhabit the host tree, *Cinnamomum kanehirai*, which is an endemic tree in Taiwan [48]. This is an example of a specialized ecological relationship that also makes *A. cinnamomea* a highly vulnerable species, thus making conservation of its host tree crucial. Some species prefer old-growth forests [49], making them indicators of forest health and biodiversity. *Antrodia crassa* has been recognized as an indicator species for natural conifer forests in Northern Europe [39]. However, the overall effectiveness of *Antrodia* species as indicator species remains unclear [46] due to limited data on comparative ecology and taxonomic studies [39]. Some *Antrodia* species are on the IUCN red list in certain countries. For example, *A. neotropica* has been categorized as vulnerable in Brazil [50], and *A. albobrunnea* is red-listed in Finland, Norway, Poland, and Sweden [51]. Habitat loss and restricted distribution contribute to their conservation concerns.

2.1.3 Chemical Properties and Bioactive Compounds

Antrodia has gained significant attention due to its potential medicinal properties [52]. *Antrodia cinnamomea*, known as "Niu-Chang-Chih," is a traditional medicinal mushroom native to Taiwan [53]. This species has been used to treat ailments such as twisted tendons, muscle damage, mental distress, influenza, headaches, and fever in Taiwanese culture [54]. Due to its slow growth rate and the rarity of its natural host, this mushroom is referred to as the "Ruby of Taiwanese forests," highlighting its high value [53]. Over 78 compounds, including terpenoids, polysaccharides, benzoquinone derivatives, and maleic and succinic acid derivatives were identified in *A. cinnamomea* [54, 55] with terpenoids being the most abundant. Tong et al. [56] reported isolation of a new naphthalenecarboxaldehyde and seven other compounds for the first time from *A. cinnamomea*. As a medicinal source, this species has significant hepatoprotective, anti-inflammatory, cytotoxic, anticancer, and other miscellaneous activity [53, 57]. It targets multiple signalling pathways to alter hepatocellular carcinoma migration [57]. A ubiquinone derivative, antroquinonol, has been isolated from *A. cinnamomea* culture, which shows potential in treating cancer, male infertility, periodontal diseases, Parkinsonism, and cardiovascular diseases [53]. Studies have also demonstrated that *A. cinnamomea* can reduce obesity and modulate the gut microbiota in high-fat diet-fed mice, suggesting potential anti-obesogenic, anti-inflammatory, and antidiabetic effects [58].

Antrodia salmonea is another medicinal species native to Taiwan, known for its antioxidant, antimicrobial, and neuroprotective properties. A key compound in *A. salmonea* is antcin, which possesses anticancer, anti-inflammatory, antioxidant, antidiabetic, antiaging, immunomodulatory, hepatoprotective, and cardioprotective effects [59]. *Antrodia salmonea* also induces apoptosis and enhances cytoprotective effects in human colon cancer cells [60]. Its anti-inflammatory activity is based on its induction of Nrf2-mediated genes that eventually inhibit lipopolysaccharide-induced inflammation [61]. While *A. cinnamomea* has been the primary focus of medicinal research, information of other *Antrodia* species is limited.

2.1.4 Emerging Applications

Recent research into *Antrodia* species, particularly *A. cinnamomea*, has revealed several emerging applications. Li et al. [62] highlighted the neuroprotective activity of *A. cinnamomea* suggesting its use in managing neurological disorders. On the other hand, renal protective activity was reported by Tain et al. [63] based on animal models with chronic kidney diseases. Treatment with extracts of this species was reported to improve kidney function, reduce proteinuria, and lower blood pressure. Additionally, its anti-melanogenic, antioxidant, and antiproliferative effects in melanoma cells showed its potential applications in skin health

[64]. Furthermore, supplementation with *A. cinnamomea* extracts were found to reduce obesity and modulate the gut microbiota.

Ongoing research continues to explore the applications of *Antrodia*, suggesting the need for its conservation and advancement to identify its full potential. However, taxonomic accuracy and molecular data are essential for correctly identifying the species with medicinal potential.

One of the obstacles is finding the legitimate name of the source mushroom. For instance, three names have been associated with *A. cinnamomea*: *Ganoderma camphoratum*, *Antrodia camphorata*, and *Taiwanofungus camphoratus* [53]. Though the legitimate name is *A. cinnamomea* based on the holotype specimen, some research has used the name *A. camphorata* [54, 65, 66]. This leads to confusion about which species possess significant medicinal properties. Therefore, taxonomic information should be accurate and integrated with molecular data ensuring correct identification and naming of the source material.

2.2 Auricularia *Bull., Herb. Fr. (Paris) 3: pl. 290 (1780)*

2.2.1 Distribution and Taxonomy

Auricularia (*Auriculariaceae, Auriculariales, Auriculariomycetidae, Agaricomycetes, Agaricomycotina, Basidiomycota*) was introduced in 1780, with *A. mesenterica* designated as the type species. Currently, there are 37 accepted species of *Auricularia* with a wide distribution [67]. *Auricularia* species are known by several names, such as ear mushrooms, jelly fungus, and *Hed-hoo-noo* in Thai [68]. They are particularly found in tropical forests, growing on dead trees, stumps, fallen trunks, branches, and rotting wood. These fungi can also be found in temperate, tropical, and subtropical regions [68].

Morphologically, *Auricularia* spp. can be identified by gelatinous, resupinate to substipitate basidiomata with hairs on the upper surface, cylindrical to clavate, and transversely three-septate basidia with oil guttules, and hyaline, thin-walled, and allantoid basidiospores [69, 70]. Macro-morphological traits like the color and size of the basidiomata, and length of the hairs have been widely used to identify species in this genus [71].

2.2.2 Importance and Ecological Roles

Auricularia is a genus of edible fungi with significant nutritional, medicinal, and economic importance. Nutritionally it is rich in polysaccharides, dietary fibers, essential amino acids, and minerals like potassium and iron while being low in fat, thus making it a healthy food choice [72]. Moreover, *Auricularia* exhibits anticancer, antioxidant, antidiabetic, anti-hypercholesterolemic, anti-inflammatory, antiviral, and antimicrobial properties, largely attributed to its bioactive polysaccharides [73]. Traditionally, it has been used in Asian and indigenous medicine to treat ailments such as poisoning, anxiety, and gastric disorders [74].

These mushrooms are the fourth most farmed after *Agaricus, Lentinula,* and *Pleurotus* [68, 75]. *Auricularia* species possess nutritional and medicinal properties [76]. Several studies have demonstrated the antioxidant activity of *A. auricula-judae* [77–79], while others have reported the antioxidant, antibacterial, and anti-hypercholesterolemic properties of *A. polytricha* [80].

Auricularia is widely cultivated globally, especially in China and Southeast Asia, thriving on agricultural waste, eventually contributing to sustainable farming practices [81]. Interestingly, one report mentioned the formation of endomycorrhizas between an achlorophyllous orchid, *Erythrorchis ochobiensis*, and *Auricularia polytricha*, illustrating a unique symbiotic relationship [82].

2.2.3 Chemical Properties and Bioactive Compounds

Auricularia contains a variety of bioactive compounds that contribute to its medicinal properties. The most studied are polysaccharides, which exhibit anticancer, antioxidant, antidiabetic, and immunomodulatory effects [73]. These polysaccharides, such as β-glucans, are known to induce cancer cell apoptosis and reduce oxidative stress [83]. *Auricularia* polysaccharides offer health benefits such as antiaging, intestinal regulation, and cardiovascular support, among others [84].

Proteins from *Auricularia*, particularly immunomodulatory proteins, have shown potential as stable immune stimulants with resistance to heat, acids, and dehydration [85]. In addition, exo-biopolymers composed of carbohydrates and proteins have hypolipidemic effects, significantly reducing plasma LDL cholesterol in animal models [86]. *Auricularia* is also a source of melanin, which can be produced through submerged fermentation. Its main properties include antimicrobial and anti-biofilm activities, making it useful in food, cosmetics, and pharmaceutical industries [87].

Another interesting compound derived from *Auricularia* is adenosine, a nucleoside that inhibits platelet aggregation, suggesting potential applications in preventing thrombosis and cardiovascular diseases [88]. Additionally, polyphenolic compounds in *Auricularia* have been shown to lower cholesterol levels and reduce atherosclerosis in mice, highlighting their cardiovascular benefits [89].

2.2.4 Emerging Applications

Auricularia species, particularly *A. auricula-judae*, have shown potential in various non-food applications. Mushroom-based materials derived from *Auricularia* are being explored as sustainable substitutes for leather and textiles, offering biodegradable alternatives to synthetic materials [90]. Additionally, an enzyme produced by *A. auricula-judae* has been identified for its potential in plastic degradation. While the pure enzyme has not yet demonstrated significant degradation effects, ongoing research is focusing on mycelium-based degradation (whole cell degradation) as a more promising approach [90]. Additionally, cultivating *A. auricula* on walnut sawdust is a sustainable, eco-friendly, and cost-effective method which reduces deforestation and waste [91].

In the field of renewable energy, *Auricularia* biomass has been investigated for its potential in biofuel production. Specifically, *A. auricula-judae* biomass has been used as feedstock for biochar preparation, which serves as an effective carrier material for the controlled release of pesticides and fertilizers, contributing to sustainable agricultural practices [92]. Furthermore, melanin extracted from *Auricularia* species has gained attention in the cosmetic industry as a natural and healthful colorant, providing a safe alternative to synthetic ingredients [87]. These diverse applications highlight the versatility of *Auricularia* species beyond their traditional use as food, offering innovative solutions in biotechnology, environmental sustainability, and industrial applications [93].

2.3 Hericium *Pers., Neues Mag. Bot. 1: 109 (1794)*

2.3.1 Distribution and Taxonomy

Hericium (*Hericiaceae, Russulales, Incertae sedis, Agaricomycetes, Agaricomycotina, Basidiomycota*) is well known for its culinary and medicinal applications [15, 94]. These mushrooms exhibit a unique morphology, characterized by white, spiny or toothed fruiting bodies that resemble coral formations or a lion's mane [95]. *Hericium* species are globally distributed, particularly in North America, Europe, and Asia, where they primarily grow on dead or decaying hardwood trees in temperate regions [96]. Among them, *H. erinaceus* and *H. coralloides* are common and widespread, whereas other species are less frequently encountered and may have restricted distributions [97].

As per Index Fungorum [98], 73 epithets have been recorded for *Hericium*, while Species Fungorum [99] lists only 20 species worldwide. Currently, approximately 15 species are widely accepted, identified macroscopically by their branched or unbranched hymenophore structures bearing spines of varying lengths, which may occur singly or in clusters. Microscopically, they are distinguished by amyloid-ornamented basidiospores [94].

The taxonomy of *Hericium* continues to evolve, with new species being described and others undergoing reclassification. Advances in molecular genetics, particularly DNA sequencing, have provided greater clarity on evolutionary relationships within the genus [100]. Notable species within *Hericium* include *H. abietis*, *H. alpestre*, *H. americanum* (bear's head tooth fungus), *H. coralloides* (comb tooth fungus), *H. erinaceus* (lion's mane mushroom), and *H. laciniatum* [101].

2.3.2 Importance and Ecological Roles

Species of *Hericium* (e.g., *H. erinaceus*) are mostly saprotrophic, though some may exhibit symbiotic relationships [102]. They typically grow on hardwoods such as oak, beech, and maple, and occasionally on coniferous trees [96]. These mushrooms play a crucial role in nutrient cycling within forests and woodland ecosystems by decomposing dead or decaying trees and logs

[103]. Through wood decomposition, *Hericium* species break down complex organic materials into simpler compounds that other organisms can utilize. Additionally, they may contribute to the breakdown of pollutants or hazardous organic substances in contaminated environments [104].

The fruiting bodies of some *Hericium* species, particularly *H. erinaceus* and *H. coralloides*, are edible and highly valued for their texture and flavor [105]. They are often described as having a mild, seafood-like taste, making them a popular meat substitute in vegan and vegetarian diets [106]. *Hericium* mushrooms are rich in protein, fibers, and various essential vitamins and minerals, making them a nutritious addition to the diet [27].

Hericium erinaceus is particularly valued for its medicinal and culinary properties. As a result, there is growing interest in cultivating *Hericium* mushrooms to ensure sustainable use, making them available for future generations while preserving natural biodiversity [107].

2.3.3 Chemical Properties and Bioactive Compounds

Hericium mushrooms are valued for their chemical composition and numerous bioactive compounds that contribute to their therapeutic potential [108]. They are rich in polysaccharides, particularly beta-glucans (1,3-β-glucans and 1,6-β-glucans), as well as hericenones, erinacines (A, B, C), terpenoids, phenolic compounds, polypeptides, sterols, and fatty acids such as linoleic and oleic acids. They are also a source of proteins and amino acids [101].

These compounds exhibit various health-promoting effects, including neuroprotective, anti-inflammatory, antioxidant, anticancer, and immune-boosting activities [109]. Notably, hericenones and erinacines have been shown to stimulate nerve growth factor production, a protein essential for the development, maintenance, and survival of neurons [110]. Roda et al. [111] reported that *H. erinaceus* may benefit individuals with neurodegenerative diseases such as Alzheimer's, as well as enhancing cognitive function in healthy individuals.

2.3.4 Emerging Applications

Hericium is widely used across various industries, including biotechnology, sustainable practices, nutrition, and medicine [112]. Due to its diverse bioactive compounds, the mushroom's potential is still being explored in multiple fields.

The health applications of *Hericium* mushrooms are expanding rapidly as research uncovers their full potential in neuroprotection and cognitive health, immune system modulation, cancer immunotherapy, and as anti-inflammatory and antioxidant treatments for chronic diseases [59]. Emerging applications also include their use in functional foods and nutrition, particularly in plant-based diets, as well as in antiaging and skin-health cosmetics, biotechnology, drug development, and veterinary medicine for animal health [113].

As scientific research advances, the potential applications of *Hericium* mushrooms are expected to grow, solidifying their role as a valuable and versatile resource across multiple industries.

2.4 Lentinus *Fr., Syst. orb. veg. (Lundae): 77 (1825)*

2.4.1 Distribution and Taxonomy

Lentinus (*Polyporaceae, Polyporales, Incertae sedis, Agaricomycetes, Agaricomycotina, Basidiomycota*) comprises a diverse group of saprotrophic fungi with a global distribution, particularly abundant in tropical and subtropical regions. Currently, over 50 species are recognized, though this number remains subject to change as taxonomic understanding advances [114]. *Lentinus* species thrive in various habitats, including forests, grasslands, and agroecosystems, exhibiting remarkable ecological plasticity. Species richness is particularly high in biodiversity hotspots such as Southeast Asia, Africa, and South America.

Historically, the classification of *Lentinus* relied on morphological characteristics such as pileus structure, spore morphology, and substrate preference [15]. However, molecular phylogenetic studies have revolutionized the taxonomy of the genus, revealing greater diversity and resolving previously ambiguous classifications. DNA barcoding is crucial in identifying cryptic species, such as *Lignomyces vetlinianus* (syn. *Lentinus pilososquamulosus*), which was previously misidentified due to overlapping morphological traits [115]. Molecular studies have facilitated the reclassification of species such as *Lentinus giganteus* in Sri Lanka and Thailand, highlighting the effectiveness of ITS sequencing in resolving taxonomic uncertainties [116].

In Southeast Asia, *Lentinus squarrosulus* and *L. tigrinus* are prevalent, particularly in Malaysia, Thailand, and the Philippines [117, 118]. The region has also yielded new species discoveries, including three novel *Lentinus* species from northern Thailand [119, 120]. In the Brazilian Amazon and other Neotropical regions, *L. strigosus* commonly thrives on wood waste substrates [121], while *L. crinitus* is a characteristic species of South American forests [122]. Africa also harbors a diverse array of *Lentinus* species, as exemplified by studies in Cameroon, where *L. sajor-caju* has been documented in the Dja Biosphere Reserve, demonstrating adaptation to decaying hardwood [123].

Ongoing research continues to expand knowledge of *Lentinus* diversity, with recent discoveries including *L. punjabensis* from Pakistan [124]. Genetic diversity analyses, such as those conducted in India, have identified distinct clades within *Lentinus* populations, emphasizing regional speciation [125]. The genus also exhibits significant morphological diversity, with variations in pileus texture (e.g., the scaly pileus of *L. tigrinus*), stipe structure, and spore morphology [126].

2.4.2 Importance and Ecological Roles

Lentinus species play crucial ecological roles, primarily as white-rot fungi, which are essential decomposers of lignocellulosic biomass. Their ability to degrade lignin and cellulose in woody substrates [127] is vital for nutrient cycling in forest ecosystems [128]. This decomposition process is facilitated by the production of key ligninolytic enzymes, such as laccase and manganese peroxidase [129]. Notably, *Lentinula edodes* and *L. crinitus* have demonstrated efficient decomposition of agricultural residues, including rice straw and sawdust [130], highlighting their potential in waste management and composting.

Beyond their role in decomposition, *Lentinus* species contribute significantly to bioremediation. Several species exhibit the ability to degrade various pollutants, including synthetic dyes [131], polycyclic aromatic hydrocarbons [132], heavy metals [133], and phenol-rich industrial effluents [134]. These qualities position *Lentinus* species as promising candidates for environmental remediation applications.

Ecologically, *Lentinus* species establish their symbiotic contribution via decomposition rather than direct root symbiosis, enhancing soil health through organic matter turnover. Their extensive mycelial networks improve soil structure and water retention, which is particularly beneficial in agroforestry systems [121]. Furthermore, they serve as a food source for various invertebrates, playing a role in trophic dynamics. In tropical forests, species such as *L. squarrosulus* function as keystone decomposers, accelerating wood decay and contributing to carbon sequestration. Some species have even been observed associating with termite nests in Malaysian forests, further enhancing substrate decomposition [117].

Lentinus species exhibit antimicrobial activity, with potential applications in suppressing plant pathogens such as *Rigidoporus lignosus* [135], suggesting their role in biocontrol strategies for managing plant diseases. In agroecological contexts, *Lentinus* species may contribute to improved soil health through mycoremediation of pesticide residues [127]. These diverse ecological functions underscore the significance of *Lentinus* species in maintaining ecosystem health and functionality.

2.4.3 Chemical Properties and Bioactive Compounds

Lentinus species are a rich source of bioactive compounds, offering diverse medicinal and biotechnological applications. Polysaccharides, particularly β-glucans, are among the most prominent compounds. Extracts from *L. edodes* and *L. squarrosulus* have demonstrated significant immunomodulatory and antioxidant properties [136]. Notably, *L. squarrosulus* mycelia contain phenolic compounds and flavonoids that effectively scavenge free radicals [137], with some studies reporting higher antioxidant activity in mycelia compared to fruiting bodies [138]. *Lentinus crinitus* produces cytotoxic terpenoids, including phenolics such as protocatechuic acid and sterols like ergosterol, which have shown promise against cancer cell lines [139]. Additionally, lectins extracted from *L. edodes* exhibit hemagglutinating activity, which is valuable in biomedical research, and have demonstrated antiviral potential, including inhibition of HIV-1 replication [140].

Several *Lentinus* species also exhibit antimicrobial activity. For instance, *L. tigrinus* produces terpenoids and sterols, such as ergosta-7,22-dien-3β-ol, which inhibit the growth of foodborne pathogens like *Staphylococcus aureus* [141]. Laccases extracted from *Lentinus* species have demonstrated anti-inflammatory properties [142]. Polysaccharide-protein complexes from *L. polychrous* have exhibited anticancer effects, while extracts from this species have shown promise in anti-ulcer and anticancer applications [137, 143].

Metabolomic profiling has revealed strain-specific variations in bioactivity, emphasizing the importance of strain selection for targeted applications. For example, metabolomic studies of *L. strigellus* have identified novel secondary metabolites with antibacterial and anti-fungal activities [144]. Additionally, fermented *L. edodes* polysaccharides have been shown to enhance macrophage activation [136]. These diverse bioactivities underscore the potential of *Lentinus* species as valuable sources of natural compounds for pharmaceutical, nutraceutical, and biotechnological applications.

2.4.4 Emerging Applications

The biotechnological applications of *Lentinus* species are rapidly expanding, driven by their diverse enzymatic capabilities and bioactive compounds. Their ligninolytic enzymes are particularly valuable in applications such as biofuel production and wastewater treatment. For example, laccases from *Lentinus* species effectively degrade pollutants like Congo red dye [132], while *L. crinitus* mycelia have shown promise in biosorbing methylene blue [145]. Additionally, *L. strigosus* has demonstrated the ability to degrade anthraquinones [131], further highlighting its potential in bioremediation. *Lentinus squarrosulus* produces mycoflocculants that can reduce water turbidity [146], offering a bio-based solution for water purification.

In agriculture, *Lentinus* species are being cultivated on agro-industrial waste, promoting sustainable biomass production and resource utilization [130]. This approach not only reduces waste but also provides a valuable source of food and other products.

The food industry has also recognized the potential of *Lentinus* species. *Lentinula edodes* is particularly valued for its use in meat substitutes and as a source of functional ingredients [147]. *Lentinula edodes* polysaccharides are utilized as natural preservatives and functional ingredients [147], and probiotic fermentation has been shown to enhance their immunomodulatory properties [136].

Medicinally, *Lentinus* extracts are being formulated into nutraceuticals for antioxidant and anti-ulcer therapies [137, 138]. *Lentinus crinitus* polysaccharides have demonstrated antitumor activity [148], and *L. tigrinus* extracts have shown inhibitory effects against HIV-1 [140]. Additionally, *L. crinitus* exhibits probiotic properties that enhance livestock growth [149].

Emerging research is exploring the roles of *Lentinus* species in nanotechnology. For instance, pyrolysis of *Lentinus* stipes yields biochar with promising energy storage applications [150]. Furthermore, advancements in artificial intelligence (AI) are being integrated into *Lentinus* cultivation. AI-driven methods, such as YOLOv5seg-BotNet for *L. edodes* image segmentation, are being employed to optimize yields and improve cultivation efficiency [151], showcasing the intersection of mycology and cutting-edge technology.

2.5 Morchella *Dill. ex Pers., Neues Mag. Bot. 1: 116 (1794)*

2.5.1 Distribution and Taxonomy

Morchella (*Morchellaceae, Pezizales, Pezizomycetidae, Pezizomycetes, Pezizomycotina*, *Ascomycota*), commonly known as morels, is a key example of macro-ascomycetes [152]. They are seasonal ascomycetous mushrooms that are highly valued for their culinary and economic significance due to their distinctive gustatory and aromatic properties [153]. *Morchella* taxa are characterized by conical to subconical hymenophores that are sometimes ovoid or subcylindrical, consisting of a subacute apex, and feature a pointed apex with ridges and pits. The ridges are smooth, and appear yellowish or white when young, gradually turning to a pale brown color as they age. Initially, the ridges are rounded or flattened, but they become sharper or eroded with maturity. The pits are smooth, vertically elongated, and gray when young, transitioning to a pale yellowish-tan as they mature. The asci are cylindrical, hyaline, and contain eight spores. The elements on the ridges are septate, either hyaline or brown when mounted KOH, and the terminal cell gradually becomes subfusiform but rarely cylindrical [154].

In the past, extensive research has been conducted on *Morchella* to resolve its complex taxonomy, which was largely due to the presence of cryptic species and significant morphological plasticity [155]. Other plausible explanations for the unresolved taxonomic issues were attributed to inadequate sampling, taxonomic ambiguities in nomenclature, incomplete morphological descriptions, and lack of DNA sequence data for older taxa [156].

Advances in molecular phylogenetics including multiple nuclear DNA markers have significantly improved species delineation within *Morchella*, leading to a clearer understanding of its evolutionary relationships and biogeographic patterns [157]. However, it is highly recommended to employ a polyphasic approach that combines morphological, ecological, and DNA molecular data to delimit *Morchella* species [156, 157]. Index Fungorum [98] lists over 350 epithets of *Morchella*. However, only 100 species are listed in Species Fungorum [99]. Based on recent studies, 59 *Morchella* spp. are formally accepted with valid nomenclatures [157], while an additional 24 phylogenetic species without valid names are present [156].

Morchella species are predominantly distributed in temperate regions of the Northern Hemisphere, where they exhibit the highest species richness. East Asia, particularly China, is recognized as a biodiversity hotspot for these fungi [155]. Morels thrive in diverse ecological niches, including deciduous and coniferous woodlands, forest edges, riparian zones, and limed soils. Their occurrence is particularly notable in post-disturbance environments such as burn sites, where organic decomposition facilitates their fruiting. Additionally, *Morchella* species have been observed forming associations with specific tree species, often appearing near or beneath decaying elms (*Ulmus* spp.), ash (*Fraxinus* spp.), poplar (*Populus* spp.), and apple trees (*Malus* spp.) [158].

2.5.2 Importance and Ecological Roles

While some *Morchella* species exhibit saprotrophic tendencies, others display facultative mycorrhizal interactions, contributing to their ecological adaptability and widespread distribution [159].

Morchella species are economically significant. They have high nutritional and medicinal values, with a long consumption history, globally [160]. For instance, true morels are widely used in Chinese medicine as they have high contents of protein, fatty acids, vitamins, minerals, organic acids, and fibres [161]. In addition, they serve as a nutritious dietary component, characterized by their low carbohydrate, fat, and calorie content [162].

Considering the numerous benefits of morels, related research is crucial due to the increasing impact of fungal diseases, such as cobweb disease, pileus rot, and stipe rot, which significantly reduce yields, resulting in economic losses in morel cultivation [163]. Understanding and managing these fungal threats is vital for improving production and ensuring economic sustainability.

2.5.3 Chemical Properties and Bioactive Compounds

Morchella species are abundant in a variety of bioactive compounds, including polysaccharides, phenolic compounds, tocopherols, and ergosterols, which contribute to numerous health benefits [164]. These compounds have been shown to possess anti-oxidative, anti-inflammatory, immune-boosting, and gut health-supporting properties, and protective effects against cancer and cardiovascular diseases [164–166]. In addition, morels have demonstrated neuroprotective and hepatoprotective effects, making them valuable in maintaining overall health [167]. These mushrooms have also been recognized for their antidiabetic, anticancer, antimicrobial, and liver-protecting properties, underlining their potential as a natural remedy for various diseases [164, 168].

Morchella esculenta produces various primary and secondary metabolites (e.g., polyphenols, sterols, flavonoids, alkaloids) and is rich in carbohydrates, proteins, vitamins, and mineral elements [169]. Another study reported several different bioactive compounds derived from *M. tridentina*, that can be used for manufacturing drugs to treat cancer, heart diseases, and edema [170]. Additionally, other compounds derived from *Morchella* species are terpenoids, fatty acids, amino acids, sterols, peptides, alkaloids, glutamic acid, propanoic acid, coumarins, and cyclopyrrolones, reported from *M. conica*, *M. delicosa*, and *M. esculenta* [171].

2.5.4 Emerging Applications

Future research on morels could explore innovative approaches to enhancing large-scale cultivation techniques, focusing on optimizing environmental conditions for increased yields and consistent quality. Additionally, advances in biotechnology could lead to the development of morel-based bioproducts, such as biofertilizers or biodegradable materials, utilizing the mushroom's mycelium. Research into the untapped potential of morel metabolites may reveal new therapeutic compounds, particularly in the fields of cancer treatment and neurodegenerative diseases. Exploring the interaction between morel bioactive compounds and the human microbiome could also provide insights into their roles in gut health and immunity. Furthermore, integrating morel extracts into personalized nutrition could pave the way for functional foods tailored to individual health needs.

2.6 Pleurotus *(Fr.) P. Kumm., Führ. Pilzk. (Zerbst): 24 (1871)*

2.6.1 Distribution and Taxonomy

The systematics of *Pleurotus* (*Pleurotaceae, Agaricales, Agaricomycetidae, Agaricomycetes, Agaricomycotina, Basidiomycota*) mushrooms integrates traditional morphological methods with modern molecular techniques. The genus is highly diverse, with species adapted to a wide range of ecological niches. Comprehensive surveys and molecular studies are critical for documenting species diversity, understanding phylogenetic relationships, and identifying novel species. The taxonomy and phylogenetic identification of *Pleurotus* species can be complex, often leading to misidentification. Therefore, understanding their taxonomy and phylogeny is essential for their conservation, cultivation, and industrial applications.

Recent biochemical and molecular techniques, such as DNA-based identification, have been employed to validate species [172]. Sequences of the loci of the internal transcribed spacer (ITS) and large subunit (28S), and the gene for the translation elongation factor 1-alpha (*TEF1-α*), have played a pivotal role in classifying *Pleurotus* species. These molecular markers aid in resolving species boundaries and clarifying evolutionary relationships [173], which are vital for breeding programs and commercial mushroom production [174] and help to resolve misidentifications.

Pleurotus is characterized by significant diversity, with variations in morphology, ecological roles, and geographical distribution [175]. Historically, the taxonomy of *Pleurotus* species has been challenging due to their widespread distribution and the frequent association of multiple nomenclatures with a single species. Recent advancements that combine morphological analyses with molecular phylogenetic techniques have improved the accuracy of *Pleurotus* species classification [176]. As per Index Fungorum [98], there are 777 records of *Pleurotus* species, but only 25 accepted species, with *P. ostreatus* designated as the type species [17]. The genus is distributed across temperate, subtropical, and tropical regions, reflecting its ability to thrive in diverse climates and habitats [175].

2.6.2 Importance and Ecological Roles

Pleurotus represents a group of ecologically and economically important fungi within the family *Pleurotaceae* [177]. Commonly known as oyster mushrooms, these species are prized for their culinary value, medicinal properties, and roles in biotechnological applications such as bioremediation [178].

Pleurotus species are primarily saprotrophic, playing an essential role in the decomposition of lignocellulosic materials, such as wood, agricultural residues, and forest litter. This decomposition process is vital for nutrient cycling and maintaining ecosystem sustainability [179]. The ecological versatility of *Pleurotus* allows it to thrive in a wide range of habitats, contributing significantly to the recycling of organic matter in various ecosystems. Despite their adaptability, *Pleurotus* species are threatened by habitat loss, deforestation, and climate change [180]. Many species, particularly in regions such as Southeast Asia, South America, and Africa, remain underexplored [181].

2.7 Chemical Properties and Bioactive Compounds

Pleurotus comprises a diverse group of edible species known for their exceptional nutritional content, including high levels of fibers, essential vitamins (such as thiamine, riboflavin, ascorbic acid, ergosterol, and niacin), vital minerals (like phosphorus and iron), and carbohydrates [182]. The remarkable biodiversity of *Pleurotus* species stems from their complex and diverse chemical composition, which includes polysaccharides, glycoproteins, and secondary metabolites like alkaloids, flavonoids, and betalains [183]. In recent years, various *Pleurotus* species have gained increasing attention due to their excellent taste and significant medicinal value. These mushrooms are rich in bioactive compounds, with polysaccharides being particularly noteworthy [183]. Polysaccharides derived from *Pleurotus* exhibit a wide range of biological activities, including immunomodulatory [184], antitumor [185], and antioxidant [186]. They play a crucial role in eliciting important physiological responses in humans, animals, and alternative in vitro models.

In addition, *Pleurotus* possesses antihypertensive, antihyperlipidemic, anti-inflammatory, and other beneficial properties that helped to reduce the risk factors associated with COVID-19

[187]. Consistent consumption of *Pleurotus* mushrooms or their extracts, either as nutritional supplements or pharmaceutical ingredients, enhances immune health and may aid in preventing or treating infectious diseases [187]. Commonly found species with potential biological activities for health promotion include *P. djamor* [188], *P. djamor* var. *roseus* [189], *P. eryngii* [190], *P. ostreatus* [191], and other unidentified *Pleurotus* spp. [192].

2.7.1 Emerging Applications

Cultivation of *Pleurotus* has become globally significant due to its nutritional value, medicinal properties, and economic potential [193]. *Pleurotus* is widely cultivated because of its adaptability to diverse agroclimatic conditions, rapid growth, and ability to thrive on various agricultural waste materials [194]. Several species have been cultivated, including *P. citrinopileatus*, *P. djamor*, *P. djamor* var. *roseus*, *P. eryngii*, *P. flabellatus*, and *P. ostreatus* [193].

Cultivation of *Pleurotus* mushrooms requires a comprehensive understanding of their physical, chemical, biological, and enzymatic characteristics [195]. A thorough knowledge of both intrinsic factors (such as substrate type, particle size, moisture content, pH value, electrical conductivity, and oxygen availability) and extrinsic factors (including temperature, humidity, CO_2 levels, light, air exchange, and substrate treatment) is essential for achieving efficient and sustainable production. Adjusting these parameters according to the specific requirements of each species is necessary for optimal growth [195].

Cultivation of *Pleurotus* mushrooms using solid-state fermentation facilitates the recycling of agricultural waste. Raman et al. [193] reported several types of agricultural byproducts that serve as substrates for cultivation, including banana leaves, peanut shells, corn leaves, wheat and rice straw, mango fruits and seeds, and sugarcane leaves. *Pleurotus* species can grow on both composted and non-composted substrates [196]. Their cultivation is environmentally sustainable, as they aid in biodegradation and recycling of organic waste. With increasing demand for functional foods and sustainable agriculture, *Pleurotus* mushroom cultivation offers an eco-friendly and profitable solution for food security and waste management.

2.8 Schizophyllum Fr., *Observ. mycol. (Havniae) 1: 103 (1815)*

2.8.1 Distribution and Taxonomy

Schizophyllum (*Schizophyllaceae*, *Agaricales*, *Agaricomycetidae*, *Agaricomycetes*, *Agaricomycotina*, *Basidiomycota*) was established by E. M. Fries in 1815, with *S. commune* designated as the type species [197]. This genus, commonly known as the split-gill mushroom, is found naturally growing during the rainy season and is distributed globally, except in Antarctica [198]. Index Fungorum [98] lists 31 epithets of *Schizophyllum*, while Species Fungorum [99] recognizes only 11 species worldwide; these are *S. album*, *S. amplum*, *S. brasiliense*, *S. brevilamellatum*, *S. commune*, *S. fasciatum*, *S. leprieurii*, *S. murrayi*, *S. palmatum*, *S. radiatum*, and *S. variabile*. Members of this genus are known to cause white rot of wood. Among them, *S. commune* is the most-researched and well-known species based on morphology, taxonomy, phylogeny, distribution, and myco-chemical profiles [199].

Schizophyllum is characterized by small, white to gray, flabelliform basidiomata, with distinctive gills composed of two flat plates that curl away from each other [200]. These gills develop through dichotomous branching patterns [201]. When dry, the basidiomata margins curl inward but revive easily when moistened [202]. Initially, the basidiomata are inverted cuplike structures that expand radially and laterally in symmetrical patterns as they mature [201]. Basidiospores are cylindrical to allantoid, hyaline and smooth, producing a white spore print [202]. Basidia are narrowly clavate, four-spored, with a basal clamp connection. The hyphal system is dimitic, comprising nodose-septate generative hyphae and thick-walled, unbranched skeletal hyphae with clamp connections [202].

2.8.2 Importance and Ecological Roles

Schizophyllum commune is consumed in various regions, including Mexico and South Asian countries such as China, India, Malaysia, Thailand, and Vietnam [203]. However, it has also been implicated in human health issues, causing allergic diseases such as "chronic sinusitis" or "bronchopulmonary mycosis" [204]. Additionally, it can cause invasive infections affecting the lungs, brain, and palate in both immune-compromised and immune-competent patients [204].

In natural ecosystems, *Schizophyllum* taxa play a crucial role in breaking down complex wood into simpler components, thereby contributing to the nutrient cycle. This genus has been identified as a wood decomposer of over 150 genera of flowering plants [201]. *S. commune* has also exhibited pathogenic behavior in fruit and ornamental trees [205]. Plants with wounds, pruning scars, freeze injuries, or sunscald are particularly susceptible to this fungus, which causes a disease known as "Schizophyllum rot." The fungus primarily targets broadleaf and coniferous trees, leading to sap rot and decay of fresh plant tissues [205].

2.8.3 Chemical Properties and Bioactive Compounds

Schizophyllum commune is both an edible and medicinal mushroom, rich in beneficial bioactive compounds. It is widely used in traditional medicine across East Asia [206]. A variety of functional ingredients have been identified in this mushroom, including polysaccharides, phenolic compounds, terpenoids, and ergosterol [207]. Notably, it produces exopolysaccharides with potential applications in various industries, such as food, pharmaceuticals, and cosmetics. These applications include its use as emulsifiers, stabilizers, gelling agents, binders, lubricants, and thickening agents [208].

Schizophyllan, a natural biopolymer derived from glucose, features a β-1,3-linked backbone with single β-1,6-linked glucose side chains at every third residue [209]. It exhibits significant pharmaceutical properties, including antitumor, anticancer, antioxidant, antibacterial, anti-inflammatory, anti-tyrosinase, neuroprotective, antidiabetic, and immune-modulating activities [199, 206, 207].

The α- and β-glucans also play a crucial anti-oxidative role by neutralizing free radicals, thereby preventing cellular damage [210]. Currently, schizophyllan is commercially utilized in vaccines, anticancer therapies, and as an active component in cosmetic products [209].

2.8.4 Emerging Applications

Besides medical applications, *Schizophyllum commune* demonstrates a wide array of innovative uses, including biomaterial development, enhanced oil recovery, and as a component in biolubricants [209]. The beneficial properties of schizophyllan, such as high viscosity, film formation, and thermal stability, make it well-suited for biomaterial production [211]. Genetically modified strains of *S. commune*, such as those with the deletion of the hydrophobin gene sc3, exhibit superior biomaterial quality, primarily in enhanced density and mechanical properties compared to their wild-type counterparts [212]. Additionally, schizophyllan is effective in lubricants where aqueous solubility and thermal stability are important [213].

In the cosmetic industry, schizophyllan has been utilized for hydration, antiaging, and skin repair. This mushroom has been successfully tested for synthesizing silver nanoparticles, which have promising biotechnological and biomedical applications [214].

In the process of schizophyllan production, besides malt extract medium, corn fiber (the pericarp of the kernel) also serves as a cost-effective lignocellulosic substrate, enabling large-scale production [209]. This is facilitated by the enzymatic activity of *S. commune*, which includes cellulase, xylanase, pectinase, and lignin-degrading enzymes that degrade the substrate [215].

Nowadays, *S. commune* is treated as a model organism in transmission genetics, developmental biology, and population genetics, indicating its broad scientific significance [216]. Genomic analysis has identified 13,210 predicted genes in its genome, including 472 encoding transcription factors [217]. For instance, the genes fst3 and fst4 play a role in basidiomata development, with fst3 acting as an inhibitor and fst4 as an inducer.

3 Discussion

Mushroom diversity remains vastly under-explored, particularly in tropical regions, where unique ecological conditions foster a rich array of macrofungi. Despite significant advances in fungal research, many species remain undiscovered, underlining the urgent need for further surveys in diverse ecosystems, including preserved forests [19]. Notably, China has taken the lead in studying macrofungi due to their medicinal potential, leading to the discovery of novel species with potential therapeutic properties [218]. To bridge the knowledge gap, it is crucial to prioritize systematic fungal surveys to prevent extinction of undocumented species, ensuring the preservation of fungal biodiversity.

Macrofungi have diverse applications across several industries, including biotechnology, agriculture, and pharmaceuticals [2]. In the food industry, edible fungi such as *Pleurotus* species exhibit cholesterol-lowering effects, demonstrating their significance in functional foods and medicine [182]. In pharmaceuticals, macrofungi are a rich source of bioactive compounds, contributing to the development of antibiotics, antioxidants, and anticancer agents [22]. Clinical trials have highlighted the therapeutic potential of these fungi, particularly in immunomodulation and cancer treatment, where their bioactive metabolites show promise in enhancing immune responses and inhibiting tumor growth. A notable macrofungal genus that has not been discussed herein is *Ganoderma*, known for its immunomodulatory properties. This medicinal mushroom is increasingly incorporated into nutraceuticals and dietary supplements for their health-promoting properties [219].

Economically significant fungi, such as those discussed in this study, have been extensively researched for their biological activities. These fungi contribute immensely to ecosystem processes such as decomposition, nutrient cycling, and symbiotic interactions with plants [1, 2]. By secreting hydrolytic enzymes, they facilitate the breakdown of complex organic matter, making them essential for sustainable environmental management. The production of bioactive metabolites, including terpenoids, polyphenols, and polysaccharides, further enhances their value in human health, nutraceuticals, and pharmaceuticals [21].

The chemical diversity of these macrofungi is intricately linked to their ecological roles [182]. Fungi secrete bioactive compounds in response to environmental stresses, competition, and symbiotic interactions. The production of secondary metabolites, such as antibiotics, antioxidants, and anticancer agents, enables fungi to adapt to their surroundings while providing valuable compounds for human use. In agriculture, fungal enzymes contribute to soil health and plant growth promotion [220]. Additionally, in bioremediation, fungi are crucial for the degradation of pollutants, including heavy metals and petroleum hydrocarbons, thus showcasing their potential in environmental sustainability [221].

The taxonomy of macrofungi remains a major challenge in mycology. Their accurate identification is essential for ecological studies and biotechnological applications. Integrative taxonomic approaches combining morphological, molecular, and biochemical analyses are needed to resolve taxonomic ambiguities [31]. Enhanced collaboration among mycologists worldwide will lead to more accurate species identification, preventing misclassification. Chemotaxonomy and proteomics offer promising tools for identifying fungi based on specific proteins and bioactive compounds, providing an alternative approach to traditional taxonomic methods [31].

3.1 Future Directions

Future research should focus on refining fungal taxonomy to enhance species identification and classification. Advanced molecular techniques, such as high-throughput sequencing and metabolomics, will help uncover cryptic species and clarify fungal phylogenetic relationships. Expanding chemical investigations into macrofungi will provide insights into their potential applications in medicine, agriculture, and industry. Ultimately, fostering interdisciplinary collaborations between taxonomists, chemists, and biotechnologists will ensure that the vast diversity of macrofungi is adequately explored, conserved, and utilized for the benefit of humanity and the environment.

Funding remains a critical barrier to advancing fungal research. Increased financial support from governmental and private agencies, including biopharmaceutical industries, is necessary to facilitate comprehensive studies on macrofungal diversity. The economic potential of these fungi in medicine, agriculture, and biotechnology highlights the need for sustained investment in research and conservation. Establishing and maintaining fungaria and culture collections is crucial for preserving existing fungal specimens for future scientific and industrial applications. Proper documentation and long-term storage of fungal isolates will enable ongoing taxonomic and biochemical studies, ensuring the availability of valuable strains for future use.

Acknowledgments

I would like to express my sincere gratitude to all the co-authors for their invaluable collaboration and support throughout this work, with each contributing equally to the success of this chapter. We would also like to thank Dr. Rajesh Jeewon from the University of Mauritius for his assistance in revising this chapter and providing ideas for the discussion section.

Conflicts of Interest The authors declare no conflict of interest.

References

1. Hawksworth DL, Kirk PM, Sutton BC, Pegler DN (1995) Ainsworth and Bisby's dictionary of fungi. CABI, Oxon
2. Hyde KD, Xu J, Rapior S, Jeewon R, Lumyong S, Niego AGT, Abeywickrama PD, Aluthmuhandiram JVS, Brahamanage RS, Brooks S et al (2019) The amazing potential of fungi: 50 ways we can exploit fungi industrially. Fungal Divers 97:1–136. https://doi.org/10.1007/s13225-019-00430-9
3. Hyde KD, Jeewon R, Chen YJ, Bhunjun CS, Calabon MS, Jiang HB, Lin CG, Norphanphoun C, Sysouphanthong P, Pem D et al (2020) The numbers of fungi: is the descriptive curve flattening? Fungal Divers 103:219–271. https://doi.org/10.1007/s13225-020-00458-2
4. Hawksworth DL (1991) The fungal dimension of biodiversity: magnitude, significance, and conservation. Mycol Res 95:641–655
5. Blackwell M (2011) The fungi: 1, 2, 3... 5.1 million species? Am J Bot 98(3):426–438. https://doi.org/10.3732/ajb.1000298
6. Hawksworth DL, Lücking R (2017) Fungal diversity revisited: 2.2 to 3.8 million species. Microbiology. Spectrum 5(4):10–1128. https://doi.org/10.1128/microbiolspec.funk-0052-2016
7. Wu B, Hussain M, Zhang W, Stadler M, Liu X, Xiang M (2019) Current insights into fungal species diversity and perspective on naming the environmental DNA sequences of fungi. Mycology 10:127–140. https://doi.org/10.1080/21501203.2019.1614106
8. Baldrian P, Větrovský T, Lepinay C, Kohout P (2022) High-throughput sequencing view on the magnitude of global fungal diversity. Fungal Divers 114(1):539–547. https://doi.org/10.1007/s13225-021-00472-y
9. Niskanen T, Lücking R, Dahlberg A, Gaya E, Laura MS, Mikryukov V, Liimatainen K, Druzhinina I, Westrip JRS, Mueller GM et al (2023) Pushing the frontiers of biodiversity research: unveiling the global diversity, distribution, and conservation of fungi. Ann Rev Environ Resour 48:49–176. https://doi.org/10.1146/annurev-environ-112621-090937
10. Hyde KD, Saleh A, Aumentado HDR, Boekhout T, Bera I, Khyaju S, Bhunjun CS, Chethana KWT, Phukhamsakda C, Doilom M et al (2024) Fungal numbers: global needs for a realistic assessment. Fungal Divers:1–35. https://doi.org/10.1007/s13225-024-00545-8
11. Hyde KD, Noorabadi MT, Thiyagaraja V, He MQ, Johnston PR, Wijesinghe SN, Armand A, Biketova AY, Chethana KWT, Erdoğdu M et al (2024) The 2024 outline of fungi and fungus-like taxa. Mycosphere 15(1):5146–6239. https://doi.org/10.5943/mycosphere/15/1/25
12. Kinge TR, Goldman G, Jacobs A, Ndiritu GG, Gryzenhout M (2020) A first checklist of macrofungi for South Africa. MycoKeys 63. https://doi.org/10.3897/mycokeys.63.36566
13. De Meeûs T, Renaud F (2002) Parasites within the new phylogeny of eukaryotes. Trends Parasitol 18(6):247–251
14. Mueller GM, Schmit JP, Leacock PR, Buyck B, Cifuentes J, Desjardin DE, Halling RE, Hjortstam K, Iturriaga T, Larsson KH et al (2007) Global diversity and distribution of macrofungi. Biodivers Conserv 16:37–48. https://doi.org/10.1007/s10531-006-9108-8
15. Kirk PM, Cannon PF, Minter DW, Stalpers JA (2008) Ainsworth & Bisby's dictionary of the fungi, 10th edn. CAB International, Wallingford
16. Begerow D, McTaggart A, Agerer R, Frey W (2018) Syllabus of plant families-A. Engler's Syllabus der Pflanzenfamilien Part 1/3
17. He MQ, Zhao RL, Hyde KD, Begerow D, Kemler M, Yurkov A, McKenzie EHC, Raspé O, Kakishima M, Sánchez-Ramírez S et al (2019) Notes, outline and divergence times of Basidiomycota. Fungal Divers 99: 105–367. https://doi.org/10.1007/s13225-019-00435-4
18. He MQ, Cao B, Liu F, Boekhout T, Denchev TT, Schoutteten N, Denchev CM, Kemler M, Gorjón SP, Begerow D et al (2024) Phylogenomics, divergence times and notes of orders in Basidiomycota. Fungal Divers. https://doi.org/10.1007/s13225-024-00535-w
19. He MQ, Zhao RL, Liu DM, Denchev TT, Begerow D, Yurkov A, Kemler M, Millanes AM, Wedin M, McTaggart AR et al (2022) Species diversity of Basidiomycota. Fungal

Divers 114(1):281–325. https://doi.org/10.1007/s13225-021-00497-3
20. Valverde ME, Hernández-Pérez T, Paredes-López O (2015) Edible mushrooms: improving human health and promoting quality life. Int J Microbiol 2015:1–14. https://doi.org/10.1155/2015/376387
21. Dávila Giraldo LR, Pérez Jaramillo CC, Méndez Arteaga JJ, Murillo-Arango W (2023) Nutritional value and antioxidant, antimicrobial and cytotoxic activity of wild macrofungi. Microorganisms 11(5):1158. https://doi.org/10.3390/microorganisms11051158
22. Meng X, Liang H, Luo L (2016) Antitumor polysaccharides from mushrooms: a review on the structural characteristics, antitumor mechanisms and immunomodulating activities. Carbohydr Res 424:30–41. https://doi.org/10.1016/j.carres.2016.02.008
23. Sharif M, Nawaz A, Naeem S, Saeeda UH, Farid A, Zada NS, Shah AA, Badshah M, Alharbi MT, Tarabulsi MK, Selim S (2024) Pharmaceutical and nutraceutical importance of bioactive metabolites from macrofungi. J Biol Regul Homeost Agents 38(5):3621–3643
24. Ghosh S, Sett S, Saha R, Roy A, Acharya K (2021) Comparative phytochemical screening and antioxidant properties of infusion, decoction and hydroalcoholic extracts of wood ear mushrooms; *Auricularia delicata* and *Auricularia mesenterica*. Indian Phytopathol 74:113–121. https://doi.org/10.1007/s42360-020-00301-3
25. Cheung L, Cheung PC, Ooi VE (2003) Antioxidant activity and total phenolics of edible mushroom extracts. Food Chem 81:249–255
26. Li H, Tian Y, Menolli N Jr, Ye L, Karunarathna SC, Perez-Moreno J, Rahman MM, Rashid MH, Phengsintham P, Rizal L, Kasuya T (2021) Reviewing the world's edible mushroom species: a new evidence-based classification system. Compr Rev Food Sci Food Saf 20(2):1982–2014. https://doi.org/10.1111/1541-4337.12708
27. Zhang Y, Wang D, Chen Y, Liu T, Zhang S, Fan H, Liu H, Li Y (2021) Healthy function and high valued utilization of edible fungi. Food Sci Human Wellness 10(4):408–420. https://doi.org/10.1016/j.fshw.2021.04.003
28. Gafforov Y, Rašeta M, Rapior S, Yarasheva M, Wang X, Zhou L, Wan-Mohtar WAAQI, Zafar M, Lim YW, Wang M, Abdullaev B (2023) Macrofungi as medicinal resources in Uzbekistan: biodiversity, ethnomycology, and ethnomedicinal practices. J Fungi 9(9):922. https://doi.org/10.3390/jof9090922
29. Syme K, Lebel T, Hilton RN (2025) Macrofungi of two peoples bay nature reserve, Western Australia. Pac Conserv Biol 31(1). https://doi.org/10.1071/PC24091
30. Sánchez-Jardón L, del Rio-Hortega L, Cea NN, Mingarro M, Manubens P, Zambrano S, Gallo BA (2021) Fungal literature records database of the sub-Antarctic region of Aysén, Chile. Biodiver Data J 9: e75951. https://doi.org/10.3897/BDJ.9.e75951
31. Lücking R, Aime MC, Robbertse B, Miller AN, Aoki T, Ariyawansa HA, Cardinali G, Crous PW, Druzhinina IS, Geiser DM et al (2021) Fungal taxonomy and sequence-based nomenclature. Nat Microbiol 6(5):540–548. https://doi.org/10.1038/s41564-021-00888-x
32. Glamočlija J, Stojković D, Nikolić M, Ćirić A, Reis FS, Barros L, Ferreira IC, Soković M (2015) A comparative study on edible *Agaricus* mushrooms as functional foods. Food Funct 6(6):1900–1910
33. Piska K, Sułkowska-Ziaja K, Muszyńska B (2017) Edible mushroom *Pleurotus ostreatus* (*oyster mushroom*): its dietary significance and biological activity. Acta Scientiarum Polonorum. Hortorum Cultus 16(1):1–12
34. Bulam S, Ütstün NŞ, Pekşen A (2019) Health benefits of *Ganoderma lucidum* as a medicinal mushroom. Turkish J Agric Food Sci Technol 7:84–93. https://doi.org/10.24925/turjaf.v7isp1.84-93.2728
35. Li Q, Liu J, Shang X, Li Y, Zhang L, Li Z, Jiang N, Tan Q, Yu H, Song C (2021) Characterizing diversity based on the chemical and nutritional composition of shiitake culinary-medicinal mushroom *Lentinula edodes* (Agaricomycetes) commonly cultivated in China. Int J Med Mushrooms 23(8):51–64. https://doi.org/10.1615/IntJMedMushrooms.2021039240
36. Bazzicalupo AL, Buyck B, Saar I, Vauras J, Carmean D, Berbee ML (2017) Troubles with mycorrhizal mushroom identification where morphological differentiation lags behind barcode sequence divergence. Taxon 66(4):791–810. https://doi.org/10.12705/664.1
37. Zabin DA, Spirin V, Vlasák J, Coelho-Nascimento C, Menolli N Jr (2024) Taxonomic reinvestigation of *Favolus* in the Neotropics utilizing morphological and multigene phylogenetic analyses. Mycol Prog 23(1):44. https://doi.org/10.1007/s11557-024-01979-9
38. Spirin V, Vlasák J, Niemelä T, Miettinen O (2013) What is *Antrodia* sensu stricto?

Mycologia 105(6):1555–1576. https://doi.org/10.3852/13-039

39. Spirin V, Runnel K, Vlasák J, Miettinen O, Põldmaa K (2015) Species diversity in the *Antrodia crassa* group (Polyporales, Basidiomycota). Fungal Biol 119(12):1291–1310. https://doi.org/10.1016/j.funbio.2015.09.008
40. Donk MA (1960) The generic names proposed for *Polyporaceae*. Persoonia 1:173–302
41. Donk MA (1966) Notes on European polypores 1. Persoonia 4:337–343
42. Hibbett DS, Donoghue MJ (2001) Analysis of character correlations among wood decay mechanisms, mating systems, and substrate ranges in homobasidiomycetes. Syst Biol 50(2):215–242. https://doi.org/10.1080/10635150121079
43. Ortiz-Santana B, Lindner DL, Miettinen O, Justo A, Hibbett DS (2013) A phylogenetic overview of the *Antrodia* clade (Basidiomycota, Polyporales). Mycologia 105(6):1391–1411. https://doi.org/10.3852/13-051
44. Spirin V, Vlasak J, Rivoire B, Kotiranta H, Miettinen O (2016) Hidden diversity in the *Antrodia malicola* group (Polyporales, Basidiomycota). Mycol Prog 15:51. https://doi.org/10.1007/s11557-016-1193-9
45. Spirin V, Vlasak J, Miettinen O (2017) Studies in the *Antrodia serialis* group (Polyporales, Basidiomycota). Mycologia 109:217–230. https://doi.org/10.1080/00275514.2017.1300087
46. Runnel K, Spirin V, Miettinen O, Vlasák J, Dai YC, Ryvarden L, Larsson KH (2019) Morphological plasticity in brown rot fungi: *Antrodia* is redefined to encompass both poroid and corticioid species. Mycologia 111(5):871–883. https://doi.org/10.1080/00275514.2019.1640532
47. Justo A, Miettinen O, Floudas D, Ortiz-Santana B, Sjökvist E, Lindner D, Nakasone K, Niemelä T, Larsson KH, Ryvarden L, Hibbett DS (2017) A revised family-level classification of the *Polyporales* (*Basidiomycota*). Fungal Biol 121(9):798–824. https://doi.org/10.1016/j.funbio.2017.05.010
48. Chang TT, Chou WN (1995) Antrodia cinnamomea sp. nov. on Cinnamomum kanehirai in Taiwan. Mycol Res 99(6):756–758
49. Kotiranta H, Junninen K, Saarenoksa R, Kinnunen J, Kytövuori I (2010) Aphyllophorales & heterobasidiomycetes. The, 249–263
50. Drechsler-Santos ER, Kossmann T, da Cunha KM, Torres D, Vasco-Palacios A (2020) Antrodia neotropica. The IUCN red list of threatened species 2020:e.T172817995A172861297
51. The Global Fungal Red List (2025) Available online: https://redlist.info/en/iucn/welcome. Accessed on 25 Apr 2025
52. Li HX, Wang JJ, Lu CL, Gao YJ, Gao L, Yang ZQ (2022) Review of bioactivity, isolation, and identification of active compounds from *Antrodia cinnamomea*. Bioengineering 9(10):494. https://doi.org/10.3390/bioengineering9100494
53. Lu MC, El-Shazly M, Wu TY, Du YC, Chang TT, Chen CF, Hsu YM, Lai KH, Chiu CP, Chang FR, Wu YC (2013) Recent research and development of *Antrodia cinnamomea*. Pharmacol Ther 139(2):124–156. https://doi.org/10.1016/j.pharmthera.2013.04.001
54. Geethangili M, Tzeng YM (2011) Review of pharmacological effects of *Antrodia camphorata* and its bioactive compounds. Evid Based Complement Alternat Med 1:212641. https://doi.org/10.1093/ecam/nep108
55. Cherng IH, Chiang HC, Cheng MC, Wang YU (1995) Three new triterpenoids from *Antrodia cinnamomea*. J Nat Prod 58(3):365–371
56. Tong ZB, Cui XH, Wang J, Zhang CL, Zhang YY, Ren ZJ (2017) Constituents from solid-cultured *Antrodia camphorata*. Nat Prod Res 31(21):2564–2567. https://doi.org/10.1080/14786419.2017.1320785
57. Ganesan N, Baskaran R, Velmurugan BK, Thanh NC (2019) *Antrodia cinnamomea* – an updated minireview of its bioactive components and biological activity. J Food Biochem 43(8):e12936. https://doi.org/10.1111/jfbc.12936
58. Chang CJ, Lu CC, Lin CS, Martel J, Ko YF, Ojcius DM, Wu TR, Tsai YH, Yeh TS, Lu JJ, Lai HC, Young JD (2018) *Antrodia cinnamomea* reduces obesity and modulates the gut microbiota in high-fat diet-fed mice. Int J Obes 42(2):231–243. https://doi.org/10.1038/ijo.2017.149
59. Kumar K, Mehra R, Guiné RP, Lima MJ, Kumar N, Kaushik R, Ahmed N, Yadav AN, Kumar H (2021) Edible mushrooms: a comprehensive review on bioactive compounds with health benefits and processing aspects. Foods 10(12):2996. https://doi.org/10.3390/foods10122996
60. Yang HL, Liu HW, Shrestha S, Thiyagarajan V, Huang HC, Hseu YC (2021) *Antrodia salmonea* induces apoptosis and enhances cytoprotective autophagy in

colon cancer cells. Aging (Albany NY) 13(12):15964. https://doi.org/10.18632/aging.203019
61. Yang HL, Lin SW, Lee CC, Lin KY, Liao CH, Yang TY, Hseu YC (2015) Induction of Nrf2-mediated genes by *Antrodia salmonea* inhibits ROS generation and inflammatory effects in lipopolysaccharide-stimulated RAW264.7 macrophages. Food Funct 6(1):229–240
62. Li W, Wan P, Qiao J, Liu Y, Peng Q, Zhang Z, Sun B (2024) Current and further outlook on the protective potential of *Antrodia camphorata* against neurological disorders. Front Pharmacol 15:1372110. https://doi.org/10.3389/fphar.2024.1372110
63. Tain YL, Chang-Chien GP, Lin S, Hou CY, Hsu CN (2023) Renoprotective effects of solid-state cultivated *Antrodia cinnamomea* in juvenile rats with chronic kidney disease. Nutrients 15(21):4626. https://doi.org/10.3390/nu15214626
64. Wang JJ, Wu CC, Lee CL, Hsieh SL, Chen JB, Lee CI (2017) Antimelanogenic, antioxidant and antiproliferative effects of *Antrodia camphorata* fruiting bodies on B16-F0 melanoma cells. PLoS One 12(1):e0170924. https://doi.org/10.1371/journal.pone.0170924
65. Wu SH, Ryvarden L (1997) *Antrodia camphorata* ("niu-chang-chih"), new combination of a medicinal fungus in Taiwan. Bot Bull Acad Sinica Taipei 38:273–276
66. Huang GJ, Huang SS, Lin SS, Shao YY, Chen CC, Hou WC, Kuo YH (2010) Analgesic effects and the mechanisms of anti-inflammation of ergostatrien-3β-ol from *Antrodia camphorata* submerged whole broth in mice. J Agric Food Chem 58(12): 7445–7452
67. Wu F, Tohtirjap A, Fan LF, Zhou LW, Alvarenga RL, Gibertoni TB, Dai YC (2021) Global diversity and updated phylogeny of *Auricularia* (Auriculariales, Basidiomycota). Journal of Fungi 7(11):933. https://doi.org/10.3390/jof7110933
68. Bandara AR, Rapior S, Mortimer PE, Hyde KD (2015) *Auricularia* – a review of its traditional uses, bioactive compounds, and therapeutic potential. Mycosphere 6(5):621–637
69. Duncan E (1967) Morphological variation and taxonomy of *Auricularia* species. Mycologia 59(2):255–267
70. Montoya-Alvarez D (2011) *Auricularia* taxonomy and systematics. Mycol Prog 10(4): 421–432
71. Li H (2015) Species differentiation in *Auricularia* using morphological and molecular markers. Fungal Biol 119(5):467–478
72. Cheung PCK (2013) Mini-review on edible mushroom-derived β-glucans and their biological activities. J Agric Food Chem 61(1):7595–7601. https://doi.org/10.1016/j.fshw.2013.08.001
73. Ma Y, Wang C, Zhang Q, Peng X, Feng Y, Meng X (2018) Polysaccharides from *Auricularia* and their pharmacological effects. Int J Biol Macromol 118(C):1501–1510. https://doi.org/10.1016/j.phrs.2018.04.010
74. Fuller P, Guzmán G (2004) Traditional medicinal uses of *Auricularia* species. J Ethnopharmacol 94(1):231–238
75. Chang ST (1996) Mushroom research and development – equality and mutual benefit. Mushroom Biol Mushroom Products 16(1): 1–10
76. De Silva DD, Rapior S, Hyde KD et al (2013) Nutritional and therapeutic potential of *Auricularia* species. Int J Med Mushrooms 15(3): 219–237
77. Fan L, Ma R, Zhao M et al (2006) Antioxidant activity of *Auricularia auricula-judae* polysaccharides. Carbohydr Polym 65(2): 299–305
78. Kho HP, Chai TT, Wong FC (2009) Evaluation of antioxidant properties of *Auricularia auricula-judae*. Asian J Tradit Med 4(3): 75–83
79. Choi SW, Lee SS, Lee JH (2018) Antioxidant activities of *Auricularia auricula-judae* extracts. Food Sci Biotechnol 27(3):789–795
80. Sun L, Wang L, Li J (2010) Antioxidant properties of *Auricularia polytricha* extracts. Food Chem 119(3):1172–1178
81. Chang ST, Wasser SP (2017) The cultivation and environmental impact of mushrooms. Appl Microbiol Biotechnol 102(21): 8623–8630. https://doi.org/10.1093/acrefore/9780199389414.013.231
82. Umata H (1997) Symbiosis of *Auricularia polytricha* with *Erythrorchis ochobiensis*. New Phytol 136(2):303–312. https://doi.org/10.1007/BF02464092
83. Misaki A, Kakuta M, Sasaki T et al (1981) Structure of polysaccharides from *Auricularia* and their biological activities. Carbohydr Res 89(2):257–271
84. Yu T, Wu Q, Liang B, Wang J, Wu D, Shang X (2023) The current state and future prospects of *Auricularia auricula*'s polysaccharide processing technology portfolio. Molecules 28(2):582. https://doi.org/10.3390/molecules28020582

85. Sheu F, Chien PJ, Chien A (2004) Stability of *Auricularia* immunomodulatory proteins. Biochem Pharmacol 67(2):333–342
86. Yang BK, Ha JY, Jeong SC, Jeon YJ et al (2002) Hypolipidemic effect of an exo-biopolymer produced from submerged mycelial culture of *Auricularia polytricha* in rats. Biotechnol Lett 24:1319–1325
87. Wu Z, Zhang M, Yang H, Zhou H, Yang H (2018) Food bioscience production, physicochemical characterization and antioxidant activity of natural melanin from submerged cultures of the mushroom *Auricularia auricula*. Food Biosci 26:49–56. https://doi.org/10.1016/j.fbio.2018.09.008
88. Murata T, Suzuki Y, Ohta Y (2004) Adenosine from *Auricularia* and its effects on platelet aggregation. J Nat Prod 67(12): 2025–2029
89. Chen X, Wu G, Huang Z et al (2011) Polyphenolic compounds from *Auricularia* and their effects on cholesterol metabolism. Phytother Res 25(12):1876–1882
90. Meyer V, Basenko EY, Benz JP, Braus GH, Caddick MX, Csukai M, de Vries RP, Endy D, Frisvad JC, Cimerman NG, Haarmann T, Hadar Y, Hansen K, Johnson RI, Keller NP (2020) Growing a circular economy with fungal biotechnology: a white paper. Fungal Biol Biotechnol 7(5):1–23. https://doi.org/10.1186/s40694-020-00095-z
91. Hao Z, Zhang WE, Tian F, Wei R, Pan X (2022) Enhancing the nutritional and functional properties of *Auricularia auricula* through the exploitation of walnut branch waste. Foods 11(20):3242. https://doi.org/10.3390/foods11203242
92. Wang Y, Peng Z, Yang Y, Li Z, Wen Y, Liu M, Li S, Su L, Zhou Z, Zhu Y, Zhou N (2022) *Auricularia auricula* biochar supported γ-FeOOH nanoarrays for electrostatic self-assembly and pH-responsive controlled release of herbicide and fertilizer. Chem Eng J 437:1–12. https://doi.org/10.1016/j.cej.2022.134984
93. Regis MAJ, Geösel A, Geösel RM (2024) Cultivation of *Auricularia* species: a review of the history, health benefits, principles, practices, environmental conditions, research methods, and recent trends. Sydowia 76:21
94. Thongbai B, Rapior S, Hyde KD, Wittstein K, Stadler M (2015) *Hericium erinaceus*, an amazing medicinal mushroom. Mycol Prog 14:1–23. https://doi.org/10.1007/s11557-015-1105-4
95. Sholyavei S (2020) Study on different strains of *Hericium erinaceus* (Bull.: Fr) Pers. for their qualitative & quantitative traits. Doctoral dissertation, GB Pant University of Agriculture and Technology, Pantnagar-263145, Uttarakhand
96. Stamets P (2011) Growing gourmet and medicinal mushrooms. Ten Speed Press
97. Boddy L, Crockatt ME, Ainsworth AM (2011) Ecology of *Hericium cirrhatum*, *H. coralloides* and *H. erinaceus* in the UK. Fungal Ecol 4(2):163–173. https://doi.org/10.1016/j.funeco.2010.10.001
98. Index Fungorum (2025) Available online: http://www.indexfungorum.org/Names/Names.asp. Accessed on 15 Feb 2025
99. Species Fungorum (2025) Available online: http://www.speciesfungorum.org/Names/Names.asp. Accessed on 15 Feb 2025
100. Lu L, Li J, Cang Y (2002) PCR-based sensitive detection of medicinal fungi *Hericium* species from ribosomal internal transcribed spacer (ITS) sequences. Biol Pharm Bull 25(8):975–980. https://doi.org/10.1248/bpb.25.975
101. Kostanda E, Musa S, Pereman I (2024) Unveiling the chemical composition and biofunctionality of *Hericium* spp. Fungi: a comprehensive overview. Int J Mol Sci 25(11): 5949. https://doi.org/10.3390/ijms25115949
102. Van der Berg C (2018) Industrial and medicinal application of Reishi and Lion's Mane mushrooms. Doctoral dissertation, University of the Free State
103. Shuting CH (2024) Mushrooms: fungi for all seasons. Acta Edulis Fungi 31(1)
104. Kousar A, Khan HA, Farid S, Zhao Q, Zeb I (2024) Recent advances on environmentally sustainable valorization of spent mushroom substrate: a review. Biofuels Bioprod Biorefin 18(2):639–651. https://doi.org/10.1002/bbb.2559
105. Díaz-Godínez G, Téllez-Téllez M (2021) Mushrooms as edible foods. In: Fungi in sustainable food production. Springer International Publishing, Cham, pp 143–164
106. Singh A, Sit N (2024) Fungi-based meat analogs. In: Handbook of plant-based meat analogs. Academic Press, pp 99–119. https://doi.org/10.1016/B978-0-443-21846-0.00018-6
107. Loc NQ, Bui TK, Tu NM, Nguyen TT, Hoang ND, Nguyen TB, Minh LB, Ta TY, Quan NH, Ghosh SK (2024) Enhancing circular economy in the mushroom production chain: systematic literature review and field study in the central highlands of Vietnam. J Solid Waste Technol Manag 50(4):689–710.

https://doi.org/10.5276/jswtm/iswmaw/50S1/2024.689

108. Friedman M (2015) Chemistry, nutrition, and health-promoting properties of *Hericium erinaceus* (lion's mane) mushroom fruiting bodies and mycelia and their bioactive compounds. J Agric Food Chem 63(32): 7108–7123
109. Hobbs C (2023) The health and clinical benefits of medicinal fungi. In: Biochemical engineering and biotechnology of medicinal mushrooms. Advances in biochemical engineering/biotechnology, vol 184, pp 285–356. https://doi.org/10.1007/10_2023_230
110. Szućko-Kociuba I, Trzeciak-Ryczek A, Kupnicka P, Chlubek D (2023) Neurotrophic and neuroprotective effects of *Hericium erinaceus*. Int J Mol Sci 24(21):15960. https://doi.org/10.3390/ijms242115960
111. Roda E, Priori EC, Ratto D, De Luca F, Di Iorio C, Angelone P, Locatelli CA, Desiderio A, Goppa L, Savino E, Bottone MG (2021) Neuroprotective metabolites of *Hericium erinaceus* promote neuro-healthy aging. Int J Mol Sci 22(12):6379. https://doi.org/10.3390/ijms22126379
112. Llanaj X, Törös G, Hajdú P, Abdalla N, El-Ramady H, Kiss A, Solberg SØ, Prokisch J (2023) Biotechnological applications of mushrooms under the water-energy-food nexus: crucial aspects and prospects from farm to pharmacy. Foods 12(14):2671. https://doi.org/10.3390/foods12142671
113. You SW, Hoskin RT, Komarnytsky S, Moncada M (2022) Mushrooms as functional and nutritious food ingredients for multiple applications. ACS Food Sci Technol 2(8): 1184–1195
114. Kalichman J, Kirk PM, Matheny PB (2020) A compendium of generic names of agarics and *Agaricales*. Taxon 69(3):425–447. https://doi.org/10.1002/tax.12240
115. Psurtseva NV, Zmitrovich IV, Seelan JSS, Bulakh EM, Hughes KW, Petersen RH (2021) New data on morphology, physiology, and geographical distribution of *Lignomyces vetlinianus*, its identity with *Lentinus pilososquamulosus*, and sufficient phylogenetic distance from *Le. martianoffianus*. Mycol Prog 20(6):809–821. https://doi.org/10.1007/s11557-021-01701-z
116. Karunarathna SC, Yang ZL, Raspé O, Ko TK, Vellinga EC, Zhao RL, Bahkali AH, Chukeatirote E, Degreef J, Callac P, Hyde KD (2012) *Lentinus giganteus* revisited: new collections from Sri Lanka and Thailand. Mycotaxon 118(1):57–71. https://doi.org/10.5248/118.57
117. Bolhassan MH, Abdullah N, Sabaratnam V, Tsutomu H, Abdullah S, Rashid NN, Musa MY (2012) Diversity and distribution of Polyporales in peninsular Malaysia. Sains Malaysiana 41(2):155–161
118. Fabros JA, Dulay RMR, De Leon A, Kalaw S, Reyes R (2022) Distribution, cultivation, nutritional composition, and bioactivities of *Lentinus* (*Polyporaceae*, *Basidiomycetes*): a review. Curr Res Environ Appl Mycol 12: 170–219. https://doi.org/10.5943/cream/12/1/13
119. Karunarathna SC, Yang ZL, Zhao RL, Vellinga EC, Bahkali AH, Chukeatirote E, Hyde KD (2011) Three new species of *Lentinus* from northern Thailand. Mycol Prog 10: 389–398. https://doi.org/10.1007/s11557-010-0701-6
120. Thetsrimuang C, Khammuang S, Chiablaem K, Srisomsap C, Sarnthima R (2011) Antioxidant properties and cytotoxicity of crude polysaccharides from *Lentinus polychrous* Lév. Food Chem 128(3): 634–639. https://doi.org/10.1016/j.foodchem.2011.03.077
121. Andrade MCND, Sales-Campos C, Carvalho CSM, Aguiar LVB, Minhoni MTDA (2013) Uso de resíduos madeireiros da Amazônia brasileira no cultivo *in vitro* de *Lentinus strigosus*. Ambiência 9(1):189–196. https://doi.org/10.5777/ambiencia.2013.01.01nt
122. Bertéli MBD, Oliveira Filho OBQ, Freitas JDS, Bortolucci WC, Silva GR, Gazim ZC, Lívero FAR, Lovato ECW, Valle JS, Linde GA, Barros L, Reis FS, Ferreira ICFR, Paccola-Meirelles LD, Colauto NB (2021) *Lentinus crinitus* basidiocarp stipe and pileus: chemical composition, cytotoxicity and antioxidant activity. Eur Food Res Technol 247: 1355–1366
123. Njouonkou AL, Mossebo DC, Akoa A (2013) The genera *Lentinus* and *Panus* in the Dja biosphere reserve and its periphery, Cameroon. Kew Bull 68:517–521. https://doi.org/10.1007/s12225-013-9461-1
124. Asif M, Saba M, Maula F, Raza M (2024) *Lentinus punjabensis* (Polyporaceae), a new species from Lal Suhanra National Park, southern Punjab, Pakistan. Phytotaxa 642(2):169–181. https://doi.org/10.11646/phytotaxa.642.2.5
125. Kakoti M, Hazarika DJ, Kumar A, Barooah M, Modi MK, Bhattacharyya A, Boro RC (2021) Genetic diversity and DNA barcoding of wild mushrooms from North-east India

126. Seelan JSS, Justo A, Nagy LG, Grand EA, Redhead SA, Hibbett D (2015) Phylogenetic relationships and morphological evolution in *Lentinus*, *Polyporellus*, and *Neofavolus*, emphasizing southeastern Asian taxa. Mycologia 107(3):460–474. https://doi.org/10.3852/14-084
127. Serbent MP, Guimarães DKS, Drechsler-Santos ER, Helm CV, Giongo A, Tavares LBB (2020) Growth, enzymatic production and morphology of the white-rot fungi *Lentinus crinitus* (L.) Fr. upon 2, 4-D herbicide exposition. Int J Environ Sci Technol 17: 2995–3012. https://doi.org/10.1007/s13762-020-02693-1
128. Valle JS, Vandenberghe LPS, Santana TT, Almeida PH, Pereira AM, Linde GA, Colauto NB, Soccol CR (2014) Optimum conditions for inducing laccase production in *Lentinus crinitus*. Genet Mol Res 13(4):8544–8551
129. Conceição TA, Koblitz MGB, Kamida HM, Góes-Neto A (2017) Study of the production of *Lentinus crinitus* (L.) Fr. Lignolytic enzymes grown on agro-industrial waste. Adv Biosci Biotechnol 8:259–272. https://doi.org/10.4236/abb.2017.88019
130. Kalaw SP, Dulay RMR, Damaso EJ, Ramos JC, del Rosario MAG, Abon MD, De Leon AM, Undan JR, Reyes RG (2021) Optimization of mycelial culture conditions and fructification of *Lentinus* species using rice straw and sawdust-based substrates. Stud Fungi 6(1):519–530. https://doi.org/10.5943/sif/6/1/42
131. Hsu CA, Wen TN, Su YC, Jiang ZB, Chen CW, Shyur LF (2012) Biological degradation of anthroquinone and azo dyes by a novel laccase from *Lentinus* sp. Environ Sci Technol 46(9):5109–5117
132. Coelho GD, Silva MA, de Melo Pinheiro MA, Nadvorny D, Costa Amador V, Maia RT (2024) *In silico* and *in vitro* assays suggests Congo red dye degradation by a *Lentinus* sp. laccase enzyme. J Biomol Struct Dyn 42(7):3802–3813. https://doi.org/10.1080/07391102.2023.2216282
133. Bayramoğlu G, Arıca MY (2008) Removal of heavy mercury (II), cadmium (II) and zinc (II) metal ions by live and heat-inactivated *Lentinus edodes* pellets. Chem Eng J 143(1–3):133–140. https://doi.org/10.1016/j.cej.2008.01.002
134. Kadimaliev DA, Revin VV, Atykyan NA, Nadezhina OS, Parshin AA (2011) The role of laccase and peroxidase of *Lentinus* (*Panus*) tigrinus fungus in biodegradation of high phenol concentrations in liquid medium. Appl Biochem Microbiol 47:66–71. https://doi.org/10.1134/S0003683811010066
135. Sudirman LI, Lefebvre G, Kiffer E, Botton B (1994) Purification of antibiotics produced by *Lentinus squarrosulus* and preliminary characterization of a compound active against *Rigidoporus lignosus*. Curr Microbiol 29:1–6. https://doi.org/10.1007/BF01570183
136. Liang J, Zhang M, Li X, Yue Y, Wang X, Han M, Yue T, Wang Z, Gao Z (2024) Structure and immunomodulatory activity of *Lentinus edodes* polysaccharides modified by probiotic fermentation. Food Sci Human Wellness 13(1):421–433. https://doi.org/10.26599/FSHW.2022.9250036
137. Omar NA, Abdullah N, Kuppusamy UR, Abdulla MA, Sabaratnam V (2011) Nutritional composition, antioxidant activities, and antiulcer potential of *Lentinus squarrosulus* (Mont.) mycelia extract. Evid Based Complement Alternat Med 2011(1):539356. https://doi.org/10.1155/2011/539356
138. Thetsrimuang C, Khammuang S, Sarnthima R (2011) Antioxidant activity of crude polysaccharides from edible fresh and dry mushroom fruiting bodies of *Lentinus* sp. strain RJ-2. Int J Pharmacol 7(1):58–65. https://doi.org/10.3923/ijp.2011.58.65
139. dos Santos Filho JR, da Silva SÉ, Linde GA, Colauto NB, Gonçalves RAC, de Oliveira AJB (2023) *Lentinus crinitus*: traditional use, phytochemical and pharmacological activities, and industrial and biotechnological applications. Food Res Int 172:113179. https://doi.org/10.1016/j.foodres.2023.113179
140. Flórez-Sampedro L, Zapata W, Orozco LP, Mejía AI, Arboleda C, Rugeles MT (2016) *In vitro* anti-HIV-1 activity of the enzymatic extract enriched with laccase produced by the fungi *Ganoderma* sp. and *Lentinus* sp. Vitae 23(2):109–118
141. Bertéli MBD, Barros L, Reis FS, Ferreira ICFR, Glamočlija J, Soković M et al (2021) Antimicrobial activity, chemical composition and cytotoxicity of *Lentinus crinitus* basidiocarp. Food Funct 12(15):6780–6792. https://doi.org/10.1007/s00217-021-03713-1
142. Maestre-Reyna M, Liu WC, Jeng WY, Lee CC, Hsu CA, Wen TN, Wang AHJ, Shyur LF (2015) Structural and functional roles of glycosylation in fungal laccase from *Lentinus* sp. PLoS One 10(4):e0120601. https://doi.org/10.1371/journal.pone.0120601
143. Naskar A, Saha R, Basak G, Acharya K (2025) Therapeutic scrutiny of *Lentinus polychrous* with attention to its antioxidant, antimicrobial, and anticancer attributes. Appl Biochem

Biotechnol 1–28. https://doi.org/10.1007/s12010-024-05146-4

144. Julca-Canto M, Aguilar-Pérez MM, Rios N, Sousa JPB, Cubilla-Rios L (2016) Additional new natural products produced by *Lentinus strigellus*: a biotechnological approach. Tetrahedron Lett 57(6):650–653. https://doi.org/10.1016/j.tetlet.2015.12.104
145. De Freitas JDS, Bertéli MBD, Neto JC, Gasparotto ES, Gonçalves AC Jr et al (2023) Biosorption of methylene blue by residue from *Lentinus crinitus* mushroom cultivation. World J Microbiol Biotechnol 39:110. https://doi.org/10.1007/s11274-023-03562-y
146. Jebun NX, Al Mamun A, Alam MZ, Raus RA (2018) Production and stability of myco-flocculants from *Lentinus squarrosulus* RWF5 and *Simplicillium obclavatum* RWF6 for reduction of water turbidity. IIUM. Eng J 19(1):48–58. https://doi.org/10.31436/iiumej.v19i1.843
147. Gaitán-Hernández R, López-Peña D, Esqueda M, Gutiérrez A (2019) Review of bioactive molecules production, biomass, and basidiomata of shiitake culinary-medicinal mushrooms, *Lentinus edodes* (Agaricomycetes). Int J Med Mushrooms 21(9): 841–850. https://doi.org/10.1615/IntJMedMushrooms.2019031849
148. López-Legarda X, Arboleda-Echavarría C, Parra-Saldívar R, Rostro-Alanis M, Alzate JF, Villa-Pulgarín JA, Segura-Sánchez F (2020) Biotechnological production, characterization and *in vitro* antitumor activity of polysaccharides from a native strain of *Lentinus crinitus*. Int J Biol Macromol 164:3133–3144. https://doi.org/10.1016/j.ijbiomac.2020.08.191
149. Vaquiro HA, Suárez H, Murillo W (2023) Probiotic growth-stimulating capacity and antimicrobial activities of aqueous extracts of *Lentinus crinitus* (L.) Fr (polyporales, basidiomycota). Heliyon 9(8):e18738
150. Zou H, Zhang J, Liu J, Buyukada M, Evrendilek F, Liang G (2020) Pyrolytic behaviors, kinetics, decomposition mechanisms, product distributions and joint optimization of *Lentinus edodes* stipe. Energy Convers Manag 213:112858. https://doi.org/10.1016/j.enconman.2020.112858
151. Xu X, Su X, Zhou L, Yu H, Zhang J (2024) Instance segmentation of *Lentinus edodes* images based on YOLOv5seg-BotNet. Agronomy 14(8):1808. https://doi.org/10.3390/agronomy14081808
152. Du XH, Yang ZL (2021) Mating systems in true morels (*Morchella*). Microbiol Mol Biol Rev 85:e00220–e0e320. https://doi.org/10.1128/MMBR.00220-20
153. Liu Q, Ma H, Zhang Y, Dong C (2018) Artificial cultivation of true morels: current state, issues and perspectives. Crit Rev Biotechnol 38(2):259–271. https://doi.org/10.1080/07388551.2017.1333082
154. Kuo M, Dewsbury DR, O'Donnell K, Carter MC, Rehner SA, Moore JD, Volk TJ (2012) Taxonomic revision of true morels (*Morchella*) in Canada and the United States. Mycologia 104(5):1159–1177. https://doi.org/10.3852/11-375
155. Du XH, Zhao Q, O'Donnell K, Rooney AP, Yang ZL (2012) Multigene molecular phylogenetics reveals true morels (*Morchella*) are especially species-rich in China. Fungal Genet Biol 49:455–469. https://doi.org/10.1016/j.fgb.2012.03.006
156. Loizides M, Alvarado PA, Moreau PA, Assyov B, Halasů V, Stadler M, Rinaldi A, Marques G, Zervakis GI, Borovička J, Van Vooren N (2022) Has taxonomic vandalism gone too far? A case study, the rise of the pay-to-publish model and the pitfalls of *Morchella* systematics. Mycol Prog 21(1):7–38. https://doi.org/10.1007/s11557-021-01755-z
157. Cravero M, Bonito G, Chain PS, Bindschedler S, Junier P (2024) A new species of true morel from Switzerland: *Morchella helvetica* sp. nov. Mycologia 116(6): 1101–1109. https://doi.org/10.1080/00275514.2024.2397932
158. Pilz D, Rebecca ML, Susan A, Luis VR, Shannon B, Tricia W, Parks CG, Erika MF, Blaze B (2007) Ecology and management of morels harvested from the forests of western North America. General Technical Report PNW-GTR-710 Portland. U.S. Department of Agriculture, Forest Service: Portland, OR, USA
159. Pilz D, Weber NS, Carter MC, Parks CG, Molina R (2004) Productivity and diversity of morel mushrooms in healthy, burned, and insect-damaged forests of northeastern Oregon. For Ecol Manag 198:367–386. https://doi.org/10.1016/j.foreco.2004.05.028
160. Sa W, Qiao J, Gao Q, Li Z, Shang Q (2022) DNA barcoding and species classification of *Morchella*. Genes 13(10):1806. https://doi.org/10.3390/genes13101806
161. Li Y, Chen H, Zhang X (2023) Cultivation, nutritional value, bioactive compounds of *Morchella*, and their health benefits: a systematic review. Front Nutr 10:1159029. https://doi.org/10.3389/fnut.2023.1159029

162. Soylu MK (2024) Isolation and molecular identification of the pure culture of *Morchella* collected from Türkiye and its characteristics. Horticulturae 10(10):1020. https://doi.org/10.3390/horticulturae10101020
163. Yu FM, Jayawardena RS, Luangharn T, Zeng XY, Li CJY, Bao SX, Ba H, Zhou DQ, Tang SM, Hyde KD, Zhao Q (2024) Species diversity of fungal pathogens on cultivated mushrooms: a case study on morels (*Morchella*, *Pezizales*). Fungal Divers 125(1):157–220. https://doi.org/10.1007/s13225-023-00531-6
164. Tietel Z, Masaphy S (2018) True morels (*Morchella*) – nutritional and phytochemical composition, health benefits and flavor: a review. Crit Rev Food Sci Nutr 58(11): 1888–1901. https://doi.org/10.1080/10408398.2017.1285269
165. Taşkin H, Süfer Ö, Attar SH, Bozok F, Baktemur G, Büyükalaca S, Kafkas NE (2021) Total phenolics, antioxidant activities and fatty acid profiles of six *Morchella* species. J Food Sci Technol 58:692–700. https://doi.org/10.1007/s13197-020-04583-3
166. Chen Q, Che C, Yang S, Ding P, Si M, Yang G (2022) Anti-inflammatory effects of extracellular vesicles from *Morchella* on LPS-stimulated RAW2647 cells via the ROS-mediated p38 MAPK signaling pathway. Mol Cell Biochem 478:317–327. https://doi.org/10.1007/s11010-022-04508-y
167. Xiong C, Li Q, Chen C, Chen Z, Huang W (2016) Neuroprotective effect of crude polysaccharide isolated from the fruiting bodies of *Morchella importuna* against H2O2-induced PC12 cell cytotoxicity by reducing oxidative stress. Biomed Pharmacother 83:569–576. https://doi.org/10.1016/j.biopha.2016.07.016
168. Zhang J, Zhao J, Liu G, Li Y, Liang L, Liu X, Xu X, Wen C (2023) Advance in *Morchella* sp. polysaccharides: isolation, structural characterization and structure-activity relationship: a review. Int J Biol Macromol 247: 125819. https://doi.org/10.1016/j.ijbiomac.2023.125819
169. Wu H, Chen J, Li J, Liu Y, Park HJ, Yang L (2021) Recent advances on bioactive ingredients of *Morchella esculenta*. Appl Biochem Biotechnol 193:4197–4213. https://doi.org/10.1007/s12010-021-03670-1
170. Ullah TS, Firdous SS, Mehmood A, Swati JQ, Usman M, Khalid AN (2022) Detection of bioactive compounds and amino acids from fruiting bodies of *Morchella tridentina*. Acta Sci Polonorum Hortorum Cultus 21(1): 103–114. https://doi.org/10.24326/asphc.2022.1.9
171. Haq FU, Imran M, Saleem S, Aftab U, Muazzam A, Rafi A, Jamal M, Safi SZ (2025) Chemical characterization and cytotoxic effect of three edible fungi (*Morchella*) against breast cancer cells: a therapeutic approach. Kuwait J Sci 52(1):100285. https://doi.org/10.1016/j.kjs.2024.100285
172. Menolli N Jr, Breternitz BS, Capelari M (2014) The genus *Pleurotus* in Brazil: a molecular and taxonomic overview. Mycoscience 55:378–389. https://doi.org/10.1016/j.myc.2013.12.001
173. Suwannarach N, Kumla J, Satienperakul K, Sungpalee W, Sri-Ngernyuang K, Lumyong S (2020) *Pleurotus sirindhorniae* (Pleurotaceae, Agaricales), a new species from northern Thailand. Phytotaxa 460:285–295. https://doi.org/10.11646/phytotaxa.460.4.6
174. Barh A, Sharma VP, Annepu SK, Kamal S, Sharma S, Bhatt P (2019) Genetic improvement in *Pleurotus* (oyster mushroom): a review. 3 Biotech 9:322. https://doi.org/10.1007/s13205-019-1854-x
175. Li J, Han LH, Liu XB, Zhao ZW, Yang ZL (2020) The saprotrophic *Pleurotus ostreatus* species complex: late Eocene origin in East Asia, multiple dispersal, and complex speciation. IMA Fungus 11:1–21. https://doi.org/10.1186/s43008-020-00031-1
176. Zervakis GI, Venturella G, Fryssouli V, Inglese P, Polemis E, Gargano ML (2019) *Pleurotus opuntiae* revisited – an insight to the phylogeny of dimitic *Pleurotus* species with emphasis on the *P. djamor* complex. Fungal Biol 123:1–11. https://doi.org/10.1016/j.funbio.2018.12.005
177. Torres-Martínez BDM, Vargas-Sánchez RD, Torrescano-Urrutia GR, Esqueda M, Rodríguez-Carpena JG, Fernández-López J et al (2022) *Pleurotus* genus as a potential ingredient for meat products. Foods 11:779. https://doi.org/10.3390/foods11060779
178. Buratti S, Girometta CE, Savino E, Gorjón SP (2023) An example of the conservation of wood decay fungi: the new research culture collection of corticioid and polyporoid strains of the University of Salamanca (Spain). Forests 14(10):2029. https://doi.org/10.3390/f14102029
179. Grabarczyk M, Mączka W, Wińska K, Uklańska-Pusz C (2019) Mushrooms of the *Pleurotus* genus – properties and application. Biotechnol Food Sci 83:13–30. https://doi.org/10.34658/bfs.2019.83.1.13-30

180. Kewessa G, Tolera M, Dejene T, Martín-Pinto P (2023) Exploring edible mushrooms in Ethiopian forests: an insight for ecological resilience and climate change adaptation. In: Handbook of nature-based solutions to mitigation and adaptation to climate change. Springer International Publishing, Cham, pp 1–17. https://doi.org/10.1007/978-3-030-98067-2_138-1

181. Ghafoor A, Niazi AR (2024) *Pleurotus* spp: an ultimate solution to the emerging calamities of the world. N Z J Bot:1–38. https://doi.org/10.1080/0028825X.2024.2387185

182. Venturella G, Ferraro V, Cirlincione F, Gargano ML (2021) Medicinal mushrooms: bioactive compounds, use, and clinical trials. Int J Mol Sci 22(2):634. https://doi.org/10.3390/ijms22020634

183. Barbosa JR, dos Santos Freitas MM, da Silva Martins LH, de Carvalho Junior RN (2020) Polysaccharides of mushroom *Pleurotus* spp.: new extraction techniques, biological activities and development of new technologies. Carbohydr Polym 229:115550. https://doi.org/10.1016/j.carbpol.2019.115550

184. Vetvicka V, Gover O, Karpovsky M, Hayby H, Danay O, Ezov N et al (2019) Immune-modulating activities of glucans extracted from *Pleurotus ostreatus* and *Pleurotus eryngii*. J Funct Foods 54:81–91. https://doi.org/10.1016/j.jff.2018.12.034

185. Zhang Y, Yang X, Jin G, Yang X, Zhang Y (2016) Polysaccharides from *Pleurotus ostreatus* alleviate cognitive impairment in a rat model of Alzheimer's disease. Int J Biol Macromol 92:935–941. https://doi.org/10.1016/j.ijbiomac.2016.08.008

186. Xu D, Wang H, Zheng W, Gao Y, Wang M, Zhang Y, Gao Q (2016) Characterization and immunomodulatory activities of polysaccharide isolated from *Pleurotus eryngii*. Int J Biol Macromol 92:30–36. https://doi.org/10.1016/j.ijbiomac.2016.07.016

187. Dos Reis EE, Schenkel PC, Camassola M (2022) Effects of bioactive compounds from *Pleurotus* mushrooms on COVID-19 risk factors associated with the cardiovascular system. J Integr Med 20:385. https://doi.org/10.1016/j.joim.2022.07.002

188. Sasidhara R, Thirunalasundari T (2014) Phytochemicals and antioxidant potentials of *Pleurotus djamor*. J Chem Pharm Res 6(4): 950–953

189. Jagadeesh R, Babu G, Lakshmanan H, Oh OM, Jang JK, Kong KW, Raaman N (2020) Bioactive sterol derivatives isolated from the *Pleurotus djamor* var. *roseus* induced apoptosis in cancer cell lines. Cardiovasc Hematol Agents Med Chem 18(2):124–134. https://doi.org/10.2174/1871525718666200303123557

190. Calabretti A, Mang SM, Becce A, Castronuovo D, Cardone L, Candido V, Camele I (2021) Comparison of bioactive substances content between commercial and wild-type isolates of *Pleurotus eryngii*. Sustainability 13(7):3777. https://doi.org/10.3390/su13073777

191. Oda O, Muhammed A (2024) Antibacterial effect of *Pleurotus ostreatus* on selected bacterial isolates. Sahel J Life Sci FUDMA 2(1): 132–136. https://doi.org/10.33003/sajols-2024-0201-015

192. Mkhize SS, Machaba KE, Simelane MB, Pooe OJ (2022) Mushroom-derived products as alternative antimalarial therapeutics: a review. In: Drug development for malaria: novel approaches for prevention and treatment, vol 24, pp 235–249

193. Raman J, Jang KY, Oh YL, Oh M, Im JH, Lakshmanan H, Sabaratnam V (2021) Cultivation and nutritional value of prominent *Pleurotus* spp.: an overview. Mycobiology 49:1–14. https://doi.org/10.1080/12298093.2020.1835142

194. Muswati C, Simango K, Tapfumaneyi L, Mutetwa M, Ngezimana W (2021) The effects of different substrate combinations on growth and yield of oyster mushroom (*Pleurotus ostreatus*). Int J Agron 2021(1): 9962285. https://doi.org/10.1155/2021/9962285

195. Bellettini MB, Fiorda FA, Maieves HA, Teixeira GL, Ávila S, Hornung PS, Maccari Júnior A, Ribani RH (2019) Factors affecting mushroom *Pleurotus* spp. Saudi J Biol Sci 26: 633–646. https://doi.org/10.1016/j.sjbs.2016.12.005

196. Jatwa TK, Apet KT, Wagh SS, Sayyed KS, Rudrappa KB, Sornapriya SP (2016) Evaluation of various agro-wastes for production of *Pleurotus* spp. (*P. florida*, *P. sajor-caju* and *P. eous*). J Pure Appl Microbiol 10:2783–2792. https://doi.org/10.22207/JPAM.10.4.37

197. Fries E (1815) Observationes mycologicae. G. Bonnieri

198. Khatua S, Paul S, Acharya K (2013) Mushroom as the potential source of new generation of antioxidant: a review. Res J Pharm Technol 6(5):496–505

199. Mišković J, Rašeta M, Krsmanović N, Karaman M (2023) Update on mycochemical profile and selected biological activities of genus *Schizophyllum* Fr. 1815. Microbiol Res 14(1):

409–429. https://doi.org/10.3390/microbiolres14010031

200. Raper JR, Miles PG (1958) The genetics of *Schizophyllum commune.* Genetics 43(3): 530–546. https://doi.org/10.1093/genetics/43.3.530
201. Cooke WB (1961) The genus *Schizophyllum.* Mycologia 53(6):575–599. https://doi.org/10.1080/00275514.1961.12017987
202. Nakasone KK (1996) Morphological and molecular studies on *Auriculariopsis albomellea* and *Phlebia albida* and a reassessment of *A. ampla.* Mycologia 88(5):762–775. https://doi.org/10.1080/00275514.1996.12026714
203. Olivo-Aranda F, Herrera-Suárez T (1994) The species of *Schizophyllum* in Mexico, their ecological distribution and ethnomycological importance. Phytotaxa 413(2):137–148
204. Chowdhary A, Randhawa HS, Gaur SN, Agarwal K, Kathuria S, Roy P, Meis JF (2013) *Schizophyllum commune* as an emerging fungal pathogen: a review and report of two cases. Mycoses 56(1):1–10. https://doi.org/10.1111/j.1439-0507.2012.02190.x
205. Takemoto S, Nakamura H, Imamura Y, Shimane T (2010) *Schizophyllum commune* as a ubiquitous plant parasite. Jpn Agric Res Quar 44(4):357–364. https://doi.org/10.6090/jarq.44.357
206. Ansari E, Alvandi H, Kianirad S, Hatamian-Zarmi A, Mokhtari-Hosseini ZB (2024) Research progress on production and biomedical applications of Schizophyllan as a tailor-made polysaccharide: a review. Carbohydr Polym 348(A):122770. https://doi.org/10.1016/j.carbpol.2024.122770
207. Yelithao K, Surayot U, Lee C, Palanisamy S, Prabhu NM, Lee J, You S (2019) Studies on structural properties and immune-enhancing activities of glycomannans from *Schizophyllum commune.* Carbohydr Polym 218:37–45. https://doi.org/10.1016/j.carbpol.2019.04.057
208. Du B, Yang Y, Bian Z, Xu B (2017) Characterization and anti-inflammatory potential of an exopolysaccharide from submerged mycelial culture of *Schizophyllum commune.* Front Pharmacol 8:252. https://doi.org/10.3389/fphar.2017.00252
209. Leathers TD, Nunnally MS, Stanley AM, Rich JO (2016) Utilization of corn fiber for production of schizophyllan. Biomass Bioenergy 95:132–136. https://doi.org/10.1016/j.biombioe.2016.10.001
210. Klaus A, Kozarski M, Niksic M, Jakovljevic D, Todorovic N, Van Griensven LJ (2011) Antioxidative activities and chemical characterization of polysaccharides extracted from the basidiomycete *Schizophyllum commune.* LWT Food Sci Technol 44(10):2005–2011. https://doi.org/10.1016/j.lwt.2011.05.010
211. Schulz D, Rau U, Wagner F (1992) Characteristics of films prepared from native and modified branched β-1, 3-D-glucans. Carbohydr Polym 18(4):295–299. https://doi.org/10.1016/0144-8617(92)90095-8
212. Appels FV, Dijksterhuis J, Lukasiewicz CE, Jansen KM, Wösten HA, Krijgsheld P (2018) Hydrophobin gene deletion and environmental growth conditions impact mechanical properties of mycelium by affecting the density of the material. Sci Rep 8(1): 4703. https://doi.org/10.1038/s41598-018-23171-2
213. Schey JA (1983) Tribology in metalworking friction, lubrication and Wear. American Society of Metals, Metals Park
214. Arun G, Eyini M, Gunasekaran P (2014) Green synthesis of silver nanoparticles using the mushroom fungus *Schizophyllum commune* and its biomedical applications. Biotechnol Bioprocess Eng 19:1083–1090. https://doi.org/10.1007/s12257-014-0071-z
215. Tovar-Herrera OE, Martha-Paz AM, Pérez-LLano Y, Aranda E, Tacoronte-Morales JE, Pedroso-Cabrera MT et al (2018) *Schizophyllum commune*: an unexploited source for lignocellulose degrading enzymes. MicrobiologyOpen 7(3):e00637. https://doi.org/10.1002/mbo3.637
216. Raper CA (1988) *Schizophyllum commune*, a model for genetic studies of the Basidiomycotina. Adv Plant Pathol 6:511–522. https://doi.org/10.1016/B978-0-12-033706-4.50038-4
217. Ohm RA, de Jong JF, Lugones LG, Aerts A, Kothe E, Stajich JE, de Vries RP, Record E, Levasseur A, Baker SE, Bartholomew KA, Coutinho PM, Erdmann S, Fowler TJ, Gathman AC, Lombard V, Henrissat B, Knabe N, Kües U, Lilly WW, Lindquist E, Lucas S, Magnuson JK, Piumi F, Raudaskoski M, Salamov A, Schmutz J, Schwarze FWMR, van Kuyk PA, Horton JS, Grigoriev IV, Wösten HAB (2010) Genome sequence of the model mushroom *Schizophyllum commune.* Nat Biotechnol 28(9):957–963. https://doi.org/10.1038/nbt.1643
218. Lu H, Lou H, Hu J, Liu Z, Chen Q (2020) Macrofungi: a review of cultivation strategies,

bioactivity, and application of mushrooms. Compr Rev Food Sci Food Saf 19(5): 2333–2356. https://doi.org/10.1111/1541-4337.12602

219. Karunarathna SC, Prasannath K, Lu W, Hapuarachchi KK (2024) Ganoderma: bridging traditional wisdom with modern innovation in medicinal mushroom and dietary supplement industry. N Z J Bot:1–60. https://doi.org/10.1080/0028825X.2024.2410486

220. Díaz R, Díaz-Godínez G (2022) Substrates for mushroom, enzyme and metabolites production: a review. J Environ Biol 43(3): 350–359

221. Li Q, Liu J, Gadd GM (2020) Fungal bioremediation of soil co-contaminated with petroleum hydrocarbons and toxic metals. Appl Microbiol Biotechnol 104(21): 8999–9008. https://doi.org/10.1007/s00253-020-10854-y

Chapter 2

Basic Mushroom Extracts

Pablo Navarro Simarro

Abstract

Mushrooms are a rich source of bioactive compounds, including phenolics, terpenes, alkaloids, and polysaccharides like β-glucans and chitin, which hold significant potential for the food industry. Understanding the fundamental extraction techniques using food-grade solvents is crucial for producing extracts enriched with these bioactive compounds. This chapter provides a detailed guide on preparing various mushroom extracts: an aqueous extract rich in β-glucans with prebiotic effects, an ethanol extract for isolating phenolic compounds, and an alkaline extract for recovering polysaccharides like chitin. These extraction methods are essential for subsequent compound identification, evaluating their bioactive properties, and exploring their applications in food, nanotechnology, and pharmaceuticals.

Key words Mushroom extracts, Polysaccharides, Bioactive compounds, Chitin, β-glucan, Prebiotic

1 Introduction

Edible mushrooms are a widely consumed food all over the world due to their taste, versatility, and the new cultivation techniques developed in the last century that have allowed for scaling up the world's production. Despite the existence of approximately 2800 useful wild mushrooms (those that are edible or have medicinal, cosmetic, or other uses) [1], the genera *Lentinula*, *Pleurotus*, *Auricularia*, *Agaricus*, and *Flammulina* dominate the industry, accounting for about 85% of worldwide mushroom production [2]. These crops generate a large amount of by-products, such as stems and unmarketable mushrooms that have a nutritional profile almost identical to commercialized fungi [3]. As a way to exploit these by-products, numerous extraction methods have been developed capable of concentrating different fractions of polysaccharides with prebiotic effect, extracting phenolic compounds with antioxidant power or obtaining peptides with high flavoring properties. These methods, which often involve sequential extractions with different solvents, also pave the way for the valorization of wild

Pablo Navarro Simarro (ed.), *Fungi*, Methods and Protocols in Food Science, https://doi.org/10.1007/978-1-0716-4989-3_2,
© The Author(s), under exclusive license to Springer Science+Business Media, LLC, part of Springer Nature 2026

Fig. 1 Lyophilized extracts of mushroom (*Agaricus bisporus*) prepared by different methods. The display includes lyophilized and powdered mushroom (**a**), aqueous extract (**b**), ethanolic extract (**c**) and alkaline extract (**d**). The aqueous extract shows darkening due to the action of polyphenol oxidases

mushrooms, which are rich in therapeutic compounds such as polysaccharides, phenols, steroids, and alkaloids.

In the human food field, commercial mushroom extracts are often used for the formulation of functional foods due to their safety and wide availability. Among the easiest are the aqueous extracts, which stand out for their affinity for prebiotic polysaccharides such as various glucans [4, 5] (Fig. 1). These extracts can undergo further treatment with other solvents to remove proteins or separate lower molecular weight compounds through dialysis. The most commonly used aqueous fungal extract in the food industry is derived from *Pleurotus ostreatus*, either from the caps or stems, to enhance products such as yogurts to improve its syneresis and to enhance the viability of its probiotic bacteria [6, 7].

Another efficient method for purifying polysaccharides from edible fungi is alkaline extraction using sodium hydroxide. This technique allows for the extraction of both water-soluble and hydrophobic polysaccharides, such as α and β glucans with varying degrees of polymerization and branching, as well as chitin—an important polysaccharide for biofilm production [8]. While this method is straightforward and effective for obtaining different polysaccharide fractions for analysis or use in the food industry, its primary drawback is the significant environmental impact of sodium hydroxide as a contaminant.

In addition to polysaccharides, extracting bioactive compounds like phenols is crucial for the full utilization of mushrooms. To achieve this, varying concentrations of ethanol-water mixtures have proven effective in obtaining extracts rich in antioxidant compounds such as gallic acid, caffeic acid, catechins, and myricetin, among others [9]. These compounds are found in greater abundance and diversity in wild mushrooms, which hold potential for domestication and large-scale valorization.

More advanced techniques, such as ultrasonic-assisted extraction, microwave-assisted extraction, subcritical water extraction, and enzyme-assisted extraction, can significantly enhance yield or refine the purification process [10]. However, methods described in the literature often vary depending on the intended use of the extract—whether for supplementation in food products or purely for chemical analysis. Despite this variability, this chapter will focus on straightforward, cost-effective, and efficient methods to obtain extracts rich in bioactive compounds for analysis, evaluation of bioactive properties, or application in food formulation.

2 Materials

1. Distilled or ultrapure water.
2. Absolute ethanol (*see* **Note 1**). Mix with water to prepare a 4:1 ethanol-to-water stock solution for the ethanolic extract.
3. A 1 M solution of sodium hydroxide (NaOH).
4. Fresh, dried, or lyophilized mushrooms.

3 Methods

3.1 Pretreatment

To optimize the extraction process, the mushrooms should be chopped to increase surface area, then freeze-dried, and finally ground using a hand mill. The resulting mushroom powder can be sieved to homogenize it and remove larger particles. This is particularly important for extracts where we work with pellets, such as in the alkaline extract. Finally, the freeze-dried mushroom powders can be stored at room temperature in a dry, light-protected place for several months. If a freeze dryer is not available (*see* **Note 2**).

3.2 Aqueous Extract

1. Mix between 2 and 10 g of freeze-dried mushroom powder in 250 mL of water (see **Note 3**).
2. Heat the mixture on a hot plate with magnetic stirring for 3 h at 80 °C (*see* **Note 4**).
3. Centrifuge the mixture at 8000 rpm for 10 min and collect the supernatant, which contains the water-soluble fraction (*see* **Note 5**).
4. The extract can be used immediately or stored frozen and later freeze-dried. If a sterile extract is required, filter it using a vacuum filter. It is important to carefully separate the supernatant from the pellet to avoid clogging the filter.

3.3 Ethanolic Extract

1. Mix between 2 and 10 g of freeze-dried mushroom powder in 250 mL of a 4:1 ethanol-water solution.
2. Incubate the mixture with stirring for 24 h at room temperature to minimize excessive degradation of the phenolic compounds.
3. Centrifuge the mixture at 8000 rpm for 10 min and collect the supernatants.
4. Concentrate the extract using a rotary evaporator until the ethanol is completely removed. Set the water bath to 40 °C.
5. Store the extract in the refrigerator if it will be used shortly. For long-term storage, freeze and then freeze-dry the extract.

3.4 Alkaline Extract

1. Add between 5 and 10 g of freeze-dried mushroom powder to 200 mL of a 1 M NaOH solution. The mixture needs to be stirred magnetically for 3 h at room temperature (see **Note 6**).
2. Centrifuge the mixture at 10,000 rpm for 10 min and discard the supernatant. The pellets should be combined and dissolved in another 200 mL of NaOH. Maintain this solution for 3 h under the same conditions as described in **step 1**.
3. Add the solution to a 1 L graduated cylinder containing absolute ethanol (v/v) to precipitate the polysaccharides. Centrifuge the resulting pellet and dialyze against water using membranes with a 10,000 Da cutoff.
4. The dialyzed polysaccharides can be frozen, freeze-dried, and stored at room temperature.

3.5 Chitin-Rich Extract

1. Use fresh mushrooms or lyophilized mushrooms, depending on availability. If working with fresh mushrooms, grind the material using an immersion blender.
2. For every 100 g of fresh mushrooms, add 0.3 L of distilled water. In the case of lyophilized mushrooms, add 20 g per 0.3 L of distilled water.

3. Heat the mixture for 30 min with stirring at 80 °C. Then, centrifuge at 8000 rpm for 12 min and discard the supernatant to remove the water-soluble fraction.
4. Resuspend the pellet in 0.3 L of 1 M NaOH and heat the mixture for 3 h with stirring at 60 °C. Then, centrifuge the mixture again at 8000 rpm for 12 min and discard the supernatant. This extract can be stored for further analysis.
5. If preparing chitosan-rich films, resuspend the pellet in distilled water at 1% w/v and homogenize. Then, perform vacuum filtration using filter paper to concentrate the dissolved particles. Finally, dry the films at 120 °C for 3 h with 5 kg of weight on top to prevent shrinkage [11].

4 Notes

1. It is important to use absolute ethanol or other dilutions that do not contain denatonium benzoate. This compound has a very bitter taste and is added to commercially available ethanol to prevent its consumption.
2. The advantages of freeze drying are that the product can be stored for months or even years without any change in its composition. However, if a freeze dryer is not available, or if you prefer to work with fresh material due to availability or cost savings, extractions can be performed using finely chopped product (for aqueous and ethanolic extracts) or puréed with a hand blender (for polysaccharide extraction, such as chitin). It is important to note that the yield will be higher when starting with freeze-dried material compared to the same weight of fresh material.
3. The concentrations are scalable and serve as guidelines. A small amount of mushroom powder will yield a low concentration, while too much may cause solubility issues.
4. The color of the extract may darken over time. This is because certain mushrooms, such as *Agaricus bisporus*, contain polyphenol oxidases, tyrosinases, and laccases that are released when the fungal membranes are disrupted [12].
5. The insoluble portion can be reserved for additional extractions, such as for isolating other polysaccharides like chitin. Utilizing all fractions of the mushroom helps maximize the use of the material.
6. The mixture should have a viscous appearance, similar to that of pudding.

Acknowledgments

This work was supported by grant 230025CONV from the Junta de Comunidades de Castilla-La Mancha (cofinanced by European Union FEDER funds).

References

1. Martins A (2016) The numbers behind mushroom biodiversity. In: Wild plants, mushrooms and nuts. Wiley, pp 15–63
2. Royse DJ, Baars J, Tan Q (2017) Current overview of mushroom production in the world. In: Edible and medicinal mushrooms. Wiley, pp 5–13
3. Navarro-Simarro P, Gómez-Gómez L, Ahrazem O, Rubio-Moraga Á (2024) Food and human health applications of edible mushroom by-products. New Biotechnol 81:43–56. https://doi.org/10.1016/j.nbt.2024.03.003
4. Chou W-T, Sheih I-C, Fang TJ (2013) The applications of polysaccharides from various mushroom wastes as prebiotics in different systems. J Food Sci 78:M1041–M1048. https://doi.org/10.1111/1750-3841.12160
5. Zhou R, Wang Y, Li C, Jia S, Shi Y, Tang Y, Li Y (2022) A preliminary study on preparation, characterization, and prebiotic activity of a polysaccharide from the edible mushroom *Ramaria flava*. J Food Biochem 46:e14371. https://doi.org/10.1111/jfbc.14371
6. Vital ACP, Goto PA, Hanai LN, Gomes-da-Costa SM, Filho BAA, Nakamura CV, Matumoto-Pintro PT (2015) Microbiological, functional and rheological properties of low fat yogurt supplemented with *Pleurotus ostreatus* aqueous extract. LWT – Food Sci Technol 64: 1028–1035. https://doi.org/10.1016/j.lwt.2015.07.003
7. Navarro-Simarro P, Gómez-Gómez L, Rubio-Moraga Á, Moreno-Gimenez E, López-Jimenez A, Prieto A, Ahrazem O (2025) Valorization of mushroom by-products for sustainability: exploring antioxidant and prebiotic properties. J Food Biochem 2025: 3527311. https://doi.org/10.1155/jfbc/3527311
8. Ospina Álvarez SP, Ramírez Cadavid DA, Escobar Sierra DM, Ossa Orozco CP, Rojas Vahos DF, Zapata Ocampo P, Atehortúa L (2014) Comparison of extraction methods of chitin from *Ganoderma lucidum* mushroom obtained in submerged culture. Biomed Res Int 2014:169071. https://doi.org/10.1155/2014/169071
9. Liu J, Jia L, Kan J, Jin C (2013) *In vitro* and *in vivo* antioxidant activity of ethanolic extract of white button mushroom (*Agaricus bisporus*). Food Chem Toxicol 51:310–316. https://doi.org/10.1016/j.fct.2012.10.014
10. Leong YK, Yang F-C, Chang J-S (2021) Extraction of polysaccharides from edible mushrooms: emerging technologies and recent advances. Carbohydr Polym 251:117006. https://doi.org/10.1016/j.carbpol.2020.117006
11. Fazli Wan Nawawi WM, Lee K-Y, Kontturi E, Murphy RJ, Bismarck A (2019) Chitin Nanopaper from mushroom extract: natural composite of nanofibers and Glucan from a single biobased source. ACS Sustain Chem Eng 7: 6492–6496. https://doi.org/10.1021/acssuschemeng.9b00721
12. Lin X, Sun D-W (2019) Research advances in browning of button mushroom (*Agaricus bisporus*): affecting factors and controlling methods. Trends Food Sci Technol 90:63–75. https://doi.org/10.1016/j.tifs.2019.05.007

Chapter 3

Fungal Polysaccharides: Purification and Characterization

Zeinab Qazanfarzadeh, Pamela Freire de Moura Pereira, and Amparo Jiménez-Quero

Abstract

Fungi are a rich source of bioactive polysaccharides with demonstrated nutritional and health-promoting properties, making them valuable for applications in the food, pharmaceutical, cosmetic, and materials science industries. The degree of polysaccharide purity required varies depending on the intended application, necessitating a clear understanding of appropriate extraction and purification techniques. Moreover, due to the structural complexity and heterogeneity of fungal polysaccharides, it is essential to characterize their composition and structure accurately. This chapter presents a comprehensive protocol for the isolation, purification, and characterization of crude polysaccharides from fungal biomass, enabling researchers to obtain materials with varying levels of purity. Additionally, it describes analytical procedures for quantifying total carbohydrate content using the phenol-sulfuric acid method, measuring acetyl group content for chitin estimation, and determining monosaccharide composition. These protocols are critical for understanding structure-function relationships, bioactivity assessments, and tailoring their physicochemical properties for applications in biomedical, nutraceutical, and biopolymer technologies.

Key words Crude polysaccharides, Chitin, β-glucan, Fungal cell wall, Acetylation degree

1 Introduction

Fungi are widely exploited for their nutritional value and potential health benefits. They produce various bioactive compounds such as polysaccharides, phenolic compounds, terpenoids, and alkaloids. Among them, polysaccharides are important because they take part in many life-related functions [1]. These molecules are formed by the combination of sugar units and/or uronic acids through different types of glycosidic bonds, increasing the complexity of their structure [2]. Polysaccharides are known for their positive effects on health, including improving immune response, showing antioxidant properties, lowering blood cholesterol, and helping to prevent lipid peroxidation [3]. Notably, fungal polysaccharides, especially those from edible and medicinal fungi, have been developed into alternative or supplement nutraceuticals for drug

Pablo Navarro Simarro (ed.), *Fungi*, Methods and Protocols in Food Science, https://doi.org/10.1007/978-1-0716-4989-3_3,
© The Author(s), under exclusive license to Springer Science+Business Media, LLC, part of Springer Nature 2026

therapies [4]. Fungal polysaccharides have also shown a wide range of potential applications across various fields, including the food industry, cosmetics, packaging, and the development of novel materials [5].

Fungi produce a variety of polysaccharides, including acidic heteropolysaccharides containing acidic groups, such as uronic acid or sulfate, as well as neutral heteropolysaccharides composed mainly of neutral sugars like xylose, galactose, and mannose [6]. Structurally, their cell walls contain two main types of polysaccharide, including a rigid framework of chitin and a softer matrix made up of α- and β-glucans along with glycoproteins [7]. Fungi β-glucans are primarily composed of a (1 → 3)-β-D-glucose backbone, where every 2–3 glucose units may be connected via β-(1 → 6) linkages to another β-(1 → 3) chain. Their tertiary structure depends on how frequently these β-(1 → 6) bonds occur. The low branching leads to a single-helix form, while a higher degree of β-(1 → 6) linkages promotes the formation of a triple-helix structure [8].

Therefore, the extraction and purification of fungal polysaccharides are crucial for enhancing their bioactivity, ensuring consistency, and removing impurities, making them more effective and safer for pharmaceutical, nutraceutical, and food applications. The traditional methods of extracting mushroom polysaccharides mostly include extraction using alkali solution, hot water, and polyethylene glycol, among others [9, 10]. However, some new extraction methods have emerged, including ultrasonic extraction, high-pressure homogenization, and subcritical water extraction to enhance the efficiency of traditional techniques. Interestingly, the structure and biological activities of polysaccharides can be affected by the extraction method [11].

Crude polysaccharides extracted through the methods mentioned often contain significant amounts of proteins, pigments, salts, lipids, and other impurities. These can interfere with structural analysis and biological activity, making it essential to remove such contaminants thoroughly during purification [12]. Purification involves steps like removing these small molecules and proteins, decolorizing, degreasing, and using chromatography to isolate a uniform polysaccharide product [13]. Therefore, various purification methods have been reported in the literature, ranging from crude extraction with hot water to fractionation, followed by more advanced purification using a combination of techniques such as ion-exchange and gel filtration chromatography. The choice of purification method can also depend on the initial composition and structural complexity of the mushroom material, as different species and growth conditions may yield polysaccharides with diverse properties. This chapter presents a comprehensive protocol for the purification and quantification of mushroom polysaccharides, adaptable to a wide range of applications requiring different purities. Additionally, it outlines a straightforward approach for the characterization of crude polysaccharides using chromatography techniques.

2 Materials

2.1 Isolation and Purification of Mushroom Polysaccharides

1. Fresh mushroom fruiting bodies.
2. Deionized and Milli-Q water.
3. Absolute ethanol (≥99.5%).
4. Sevag reagent: mix chloroform and n-butanol in a ratio of 4:1 (v/v).
5. Hydrogen peroxide (H_2O_2) 30%, ammonia water, and active carbon (for decolorization).
6. Dialysis membrane (cutoff 3.5–10 kDa).
7. 0.025 M Tris–HCl buffer (pH 7.0): eigh 3.03 g of Tris base (molecular weight 121.14 g/mol) and dissolve it in about 800 mL of Milli-Q water in a beaker. Next, adjust the pH to 7.0 by slowly adding 1 M HCl drop by drop while stirring. Once the pH reaches 7.0, transfer the solution to a 1-liter volumetric flask and add Milli-Q water to bring the total volume to 1 liter.
8. Aqueous sodium chloride (NaCl) solutions (0.1, 0.3, 0.5, 0.7, and 1 M): weigh and dissolve 5.84, 17.53, 29.22, 40.91, and 58.44 g of NaCl, respectively, in 1 liter of Milli-Q water, stirring until fully dissolved.
9. DEAE-Cellulose-52 column (φ26 × 500 mm, for ion-exchange chromatography).
10. Sephadex G-150 column (for gel filtration chromatography).

2.2 Total Carbohydrates Measurement by Phenol-Sulfuric Acid Method

1. Glucose standard solutions: use a glucose stock solution of 1 mg/mL and dilute with Milli-Q water to concentrations of 0, 5, 10, 20, 40, 80, 160 μg/mL.
2. 2.5 M hydrochloric acid (HCl): mix 20.8 mL of concentrated HCl (12 M) with Milli-Q water to make 100 mL.
3. 2.5 M sodium hydroxide (NaOH): dissolve 10.00 g of NaOH in Milli-Q water to make 100 mL.
4. A 96-well, flat-bottomed, polystyrene microplate.
5. Phenol solution 5% w/v in Milli-Q water (*see* **Note 1**).
6. Concentrated sulfuric acid (H_2SO_4, 95–98%).

2.3 Monosaccharide Composition by Anion-Exchange Chromatography

1. MiliQ water.
2. 72% H_2SO_4: dilute 73.50 mL of concentrated H_2SO_4 (98%, d = 1.84 g/mL) to 100 mL with MiliQ water.
3. Stock solutions of standards in a concentration of 100 mg/mL, including galactose, glucose, glucosamine, N-acetylglucosamine, and mannose.
4. 200 mM NaOH: dissolve 8 g NaOH in 700 mL Mili-Q water and dilute to 1 L. Then degas the solution.
5. Nylon filters (0.2 μm).

2.4 Acetyl Content by Liquid Chromatography

1. 0.8 M NaOH: dissolve 32 g NaOH pellets in 700 mL Milli-Q water, then dilute to 1 L.
2. 37% HCl.
3. 5 mM H_2SO_4: dilute 272 μL of concentrated H_2SO_4 (98% and density 1.84 g/mL) to 1 L with Milli-Q water.
4. 1 M propionic acid ($C_3H_6O_2$): take an aliquot of 1.865 mL and transfer to a 25 mL volumetric flask previously containing 10 mL of Milli-Q water. Bring the volume up to 25 mL.
5. 1 M acetic acid (CH_3COOH): take an aliquot of 2.871 mL from acetic acid glacial in a 50 mL volumetric flask previously containing 10 mL of Milli-Q water. Bring the volume up to 50 mL.
6. Nylon filters (0.2 μm).

3 Methods

3.1 Hot Water Extraction of Crude Fungal Polysaccharides from Mushrooms

Hot water extraction of crude fungal polysaccharides from mushrooms involves immersing the fungal biomass in hot water, which capitalizes on the high solubility of hot water to cause the cells to absorb water, expand, and rupture their cell walls. Additionally, pretreating the biomass by reducing its size can enhance the extraction process by increasing the surface area.

1. Wash the mushroom's fruiting bodies with distilled water, cut them into small pieces, and lyophilize them. After drying, grind them and filter through a 20-mesh sieve.
2. Defat the mushroom powder with absolute ethanol at a ratio of 1:10 (g/mL) with constant stirring over 12 h.
3. Extract the polysaccharides using deionized water sequentially 3 times at a temperature of 90–95 °C for 2 h and biomass: solvent ratio of 1:10–20 g/mL (*see* **Note 2**). Centrifuge the mixture after each step of extraction at 2000 rpm for 15 min, and mix the supernatants obtained from 3 extractions.

3.2 Purification of Mushroom Crude Polysaccharides

After extraction, the crude polysaccharide extract undergoes a purification process that includes deproteinization, decolorization, and desalting to eliminate proteins, pigments, salts, and small molecules. Dialysis is used to separate monosaccharides, oligosaccharides, other low-molecular-weight nonpolar substances, and inorganic salts, relying on selective diffusion through a semipermeable membrane.

1. Remove proteins by mixing the crude polysaccharides with 1 volume of Sevag reagent. Shake the mixture vigorously for 15 minutes to ensure proper interaction between the solvents and proteins. Centrifuge at 5000 rpm for 10–20 min to separate the phases. The upper aqueous phase contains purified polysaccharides, while the lower organic phase contains proteins and other impurities.
2. To decolorize the crude polysaccharides, adjust the pH of the polysaccharide solution to 8.0 using ammonia water. Then, mix 50 mL of the pH-adjusted solution with 5 mL of 30% hydrogen peroxide in a 250 mL Erlenmeyer flask. Shake the mixture at 110 rpm at 55 °C for 30 min. After the reaction, centrifuge the mixture at 2000 rpm for 15 min to remove insoluble impurities (*see* **Note 3**).
3. Dialyze the deproteinized polysaccharides with dialysis membrane (cutoff 3.5–10 kDa) for 24 hours in a cold room to remove small molecules and inorganic salts. Replace the dialysis water every 3–4 h to maintain the concentration gradient.
4. Add absolute ethanol to the extract at a ratio of 3:1 (v/v), mix well, and keep overnight at 4 °C to form a precipitate. Afterward, separate the precipitates by centrifugation at 2000 rpm for 15 min. Leave the precipitate under a fume hood to evaporate the remaining ethanol, then freeze-dry the polysaccharides.

3.3 Further Purification

As a next step in the purification process, column chromatography techniques such as ion exchange and gel filtration chromatography are employed to further purify crude polysaccharide extracts. Ion exchange chromatography is highly selective and particularly effective for separating by splitting neutral and acidic polysaccharides based on their charge properties. In contrast, gel filtration chromatography, often using hydrophilic dextran gels as molecular sieves, separates polysaccharides according to their molecular size, allowing for efficient fractionation of components with different molecular weights (*see* **Note 4**).

1. Purify the polysaccharide solution using DEAE-cellulose ion-exchange chromatography. Prepare DEAE-cellulose for polysaccharide purification (*see* **Note 5**). Dissolve the polysaccharides in 0.025 Tris–HCl (pH 7.0), centrifuge, and then

inject the supernatant (10 mL) into a column of DEAE-Cellulose-52 (φ26 × 500 mm) equilibrated with 0.025 M Tris–HCl (pH 7.0). Then elute the crude polysaccharides with a gradient NaCl aqueous solution (0, 0.1, 0,3, 0.5, 0.7, and 1 M) and collect different fractions based on their charge intensity.

2. Measure the polysaccharide content of eluate fractions using the phenol-sulfuric acid microplate method according to the method described in Subheading 3.4.
3. Purify further polysaccharide-enriched eluents through gel filtration. Dissolve the purified polysaccharides in Milli-Q water (10.0 mg/mL) and filter through a 0.45 μm membrane filter. Apply the solution (3.0 mL) to the Sephadex G-150 column at a flow rate of 0.27 BV/h and elute using 0.1 M NaCl (see **Note 6**).
4. Dialyze the purified polysaccharides against Milli-Q water for 24 h and then lyophilize them. Seal and store them in a cool place or at room temperature for further analysis.

3.4 Total Carbohydrates Measurement by Phenol-Sulfuric Acid Method

This method is based on the reaction of carbohydrates with concentrated sulfuric acid, which hydrolyzes polysaccharides and oligosaccharides into monosaccharides. These monosaccharides are subsequently dehydrated to form furfural (from pentoses) or hydroxymethyl furfural (from hexoses). These furfural derivatives then react with phenol to produce a yellow-orange complex that exhibits a strong absorbance at 483–490 nm [14].

1. To release monosaccharides prior to analysis, hydrolyze 2 mg of the crude polysaccharide by adding 1 mL of 2.5 M HCl and heating the mixture at 100 °C for 3 h in a sealed tube or tightly capped vial. After the reaction, allow the sample to cool, then neutralize it with 2.5 M NaOH. Centrifuge the mixture and dilute the resulting supernatant tenfold with Milli-Q water for further analysis.
2. Prepare the calibration standard using a stock solution of glucose (1 mg/mL) by diluting it with Milli-Q water to concentrations of 0–160 μg/mL according to Table 1.
3. Add 50 μL of each standard, sample, and blank (Milli-Q water) into separate wells of a 96-well plate. Rapidly add 150 μL of concentrated sulfuric acid to each well (*see* **Note** 7).
4. Add immediately 30 μL of 5% phenol solution to each well and mix gently by pipetting or shaking the plate.
5. Incubate at 90 °C for 5 min. After that, cool the plate to room temperature (~10 min).
6. Read the absorbance at 490 nm using a microplate reader. Use the standard curve to determine carbohydrate concentration in samples.

Table 1
Preparation of glucose standards

Concentration (μg/L)	Volume of stock standard (μL of 1 mg/mL)	Water (μL)
0	0	1000
5	5	995
10	10	990
20	20	980
40	40	960
80	80	920
160	160	840

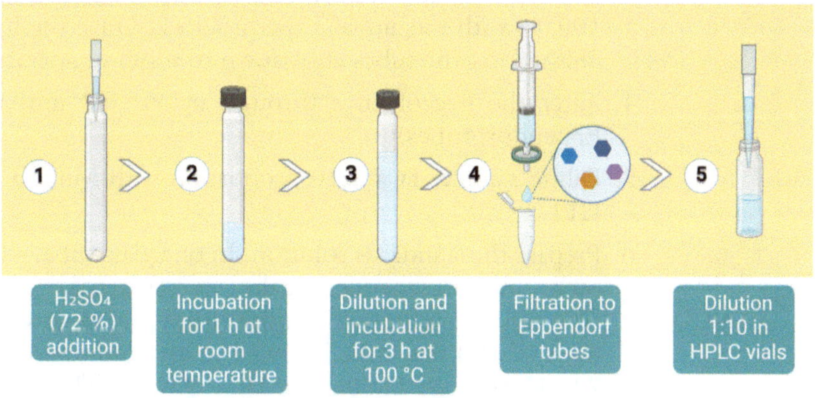

Fig. 1 Schematic representation of two-step sulfuric hydrolysis

3.5 Monosaccharide Composition by Anion-Exchange Chromatography

To understand the carbohydrate composition of the fungal cell wall, the investigation of neutral sugars is required. For this, a two-step sulfuric acid hydrolysis is adopted, which is responsible for releasing the monomers from galactans, glucans, mannans, and chitin, as galactose, glucose, mannose, and glucosamine (considering the acid-lability of N-acetyl residues in N-acetyl-glucosamine), respectively (*see* **Note 8**) (Fig. 1). The released monosaccharides are subsequently separated using high-performance anion-exchange chromatography (HPAEC) and quantified using a pulsed amperometric detector (PAD) [15].

1. Weigh approximately 1–2 mg of sample in glass tubes, add 125 μL of 72% H_2SO_4 to the samples, and soak them well with acid. Close the flasks with their caps to avoid spillage accidents.
2. Incubate the prepared glass tubes containing the samples for 1 h at room temperature.

Table 2
Preparation of monosaccharide standards from 1 g/L stock solution

Concentration (g/L)	Volume of stock standard (μL of 1 g/L)	Water (μL)
0	0	1000
0.005	5	995
0.01	10	990
0.02	20	980
0.05	50	950
0.1	100	900

3. Carefully add 1375 μL of Milli-Q water to the glass tubes. Close the tubes with the caps and hydrolyze them for 3 h at 100 °C with the aid of a thermoblock. Once the hydrolysis is finished, let the tubes cool down to room temperature.
4. Filter the hydrolysates through nylon (0.2 μm) filters into Eppendorf tubes.
5. Dilute samples at a 1:10 proportion with Milli-Q water into HPLC vials.
6. Prepare the standards solution. In one glass tube, add 15 μL of each monosaccharide (galactose, glucose, glucosamine, N-acetylglucosamine, mannose) from preprepared stock solutions of 100 mg/mL, followed by the addition of 125 μL of 72% H_2SO_4 and 1300 μL of Milli-Q water. The standard tubes will be treated following **steps 3–4**. The final concentration of each monosaccharide present in the tube will be 1 g/L, which will be used for the preparation of the different dilutions with Milli-Q water, following Table 2.
7. Inject the samples into the HPAEC-PAD system. A column CarboPac PA20 (3 × 150 mm, Thermo Fischer, USA) at a flow rate of 0.4 mL/min is used for monosaccharides separation. A gradient method is performed using Milli-Q water as eluent A and 200 mM NaOH as eluent B as follows: equilibration for 8 min with 1.2% B, and 20 min elution with eluent B. Quantification is carried out against a calibration curve built with neutral sugars (galactose, glucose, glucosamine, N-acetylglucosamine, mannose).

3.6 Acetyl Content by Liquid Chromatography

Investigation of acetyl content is crucial for chitin and derivatives, determining their chemical and physicochemical properties. It is also fundamental to distinguish chitin from chitosan, a derivative product formed by the partial deacetylation of N-acetyl-glucosamine groups. The following protocol consists of the saponification of acetyl groups and their release as free acetic acid [16].

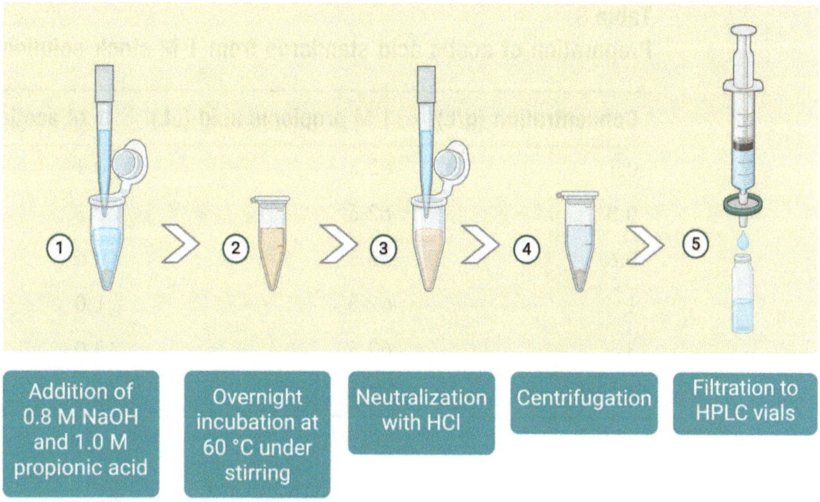

Fig. 2 Schematic representation of the determination of acetylation degree assay

1. Weigh samples at approximately 7 mg in Eppendorf tubes. Add 300 mL of Milli-Q water and vortex it. Add 1.2 mL of 0.8 M NaOH to the samples and incubate them overnight at 60 °C under continuous stirring (Fig. 2).
2. Propionic acid can be used as an internal standard during the sample preparation. Add 10 mL of 1 M propionic acid to the samples from the initial incubation step and vortex them.
3. After complete dissolution of the samples, neutralize them with 37% HCl and record the amount of acid used for further concentration calculations.
4. Centrifuge the samples for 10 min at 3000 rpm and filter the supernatant through 0.2 um nylon filters with the aid of a syringe directly into HPLC glass vials.
5. The HPLC run is performed using 5 mM H_2SO_4 as mobile phase at a flow rate of 0.7 mL/min, an organic acid column (e.g.: Aminex® HPX-87H, BioRad, Hercules, CA, USA) kept at 50 °C, and an HPLC system equipped with a refractive index (RI) detector, settled at 45 °C. The run is performed over 30 min. Sample quantification is performed against an acetic acid calibration standard curve and prepared as described in Table 3 (*see* **Note 9**).
6. Prepare different dilutions of acetic acid (from 1 M stock solution) in a 100 mL volumetric flask as suggested in Table 3, and bring the volume up to 100 mL with Milli-Q water.
7. Calculate the degree of acetylation (DA) using Eqs. (1) and (2) [17].

Table 3
Preparation of acetic acid standards from 1 M stock solution

Concentration (g/L)	1 M propionic acid (μL)	1 M acetic acid (mL)
0.1	62.5	0.1
0.3	62.5	0.3
0.6	62.5	0.6
1	62.5	1.0
3	62.5	3.0
6	62.5	6.0

$$\%\text{acetyl} = \frac{MM_{\text{acetyl}} \cdot C_{\text{acetyl}}}{C_{\text{polysaccharide}}} \cdot 100 \qquad (1)$$

Where MM_{acetyl} is 43 g/mol, C_{acetyl} is the concentration of acetic acid (M), and $C_{\text{polysaccahride}}$ is the concentration of polysaccharide present in the sample (g/L).

$$DA = \frac{161.1 \cdot \%\text{acetyl}}{(MM_{\text{acetyl}} \cdot 100) - (MM_{\text{acetyl}} - 1) \cdot \%\text{acetyl}} \qquad (2)$$

Where 161.1 is the molar mass of a GlcN monomer (g/mol), $MM_{\text{acetyl}\,-\,1}$ is the difference in molar mass between a hydrogen and an acetyl group (42 g/mol).

4 Notes

1. A fresh solution of phenol should be prepared for each experiment. After mixing, keep the solution protected from light, as it is light-sensitive.
2. To enhance the extraction yield of crude polysaccharides, a pretreatment with ultrasound at 25 °C for 30 min is recommended.
3. Decolorization can alternatively be performed using activated carbon. In this method, mix 50 mL of the crude polysaccharide solution with 0.5 g of dried activated carbon in a 50 mL centrifuge tube. Shake the mixture at 110 rpm at 55 °C for 30 minutes. Afterward, centrifuge at 2000 rpm for 10 min to remove the residual activated carbon and insoluble impurities. Besides, decolorization can be carried out using macroporous adsorbent resins (e.g., D301, NKA-9, SP-825). These resins have also shown the potential to remove proteins, eliminating the need for the Sevag method [18, 19].

4. The choice between ion-exchange and gel chromatography depends on the properties of the polysaccharides and the intended purpose. Ion-exchange chromatography is ideal when separation based on charge is needed, while gel chromatography is better suited for preserving the structural integrity and bioactivity of the polysaccharides.
5. For preparation of DEAE-cellulose for polysaccharide purification, suspend DEAE-cellulose in Milli-Q water (1:10 w/v), allow it to swell for 1–2 hours, and wash it 3–5 times. Remove fines by settling, then equilibrate with 0.5 M NaCl or 0.1 M acetic acid. Rinse until neutral pH is achieved, then wash with a low ionic strength buffer (e.g., 20 mM phosphate buffer, pH 7–8), and store at 4 °C until use.
6. The selection of an appropriate chromatography column depends on the specific requirements of the purification process. Different types of Sephadex G-series columns, such as Sephadex G-75, G-100, G-150, and G-200, can be chosen based on the molecular weight range of the target polysaccharides.
7. Sulfuric acid is highly corrosive. Always use gloves, eye protection, and handle it in a fume hood.
8. In function of the acid liability of N-acetyl-glucosamine, the separation of glucosamine and N-acetyl-glucosamine in HPAEC-PAD is challenging. Therefore, the peak must represent the sum of both monosaccharides as shown in Fig. 3.
9. Both RI and UV detectors at 210 nm can be used for the quantification of acetyl content, as shown in Fig. 4. Additionally, a column Rezex ROA 300 × 7.8 mm (Phenomenex, Torrance, CA, USA), kept at 50 °C, can be used alternatively to the column Aminex® HPX-87H.

Acknowledgments

The authors acknowledge the financial support from the European Union's Horizon 2020 Research and Innovation Programme under Grant Agreement No. 101037796, as well as from the Horizon Europe Programme under the Marie Skłodowska-Curie Actions, Grant Agreement No. 101107449.

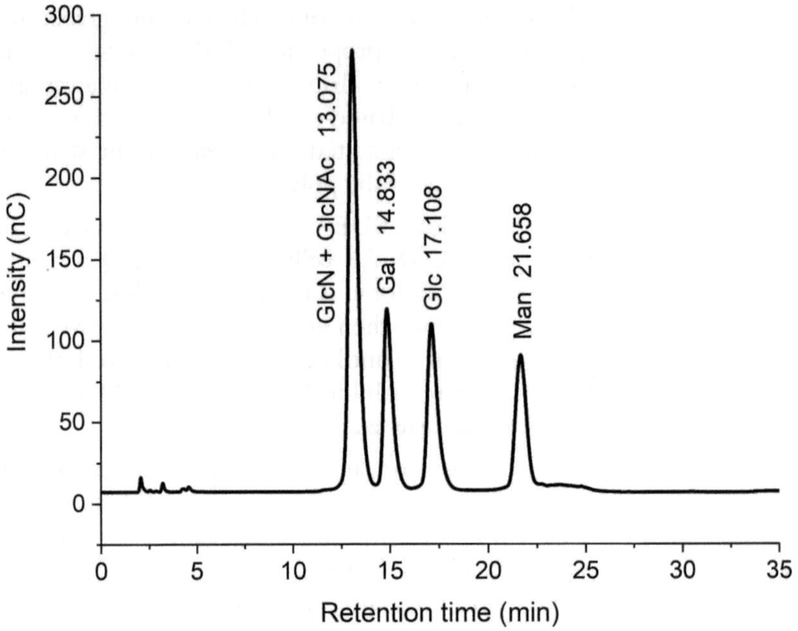

Fig. 3 Chromatogram of a mixture of standards (a) Glucosamine (GlcN), N-acetil-glucosamine (GlcNAc), galactose (Gal), glucose (Glc), and mannose (Man) prepared following the same procedure for the samples

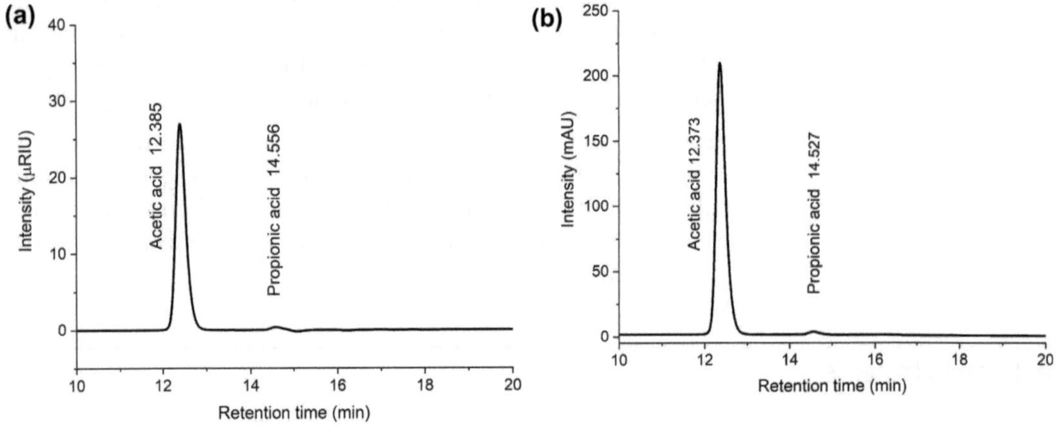

Fig. 4 Chromatograms of an acetic acid standard containing propionic acid as internal standard obtained using (**a**) a refractive index (RI) and (**b**) an ultraviolet (UV) detector

References

1. Mohammed ASA, Naveed M, Jost N (2021) Polysaccharides; classification, chemical properties, and future perspective applications in fields of pharmacology and biological medicine (a review of current applications and upcoming potentialities). J Polym Environ 29:2359–2371
2. Murphy EJ, Fehrenbach GW, Abidin IZ, Buckley C, Montgomery T, Pogue R, Murray P, Major I, Rezoagli E (2023) Polysaccharides – naturally occurring immune modulators. Polymers (Basel) 15:2373
3. Thimmaraju A, Seedevi P (2024) Purification, chemical, structural characterization and

biological activity of polysaccharide from blue oyster mushroom *Hypsizygus ulmarius*. Biomass Convers Biorefinery 14:19889–19899

4. Wang H, Yang L, Qiu Q, Rao Q, Liu L, Cui Y, Zhang L, Ma Y, Jin X, Yang D (2024) Exploring the health benefits of *Boletus aereus* polysaccharides: extraction, structural characterization, and Antiproliferative properties against non-Hodgkin's lymphomas (NHLs). J Agric Food Chem 72:16334–16346
5. Elsacker E, Vandelook S, Van Wylick A, Ruytinx J, De Laet L, Peeters E (2020) A comprehensive framework for the production of mycelium-based lignocellulosic composites. Sci Total Environ 725:138431
6. Lin B, Huang G (2022) Extraction, isolation, purification, derivatization, bioactivity, structure–activity relationship, and application of polysaccharides from white jellyfungus. Biotechnol Bioeng 119:1359–1379
7. Leong YK, Yang F-C, Chang J-S (2021) Extraction of polysaccharides from edible mushrooms: emerging technologies and recent advances. Carbohydr Polym 251:117006
8. Nitschke J, Modick H, Busch E, Von Rekowski RW, Altenbach H-J, Mölleken H (2011) A new colorimetric method to quantify β-1, 3-1, 6-glucans in comparison with total β-1, 3-glucans in edible mushrooms. Food Chem 127:791 796
9. Yongjiang W, Zhong C, Jianwei M, Minger F, Xueqian W (2009) Optimization of ultrasonic-assisted extraction process of Poria cocos polysaccharides by response surface methodology. Carbohydr Polym 77:713–717
10. Zhang L, Wang M (2016) Polyethylene glycol-based ultrasound-assisted extraction and ultrafiltration separation of polysaccharides from *Tremella fuciformis* (snow fungus). Food Bioprod Process 100:464–468
11. Zhang J, Zeng X, Liu G, Wen C, Xu X (2024) Extraction of Lentinus Edodes polysaccharides with ultrasound enhanced with deep eutectic solvent and their structural characterization and antioxidant activity. Chem Biodivers 21: e202400141
12. Zhu F, Du B, Xu B, Ramawat KG, Mérillon JM (2021) Preparation and characterization of polysaccharides from mushrooms. In: Polysaccharides: bioactivity and biotechnology, pp 1–16
13. Song W, Li M, Yue X, Zheng Y (2024) Effect of extraction and purification on the structure and activity of *Flammulina velutipes* polysaccharides: a review. Int J Food Sci Technol 59: 9571–9591
14. Masuko T, Minami A, Iwasaki N, Majima T, Nishimura S-I, Lee YC (2005) Carbohydrate analysis by a phenol–sulfuric acid method in microplate format. Anal Biochem 339:69–72
15. Saeman JF (1954) Techniques for the determination of pulp constituents by quantitative paper chromatography. Tappi 37:336–343
16. Xu C, Leppänen A-S, Eklund P, Holmlund P, Sjöholm R, Sundberg K, Willför S (2010) Acetylation and characterization of spruce (*Picea abies*) galactoglucomannans. Carbohydr Res 345:810–816
17. Laignel B, Bliard C, Massiot G, Nuzillard JM (1997) Proton NMR spectroscopy assignment of D-glucose residues in highly acetylated starch. Carbohydr Res 298:251–260
18. He B, Guo L, Zheng Q, Lin S, Lin J, Wei T, Ye Z (2019) A simple and effective method using macroporous resins for the simultaneous decoloration and deproteinisation of *Cordyceps militaris* polysaccharides. Int J Food Sci Technol 54:1741–1751
19. Shu X, Zhang Y, Jia J, Ren X, Wang Y (2019) Extraction, purification and properties of water-soluble polysaccharides from mushroom *Lepista nuda*. Int J Biol Macromol 128:858–869

Chapter 4

Extraction Strategy for Enhancing Solubility of Mushroom Polysaccharide Extracts: From Insoluble Fractions to Bioactive Compounds

Andreia Silva Fernandes and Freni K. Tavaria

Abstract

Mushroom polysaccharide solubility is influenced by several factors. Their branched architecture and large molecular weights often promote aggregation, limiting dissolution. While conventional methods focus on separating soluble from insoluble fractions, a significant part of bioactive compounds remains locked within cell walls. Harsher acid or alkaline treatments can free these compounds by loosening the rigid wall matrix. In addition, enzymatic or ethanol-based pretreatments can enhance extraction efficiency and purity. This chapter offers practical guidance on pretreatment, extraction, and purification of insoluble mushroom polysaccharides, with emphasis on reproducible protocols and improved solubility.

Key words Polysaccharides, Solubility, Alkaline extraction, Pretreatment, Purification

1 Introduction

Polysaccharides have very complex structural features that have a significant impact on their solubility. A combination of chemical, structural, and process-related factors will determine the solubility of these molecules [1, 2]. In general, these compounds contain highly branched configurations with numerous glycosidic linkages, which can block their capability to dissolve homogeneously in solvents. Molecular weight (MH) is also a key factor in polysaccharide solubility: high-MH polysaccharides tend to aggregate in solution, negatively affecting solubility [3].

The most frequently used methods for polysaccharide extraction in mushrooms consist of separating soluble components from insoluble cell walls [4]. Nevertheless, a large proportion of bioactive compounds remains locked within the insoluble fractions of the mushroom cell wall, making their extraction a key challenge. These insoluble fractions require harsher treatments (e.g., acid or alkaline solvents) that will break down or loosen cell-wall structures,

Pablo Navarro Simarro (ed.), *Fungi*, Methods and Protocols in Food Science, https://doi.org/10.1007/978-1-0716-4989-3_4,
© The Author(s), under exclusive license to Springer Science+Business Media, LLC, part of Springer Nature 2026

releasing polysaccharides into solution [5, 6]. These methods require careful optimization of the extraction time, temperature, and pH to preserve the extracted polysaccharides' bioactivity.

More than that, previous pretreatment can also be an essential step towards further extraction procedures. Some enzymes, such as cellulases or xylanases, can partially degrade the insoluble cell wall matrix, releasing more polysaccharides into a soluble form [7, 8]. In other cases, solvents such as absolute ethanol can be used as a defatting agent and to remove certain phenolic compounds, improving posterior water or solvent extraction and increasing the purity and the yield of polysaccharides [9, 10].

This chapter explores practical strategies to address the common obstacles encountered when breaking down insoluble mushroom fractions, focusing mainly on three aspects: pretreatment, extraction, and purification of insoluble fractions.

2 Materials

1. Distilled water.
2. Absolute ethanol; ethanol 96%.
3. Sodium hydroxide 0.5 M.
4. Acetic acid (30%).
5. Dried mushrooms should be stored in an airtight container at room temperature until further processing.
6. Prepare and store all the solutions and reagents at room temperature (unless indicated otherwise).

3 Methods

3.1 Sample Preparation

1. Grind the dried mushroom samples into a fine powder using a blender, grinder, or ball mill (*see* **Note 1**).
2. Sieve the powder through a 60–100 mesh to ensure uniform particle size.

3.2 Pretreatment

1. Weigh the desired amount of mushroom powder.
2. Add absolute ethanol in a ratio of 1:10 (m/v) to remove lipids are other impurities and mix well with a magnetic stirrer for 12–24 h at 4 °C (*see* **Note 2**).
3. Centrifuge at 5000 rpm for 15 min and discard the supernatant.

3.3 Hot Water Extraction

1. Add distilled water (1:20 w/v) to the pellet obtained in the last step.
2. Heat at 90 ± 5 °C with vigorous stirring for 2 h (*see* **Note 3**).
3. Cool at room temperature and Centrifuge at 5000 rpm for 15 min to obtain the two initial fractions (water-soluble and water-insoluble). (*see* **Note 4**).
4. Store the water-soluble fraction at 4 °C or freeze for long-term storage.

3.4 Alkaline Extraction

1. Add NaOH 0.5 M (1:20 w/v) to the insoluble fraction obtained in the last step.
2. Vigorously stirring the solution, certifying that most of the insoluble pellet solubilizes in the NaOH solution. (*see* **Note 5**).
3. Keep the solution at 4 °C during at least 12 h.

3.5 Neutralization

1. Centrifuge the solution at 10000 rpm for 20 min at 4 °C to obtain the two fractions (alkaline-soluble and alkaline-insoluble).
2. Carefully neutralize the alkaline-soluble fraction with a 30% acetic acid solution. (*see* **Note 6**).

3.6 Precipitation

1. Add ethanol 96% in a ratio of 1:4 (m/v) to allow the precipitation of polysaccharides and maintain the solution at 4 °C during 12 h.
2. Centrifuge at 5000 rpm for 15 min to precipitate the polysaccharides and discard the supernatant.
3. Redissolve the residue in distilled water and store at 4 °C or freeze for long-term storage.

3.7 Assessing Solubility

1. Prepare extract solutions at desire concentrations (e.g., 1, 5, 10, and 20 mg/mL) using distilled water as the solvent.
2. Mix the solutions to ensure complete dissolution.
3. Centrifuge the solutions at 3000 RPM for 10 min at room temperature to obtain clearer supernatants. (*see* **Note 7**).
4. Transfer the supernatants to pre-weighed glass Petri dishes.
5. Weigh the cooled Petri dishes.
6. Calculate the water solubility (%) using the formula:
7. Water solubility (%) = (weight of dry supernatant/weight of sample) × 100.

4 Notes

1. Choose the grinding method based on the sample size and required fineness. During grinding, excessive heat generation may degrade thermolabile compounds. To avoid this, grind in short pulses or use a cooling mechanism if available. The choice of mesh size (60–100) depends on the desired particle size. A smaller mesh (e.g., 100) will yield finer powder, which might enhance extraction efficiency.
2. Maintain the recommended 1:10 (m/v) ratio consistently. Insufficient ethanol may result in incomplete impurity removal, while excessive ethanol can dilute the sample unnecessarily.
3. Avoid exceeding the upper limit (95 °C) to prevent degradation of heat-sensitive compounds. More than that, overheating beyond 95 °C can also lead to caramelization of sugars or degradation of polysaccharides, potentially altering the chemical profile of the extract.
4. In this step, two fractions are obtained: water-soluble a water-insoluble. The water-soluble fraction can also be precipitated with ethanol, collected, and analyzed.
5. Allow sufficient time for complete solubilization; if the pellet is still not dissolving, extend the stirring duration or slightly warm the solution. However, it is important to avoid high temperatures (e.g., >60 °C). Depending on the type of polysaccharides may be necessary to adjust the pH of the solution in this step to prevent degradation. Extreme pH (e.g., >13) may lead to hydrolysis of the polysaccharides, altering their structure and bioactivity. In some cases, harsh alkaline extraction could be necessary for the optimization of these conditions [11].
6. Aim for a neutral pH (7.0) or a slightly acidic range (pH 6.5–7.5). Allow the solution to stabilize for 10–15 min after reaching the desired pH before proceeding to the next step.
7. If the polysaccharides are difficult to dissolve, gently warm the solution or use mild agitation. Note that temperature, time, and the speed of centrifugation can impact solubility. Different temperatures and pH can also be used to measure solubility in different conditions [12].

Acknowledgments

This work was supported by FCT – Fundação para a Ciência e a Tecnologia and CBQF-Centro de Biotecnologia e Química Fina

Laboratório Associado, Escola Superior de Biotecnologia through project UIDB/50016/2020 and the grant UI/BD/151390/2021.

References

1. Singh V, Indoria S, Jisha KJ, Gardas RL (2021) Structure and solubility of polysaccharides. In: Polysaccharides: properties and applications. https://doi.org/10.1002/9781119711414.ch15
2. Whistler RL (1973) Solubility of polysaccharides and their behavior in solution. ACS Publications. https://doi.org/10.1021/ba-1971-0117.ch014
3. Guo Q, Cui SW, Kang J, Ding H, Wang Q, Wang C (2015) Non-starch polysaccharides from American ginseng: physicochemical investigation and structural characterization. Food Hydrocoll 44. https://doi.org/10.1016/j.foodhyd.2014.09.031
4. Akram K, Shahbaz HM, Kim GR, Farooq U, Kwon JH (2017) Improved extraction and quality characterization of water-soluble polysaccharide from gamma-irradiated *Lentinus edodes*. J Food Sci 82(2). https://doi.org/10.1111/1750-3841.13590
5. Li S, Li J, Zhang J, Wang W, Wang X, Jing H, Ren Z, Gao Z, Song X, Gong Z, Jia L (2017) The antioxidative, antiaging, and hepatoprotective effects of alkali-extractable polysaccharides by *Agaricus bisporus*. Evid Based Complement Alternat Med 2017. https://doi.org/10.1155/2017/7298683
6. Palacios I, García-Lafuente A, Guillamón E, Villares A (2012) Novel isolation of water-soluble polysaccharides from the fruiting bodies of *Pleurotus ostreatus* mushrooms. Carbohydr Res 358. https://doi.org/10.1016/j.carres.2012.06.016
7. Zhao YM, Song JH, Wang J, Yang JM, Wang ZB, Liu YH (2016) Optimization of cellulase-assisted extraction process and antioxidant activities of polysaccharides from *Tricholoma mongolicum* Imai. J Sci Food Agric 96(13). https://doi.org/10.1002/jsfa.7662
8. Chen J, Lai P, Shen H, Zhen H, Fang R (2013) Effect of extraction methods on polysaccharide of *Clitocybe maxima* stipe. Adv J Food Sci Technol 5(3). https://doi.org/10.19026/ajfst.5.3273
9. Piskov SI, Timchenko LD, Rzhepakovsky IV, Avanesyan SS, Bondareva NI, Sizonenko MN, Areshidze DA (2019) Effect of pre-treatment conditions on the antiatherogenic potential of freeze-dried oyster mushrooms. Foods Raw Mater 7(2). https://doi.org/10.21603/2308-4057-2019-2-375-386
10. Trygg J, Beltrame G, Yang B (2019) Rupturing fungal cell walls for higher yield of polysaccharides: acid treatment of the basidiomycete prior to extraction. Innov Food Sci Emerg Technol 57. https://doi.org/10.1016/j.ifset.2019.102206
11. Liu F, Chen H, Qin L, Al-Haimi AANM, Xu J, Zhou W, Zhu S, Wang Z (2023) Effect and characterization of polysaccharides extracted from *Chlorella* sp. by hot-water and alkali extraction methods. Algal Res 70. https://doi.org/10.1016/j.algal.2023.102970
12. Hitayezu E, Kang YH (2021) Effect of particle size on the physicochemical and morphological properties of *Hypsizygus marmoreus* mushroom powder and its hot-water extracts. Korean J Food Preserv 28(4). https://doi.org/10.11002/KJFP.2021.28.4.540

Chapter 5

Mushroom Phenolic Compounds: Identification, Quantification, and Biological Properties

Gülsen Tel-Çayan, Ebru Deveci, and Fatih Çayan

Abstract

Among natural resources, mushrooms have gained fame not only for their nutritional value and attractive taste but also for their health benefits and are used as supplements, nutraceuticals, and mycotherapy products. Mushrooms are declared as the "new superfoods of the future" as a treasure trove of a wide range of bioactive compounds, such as polysaccharides, lectins, terpenes, steroids, and phenolic compounds. Phenolic compounds in particular continue to be the focus of research with their wide range of bioactive properties. Therefore, the extraction of phenolic compounds from mushrooms constitutes the most critical step, especially in identification and bioactivity studies. In this regard, this chapter will guide researchers in the extraction, identification, quantification, and examination of the biological properties of phenolic compounds.

Key words Phenolic compounds, Mushrooms, Extraction, Bioactivities, Identification

1 Introduction

Mushrooms have been a part of human life for a long time. Today, mushrooms are among the most respected functional foods valued worldwide for their nutritional and therapeutic advantages. Pronounced as the "new superfood of the future," mushrooms have become a cornerstone of the human diet in improving human health and well-being [1].

Mushrooms are generally considered food due to their long history of consumption and are now widely cultivated worldwide. It has been reported that there are currently more than 40,000 mushroom species in the world, of which 2200 are edible or therapeutic and 60 species have been commercialized [2]. In addition to their desirable taste, mushrooms continue to be the focus of consumers with their remarkable nutritional values. Mushrooms, which have been proven to be a reserve of protein, carbohydrates, elements, and vitamins, have a low lipid content. In addition, mushrooms have been reported to contain dietary fibers such as

Pablo Navarro Simarro (ed.), *Fungi*, Methods and Protocols in Food Science, https://doi.org/10.1007/978-1-0716-4989-3_5,
© The Author(s), under exclusive license to Springer Science+Business Media, LLC, part of Springer Nature 2026

chitin and β-glucans and essential amino acids [3, 4]. Mushrooms, increasingly recognized as nutritional powerhouses, have also been the subject of numerous studies examining their health-promoting properties. These studies have revealed that mushrooms are equipped with bioactive lectins, polysaccharides, phenolics, terpenoids, and steroids, and in connection with these compounds, it has been proven that mushrooms have hypoglycemic, antidiabetic, antimicrobial, hypocholesterolemic, antioxidant, anticancer, antiviral, anti-inflammatory, hepatoprotective, immunomodulatory, and neuroprotective properties [5, 6]. The fact that phenolic compounds can be used as a natural alternative for the therapy and prevention of several health issues, such as cancer, cardiovascular disorders and neurodegeneration; their potential for used in cosmetic and pharmaceutical products; their preference in the development of functional foods aimed at maintaining good health as a natural alternative to artificial food additives are effective in increasing the interest in phenolic compounds compared to other compounds [7, 8].

Compounds bearing one or more hydroxyl groups directly connected to an aromatic ring are called phenolic compounds. Phenolic compounds are the most common bioactive support compounds in the human diet. It has been informed that more than 8000 phenolic compounds with a wide range of chemical structures have been identified to date [9]. The number of phenol rings and the linkage of the rings affect the structural diversity of phenolic compounds, and thus these compounds are divided into various subclasses such as simple phenols, phenolic acids, flavonoids, stilbenes, tannins, coumarins, and lignans. Mushrooms are the source of phenolic compounds, which are of noteworthy magnitude in the food and pharmaceutical industries [10]. Phenolic compounds, which have multiple bioactive benefits for human health, have been proven to have excellent antioxidant properties and a wide spectrum of bioactive properties, including anti-inflammatory, antimicrobial, anticancer, antiviral, hepatoprotective, hypocholesterolemic, hypoglycemic, immunomodulatory, cardioprotective, antidiabetic, and neuroprotective properties [11].

Extraction is the main stage in the extraction, identification and quantification of phenolic compounds. Since various factors such as solubility, structural differences, interactions with the matrix, and different distribution ratios of phenolic compounds affect the extraction, there is no single accepted procedure for the extraction of all phenolics or phenolics of a certain group [12]. A wide variety of liquid-liquid and solid-liquid extraction methods are used in the extraction of phenolic compounds from mushrooms, with the advantages of easy use, high yield, low cost, and general applicability. Among the extraction methods used for this purpose, maceration, phenolic-rich extraction, Soxhlet, infusion, and ultrasonic-assisted extraction are at the top of the list. The most used solvents are methanol, ethanol, acetone, water, and ethyl acetate and their combinations. High-performance liquid chromatography (HPLC) is the most basic analytical technique used in the identification and quantification of phenolic compounds in mushrooms [13]. Due to the variability and difficulties in obtaining phenolic compounds from mushrooms, the subject of this section is the most effective and widely used extraction methods to obtain extracts rich in phenolic compounds in the extraction of phenolic compounds from mushrooms. In addition, the identification, quantification, and bioactive properties of phenolic compounds are also examined.

2 Materials

1. Dried and powdered mushrooms.
2. Solvents: Ethanol, methanol, acetone, acetonitrile, ethyl acetate, ultrapure water (*see* **Note 1**).
3. Acids: Formic acid, acetic acid, phosphoric acid (*see* **Note 1**).
4. Phenolic compounds: Protocatechuic acid, gallic acid, *p*-hydroxybenzoic acid, catechin, cinnamic acid, *p*-coumaric acid, caffeic acid, chlorogenic acid, syringic acid, ferulic acid, quercetin, ellagic acid, pyrogallol, esculetin, kaempferol, rutin, gentisic acid, rosmarinic acid.

3 Methods

3.1 Pretreatment

To obtain maximum phenolic compounds from mushrooms, mushroom samples should be dried and powdered in a dry and light-protected manner. Thus, the surface area of the mushrooms will be increased and the efficiency of the number of phenolic compounds to be extracted will reach the maximum yield.

It is important to macerate the dried mushroom sample with hexane 3 times at room conditions before extraction to separate the fatty acids and other apolar compounds in it. This hexane extract to be obtained may be oily and fluid. This extract can later be used to characterize the oils in the mushroom. For the extraction yield to be high, care should be taken to ensure that the temperature is high. Otherwise, if low temperatures are used, the cell wall cannot soften or decompose, which prevents the release of bound phenolics [14].

3.2 Extraction of Phenolic Compounds

3.2.1 Maceration

1. The dried and powdered mushroom sample is mixed with the solvent at least 3 times and at most 10 times its weight (*see* **Note 2**).

2. The mixture is left for extraction for at least 1 day and at most 7 days.
3. The mixture is filtered with Whatman no. 4 paper.
4. The filtrate is centrifuged at 4000 rpm for 10 min (*see* **Note 3**).
5. The mushroom residue is re-extracted two more times with the same procedure mentioned above.
6. The supernatants are collected, combined, and the solvents are evaporated in a rotary evaporator at 40 °C to obtain the extract.
7. For storage conditions of the extract (*see* **Note 4**).

3.2.2 Phenolic-Rich Extraction

1. 3 g of mushroom sample is mixed with 30 mL of acetone: water (80:20 v/v) at −18 °C for 24 h (*see* **Note 5**).
2. The mixture is brought to room temperature.
3. The mixture is kept in an ultrasonic bath for 15 min.
4. The mixture is then centrifuged at 4000 rpm for 10 min, and the supernatant, which contains phenolic compounds, were filtered through Whatman no. 4 paper (*see* **Note 3**).
5. The mushroom residue is re-extracted two more times with the same procedure mentioned above.
6. All supernatants are combined, and the solvents are evaporated at 40 °C under reduced pressure.
7. For storage conditions of the extract (*see* **Note 4**).

3.2.3 Soxhlet Extraction

1. 25 g or 100 g of the mushroom sample is placed in a thimble (porous bag) made from either a firm filter paper or cellulose.
2. Extraction solvents such as ethanol, methanol, or water are placed in the bottom flask (*see* **Note 6**).
3. The solvent is then heated at 50 or 60 °C and vaporized in the sample thimble, is next condensed in the condenser on top of the apparatus, and then drips back.
4. The duration of the extraction can vary from 3 to 24 h (*see* **Note 7**).
5. After Soxhlet extraction is completed, the mixture is cooled to room temperature and filtered through Whatman no. 4 paper.
6. The ethanol and methanol extracts are concentrated at 40 °C under reduced pressure, and the water extract is concentrated with a freeze dryer.
7. For storage conditions of the extract (*see* **Note 4**).

3.2.4 Infusion

1. The dried and powdered mushroom sample is mixed with 5 times its volume of ultrapure water heated to 80 °C.
2. This mixture is left for extraction on a shaker at room temperature for 24 h.

3. The mixture is then filtered under vacuum by filtration.
4. The resulting mixture is centrifuged at 4000 rpm for 10 min, and the supernatants are collected (*see* **Note 3**).
5. The supernatants are then completely dried with a lyophilizer to obtain the water extract.
6. For short- or long-term storage, the extract must be frozen (*see* **Note 8**).

3.2.5 Ultrasonic-Assisted Extraction

1. Dried and powdered mushroom sample is mixed with a solvent 3 or 5 times its weight.
2. The mixture is left for extraction in an ultrasonic bath at a frequency range of 20 to 100 kHz, 20–70 °C for periods ranging from 15 min to 1 h (*see* **Note 9**).
3. The mixture is then filtered.
4. The solvent is evaporated in a rotary evaporator at 40 °C, and the extract is obtained.
5. For storage conditions of the extract (*See* **Note 4**).

3.3 Identification of Phenolic Compounds

Any of the extracts described in Subheading 3.2 is dissolved with the mobile phase system or methanol:water mixture and filtered through a disposable LC filter disk. Then, it is injected into the device where chromatographic separation will be performed.

1. Reverse-phase columns containing an octadecyl group are used for chromatographic separation.
2. If phenolic compounds are to be identified by HPLC, a reversed-phase column with an internal diameter ranging from 2.1 to 5 mm is used. If UHPLC is to be used, the internal diameter of the column should be between 1.1 and 2.1 mm.
3. In the identification of phenolic compounds by HPLC, the particle size of the reversed-phase column should be between 2.6 and 5.0 μm, and for the UHPLC device, it should be between 1.6 and 2.0 μm.
4. In both HPLC and UHPLC instruments, the column length can vary between 150 mm and 250 mm.
5. The flow rate of the mobile phase varies between 0.8 and 1.5 mL/min, and a gradient system should be used (*see* **Note 10**).
6. The mobile phase is usually a mixture of water and organic solvents such as methanol and acetonitrile, and an organic acid such as acetic acid, formic acid, or phosphoric acid is usually added at 0.5 to 1.5% to acidify the mobile phase (*see* **Note 11**).

7. HPLC or UHPLC coupled with a UV-Vis or photodiode array (PDA) detector is used for the identification of phenolic compounds.
8. The UV spectrum of each phenolic compound can be obtained by scanning between 200 and 400 nm.
9. If phenolic compounds are bound to non-UV-active sugar groups, which hinders the identification of phenolic compounds by UV-Vis based detectors, then an LC system coupled with a mass spectrometer (MS) or a tandem mass spectrometer (LC-MS/MS) allows the identification of phenolic compounds in complex matrices due to its higher sensitivity and better selectivity.
10. To identify phenolic compounds, first of all, chromatographic analysis of commercially available phenolic standard compounds should be performed, and complete separation should be ensured in the column. The standards of phenolic compounds should be analyzed individually under the relevant chromatographic conditions, calibration graphs should be drawn, and optimization experiments should be completed. After the analysis of the standards is completed and optimization experiments are performed, phenolic compounds of the mushroom extract can be identified by comparing retention times and UV data with commercial standards.

3.4 Quantification of Phenolic Compounds

1. The most commonly identified phenolic compounds in mushrooms (protocatechuic acid, gallic acid, *p*-hydroxy benzoic acid, catechin, cinnamic acid, *p*-coumaric acid, caffeic acid, chlorogenic acid, syringic acid, ferulic acid, quercetin, ellagic acid, pyrogallol, esculetin, kaempferol, rutin, gentisic acid, rosmarinic acid) are analyzed chromatographically on one of the HPLC, UHPLC, or LC devices separately and at different concentrations, and a calibration graph is drawn for each phenolic standard.
2. Quantification of the phenolic compounds of the mushroom extract is performed by comparing and mapping the retention times and UV data with the abovementioned standards.
3. Phenolic compound amounts of mushroom extract can be calculated in mg/L, mg/kg, mg/g, mg/100 g, μg/g, μg/mL, ppm, μg/100 g (*see* **Note 12**).

3.5 Biological Properties of Phenolic Compounds

1. The biological properties of phenolic compounds identified in mushroom extracts can be investigated by in vitro and in vivo studies using standards or by isolating them in pure form.
2. Table 1 can be examined for the biological properties of the main phenolic compounds identified in mushrooms in previous studies.

Table 1
Biological properties of the main phenolic compounds in mushrooms

Phenolic compounds	Biological properties
Protocatechuic acid	Analgesic, antiaging, antiallergic, antibacterial, anticancer, antidiabetic, antifibrogenic, antihyperglycemic, anti-inflammatory, antimicrobial, antimutagenic, antiosteoporotic, antioxidant, antiulcer, antiviral, neuroprotective properties [15]
Gallic acid	Anthelmintic, antibacterial, anticancer, anti-dengue, antidepressant, antidiabetic, antifungal, anti-inflammatory, antimalarial, anti-obesity, antioxidant, antiplatelet, antiseptic, antitubercular, antiviral, anxiolytic, cardioprotective, diuretic, nephroprotective, neuroprotective properties [16]
p-Hydroxy benzoic acid	Analgesic, antialgal, antiallergic, antibacterial, anticancer, antidiabetic, antifungal, antimicrobial, antimutagenic, antioxidant, antisickling, cardioprotective, hepatotoxic, insecticidal, nematicidal, vasodilatory properties [17]
Catechin	Antiallergic, antibacterial, antibacterial, anticancer, antidiabetic, anti-HIV, antihypertensive, anti-inflammatory, anti-obesity, antioxidant, antiviral, cardioprotective, neuroprotective properties [18]
Cinnamic acid	Anticancer, antidiabetic, anti-inflammatory, antimalarial, antimicrobial, antioxidant, neuroprotective properties [19]
p-Coumaric acid	Analgesic, anti-arthritis, anticancer, antidiabetic, anti-inflammatory, antimicrobial, anti-obesity, antioxidant, antiplatelet, antipyretic, antiulcer, antivirus, anxiolytic, cardioprotective, neuroprotective properties [20]
Caffeic acid	Analgesic, anti-atherosclerotic, antibacterial, anticancer, antihyperglycemic, anti-inflammatory, antineoplastic, antioxidant, antiproliferative, antithrombotic, antiviral, cardioprotective, immunostimulant properties [21]
Chlorogenic acid	Antibacterial, antihypertension, anti-inflammatory, antilipidemic, antimicrobial, anti-obesity, antioxidant, antiprebiotic, antipyretic, antiviral, cardioprotective, central nervous system stimulator, hepatoprotective, neuroprotective properties [22]
Syringic acid	Antiangiogenic, antibacterial, anticancer, antidiabetic, antiendotoxic, antihypertensive, anti-inflammatory, antimicrobial, antioxidant, cardioprotective, hepatoprotective, neuroprotective properties [23]
Ferulic acid	Analgesic, antibacterial, anticancer, antidiabetic, antifibrotic, anti-inflammatory, antioxidant, antiradiation, antiviral, hepatoprotective, immune-enhancing, immunostimulant, nephroprotective, neuroprotective, vascular endothelial protective properties [24]
Quercetin	Antiallergic, antiatherogenic, antibacterial, anticancer, antidiabetic, antihyperglycemic, antihypertensive, anti-inflammatory, antimicrobial, antioxidant, antiulcer, antiviral, hepatoprotective, neuroprotective properties [25]
Ellagic acid	Antibacterial, anticancer, antidiabetic, antifungal, antihyperlipidemic, anti-inflammatory, antimicrobial, antioxidant, antiviral, cardioprotective, gastroprotective, neuroprotective, probiotic properties [26]

(continued)

Table 1 (continued)

Phenolic compounds	Biological properties
Pyrogallol	Antibacterial, anticancer, antidermatitic, antifungal, anti-inflammatory, antimutagenic, antioxidant, antiviral, cardioprotective, disinfectant properties [27]
Esculetin	Anti-atherosclerotic, antibacterial, anticancer, anticoagulant, antidiabetic, antifungal, antihyperglycemic, anti-inflammatory, antioxidant, antiviral, immunomodulatory, neuroprotective properties [28]
Kaempferol	Anti-arthritis, antibacterial, anticancer, antidiabetic, antifungal, anti-inflammatory, antimicrobial, anti-obesity, antioxidant, antiparasitic, antithrombosis, antiviral, cardioprotective, gastroprotective, hepatoprotective, immunomodulatory, neuroprotective, radioprotective, renoprotective, wound healing properties [29]
Rutin	Antiallergic, antibacterial, anticancer, anticataract, antihypertensive, anti-inflammatory, antileishmanial, anti-osteoporotic, antioxidant, antiplatelet, antiprotozoal, antispasmodic, antithyroid, antiviral, cytoprotective, hypolipidemic, vasoactive properties [30]
Gentisic acid	Analgesic, anticancer, antidiabetic, antigenotoxic, anti-inflammatory, antimicrobial, antioxidant, cardioprotective, hepatoprotective, muscle relaxant, nephroprotective, neuroprotective, skin-lightening properties [31]
Rosmarinic acid	Antiaging, antiallergic, anticancer, antidepressant, antidiabetic, anti-inflammatory, antimicrobial, antioxidant, antiviral, cardioprotective, hepatoprotective, neuroprotective, renoprotective properties [32]

4 Notes

1. It is important that all chemicals to be used are HPLC grade, so that no solvent impurities are mixed into the mushroom extracts. In addition, in the analysis of the identification of phenolic compounds, the solvents to be used must be HPLC grade, so that the column is not damaged and for more accurate and reliable quantification.
2. The most used solvents in the extraction of phenolics by maceration are ethanol, methanol, acetone, or ethyl acetate and their mixtures with water in different ratios. Maceration is preferably carried out in a stoppered vessel to minimize solvent loss through evaporation and to increase the extraction efficiency of the phenolic compounds [12].
3. During centrifugal filtration, colloidal substances passing into the extracted mixture are precipitated, and a clearer upper phase is obtained. Thus, the extracted phenolic compounds are separated from colloidal substances. Also, the solubility problem of the extract is eliminated in further analyses.

4. If the extract is to be used in a short time, it can be stored in the refrigerator at +4 °C. If the extract is to be used for a long time, it should be frozen.
5. Here, the use of acetone prevents the mixture from passing from the liquid phase to the solid phase in the freezing environment. If acetone (this can be methanol) is not used, the mixture will freeze, and extraction will not occur. In the extraction of phenolic compounds by freezing, the aim is to perform the extraction without any change in the phenolic composition.
6. The number of solvents to be used for Soxhlet extraction varies between 100 and 500 mL; the best yield is obtained with this amount of solvent [33].
7. It should be noted that the extraction period should not exceed 1 day. Because continuous heating of the extracted phenolic compounds at high temperatures leads to the decomposition of thermolabile phenolic compounds [13].
8. This extract usually has a powdery appearance. It should be noted that if the water extract is stored at +4 °C, it will absorb moisture again.
9. The solvent used should have low vapor pressure. Because the disintegration of the cavitation bubble is deeper in the liquid medium compared to solvents with high vapor pressure [34]. Ultrasonic waves cause the collapse of cell walls, the reduction of particle sizes, and the destruction of fungal matrices, thus increasing the extraction rate and eliminating the need for high temperatures in the extraction process. High temperatures also cause solvent evaporation. It should be noted that prolonged sonication will negatively affect phenolic compounds.
10. The gradient method separates most phenolic compounds well, ensuring that most of their peaks show distinct and sharp individual peaks [35].
11. The addition of sufficient amount of acid to the mobile phase is beneficial to ensure complete separation, to attenuate peak signature and to improve the resolution of compounds [36]. The presence of acid in the mobile phase is necessary for complete dissolution of phenolics present in the mixture and efficient chromatography of these compounds, especially elimination of peak tails [37].
12. The amounts of the main phenolic compounds detected in mushrooms are reported in the literature as follows: protocatechuic acid (0.01–168.46 μg/g), gallic acid (0.01–290.34 μg/g), *p*-hydroxy benzoic acid (0.01–263.0 μg/g), catechin (0.89–112.61 μg/g), cinnamic acid (0.01–16.66 μg/g), *p*-coumaric acid (0.01–37.0 μg/g), caffeic acid (0.01–18.0 μg/g), chlorogenic acid

(10.01–63.04 μg/g), syringic acid (0.05–22.0 μg/g), ferulic acid (0.01–22.0 μg/g), quercetin (5.10–52.0 μg/g), ellagic acid (0.01–17.21 μg/g), pyrogallol (2.03–464.0 μg/g), esculetin (0.24–12.25 μg/g), kaempferol (0.4–53.0 μg/g), rutin (6.2–11.7 μg/g), gentisic acid (38.55–292.62 μg/g), rosmarinic acid (0.01–18.19 μg/g) [38–57].

References

1. Kour H, Kour D, Kour S, Singh S, Hashmi SAJ, Yadav AN, Kumar K, Sharma YP, Ahluwalia AS (2022) Bioactive compounds from mushrooms: emerging bioresources of food and nutraceuticals. Food Biosci 50:102124. https://doi.org/10.1016/j.fbio.2022.102124
2. Zhou Y, Chu M, Ahmadi F, Agar OT, Barrow DFR, Suleri HAR (2024) A comprehensive review on phytochemical profiling in mushrooms: occurrence, biological activities, applications and future prospective. Food Rev Int 40:924–951. https://doi.org/10.1080/87559129.2023.2202738
3. Hamza A, Mylarapu A, Krishna KV, Kumar DS (2024) An insight into the nutritional and medicinal value of edible mushrooms: a natural treasury for human health. J Biotech 381:86–99. https://doi.org/10.1016/j.jbiotec.2023.12.014
4. Lysakowska P, Sobota A, Wirkijowska A (2023) Medicinal mushrooms: their bioactive components, nutritional value and application in functional food production-a review. Molecules 28:5393. https://doi.org/10.3390/molecules28145393
5. Fogarasi M, Nemeş SA, Farcaş A, Socaciu C, Semeniuc CA, Socaciu MI, Socaci S (2024) Bioactive secondary metabolites in mushrooms: a focus on polyphenols, their health benefits and applications. Food Biosci 62:105166. https://doi.org/10.1016/j.fbio.2024.105166
6. Venturella G, Ferraro V, Cirlincione F, Gargano ML (2021) Medicinal mushrooms: bioactive compounds, use, and clinical trials. Int J Mol Sci 22:634. https://doi.org/10.3390/ijms22020634
7. Albuquerque BR, Heleno SA, Oliveira MBPP, Barros L, Ferreira ICFR (2021) Phenolic compounds: current industrial applications, limitations and future challenges. Food Funct 12:14–29. https://doi.org/10.1039/D0FO02324H
8. El-Saadony MT, Yang T, Saad AM, Alkafaas SS, Elkafas SS, Eldeeb GS, Mohammed DM, Salem HM, Korma SA, Loutfy SA, Alshahran MY, Ahmed AE, Mosa WFA, El-Mageed TAA, Ahmed AF, Fahmy MA, El-Tarabily MK, Mahmoud RM, Qamar SFA, El-Tarabily KA, Lorenzo JM (2024) Polyphenols: chemistry, bioavailability, bioactivity, nutritional aspects and human health benefits: a review. Int J Biol Macromol 277:134223. https://doi.org/10.1016/j.ijbiomac.2024.134223
9. Rathod NB, Elabed N, Punia S, Ozogul F, Kim SK, Rocha JM (2023) Recent developments in polyphenol applications on human health: a review with current knowledge. Plants 12:1217. https://doi.org/10.3390/plants12061217
10. Mihaylova D, Dimitrova-Dimova M, Popova A (2024) Dietary phenolic compounds-wellbeing and perspective applications. Int J Mol Sci 25(9):4769. https://doi.org/10.3390/ijms25094769
11. Abdelshafy AM, Belwal T, Liang Z, Wang L, Li D, Luo Z, Li L (2022) A comprehensive review on phenolic compounds from edible mushrooms: occurrence, biological activity, application and future prospective. Crit Rev Food Sci Nutr 62(22):6204–6224. https://doi.org/10.1080/10408398.2021.1898335
12. Alara OR, Abdurahman NH, Ukaegbu CI (2021) Extraction of phenolic compounds: a review. Curr Res Food Sci 4:200–214. https://doi.org/10.1016/j.crfs.2021.03.011
13. Shi L, Zhao W, Yang Z, Subbiah V, Suleria HAR (2022) Extraction and characterization of phenolic compounds and their potential antioxidant activities. Environ Sci Pollut Res 29:81112–81129. https://doi.org/10.1007/s11356-022-23337-6
14. Huynh KNN, Nguyen NVT, Ngan DT (2024) Natural phenolic compounds: a review of extraction and analysis. In: Exploring natural phenolic compounds-recent Progress and practical application. IntechOpen. https://doi.org/10.5772/intechopen.1007974
15. Cadena-Iniguez J, Santiago-Osorio E, Sanchez-Flores N, Salazar-Aguilar S, Soto-Hernandez RM, Riviello-Flores MDIL, Macias-Zaragoza VM, Aguiniga-Sanchez I (2024) The cancer-protective potential of

protocatechuic acid: a narrative review. Molecules 29:1439. https://doi.org/10.3390/molecules29071439

16. Harwansh RK, Deshmukh R, Shukla VP, Khunt D, Prajapati BG, Rashid S, Ali N, Elossaily GM, Suryawanshi VK, Kumar A (2024) Recent advancements in gallic acid-based drug delivery: applications, clinical trials, and future directions. Pharmaceutics 16:1202. https://doi.org/10.3390/pharmaceutics16091202
17. Manuja R, Sachdeva S, Jain A, Chaudhary J (2013) A comprehensive review on biological activities of *p*-hydroxy benzoic acid and its derivatives. Int J Pharm Sci Rev Res 22(2): 109–115.
18. Shikha D, Singh A, Rangra NK, Monga V, Bhati R (2024) Insights to therapeutic potentials, pharmaceutical formulations, chemistry and analytical methods of catechin. Phytochem Rev 23:1557–1598. https://doi.org/10.1007/s11101-024-09929-9
19. Ruwizhi N, Aderibigbe BA (2020) Cinnamic acid derivatives and their biological efficacy. Int J Mol Sci 21:5712. https://doi.org/10.3390/ijms21165712
20. Pei K, Ou J, Huang J, Ou S (2016) *p*-Coumaric acid and its conjugates: dietary sources, pharmacokinetic properties and biological activities. J Sci Food Agric 96(9):2952–2962. https://doi.org/10.1002/jsfa.7578
21. Maity S, Kinra M, Nampoothiri M, Arora D, Pai KSR, Mudgal J (2022) Caffeic acid, a dietary polyphenol, as a promising candidate for combination therapy. Chem Pap 76:1271–1283. https://doi.org/10.1007/s11696-021-01947-7
22. Nguyen V, Taine EG, Meng D, Cui T, Tan W (2024) Chlorogenic acid: a systematic review on the biological functions, mechanistic actions, and therapeutic potentials. Nutrients 16:924. https://doi.org/10.3390/nu16070924
23. Srinivasulu C, Ramgopal M, Ramanjaneyulu G, Anuradha CM, Suresh Kumar C (2018) Syringic acid (SA)-a review of its occurrence, biosynthesis, pharmacological and industrial importance. Biomed Pharmacother 108:547–557. https://doi.org/10.1016/j.biopha.2018.09.069
24. Kumar M, Kaushik D, Shubham S, Kumar A, Kumar V, Oz E, Brennan C, Zeng M, Proestos C, Çadırcı K, Bayrak M, Elobeid T, Karav S, Oz F (2025) Ferulic acid: extraction, estimation, bioactivity and applications for human health and food. J Sci Food Agric 105: 4168–4177. https://doi.org/10.1002/jsfa.13931
25. Ansari P, Choudhury ST, Seidel V, Rahman AB, Aziz MA, Richi AE, Rahman A, Jafrin UH, Hannan JMA, Abdel-Wahab YHA (2022) Therapeutic potential of quercetin in the management of type-2 diabetes mellitus. Life 12:1146. https://doi.org/10.3390/life12081146
26. Golmei P, Kasna S, Roy KP, Kumar S (2024) A review on pharmacological advancement of ellagic acid. J Pharmacol Pharmacother 15(2): 93–104. https://doi.org/10.1177/0976500X241240634
27. Ozturk Sarikaya SB (2015) Acethylcholinesterase inhibitory potential and antioxidant properties of pyrogallol. J Enzyme Inhib Med Chem 30(5):761–766. https://doi.org/10.3109/14756366.2014.965700
28. Garg SS, Gupta J, Sahu D, Liu CJ (2022) Pharmacological and therapeutic applications of esculetin. Int J Mol Sci 23:12643. https://doi.org/10.3390/ijms232012643
29. Yang L, Gao Y, Bajpai VK, El-Kammar HA, Simal-Gandara J, Cao H, Cheng KW, Wang M, Arroo RRJ, Zou L, Farag MA, Zhao Y, Xiao J (2023) Advance toward isolation, extraction, metabolism and health benefits of kaempferol, a major dietary flavonoid with future perspectives. Crit Rev Food Sci Nutr 63(16):2773–2789. https://doi.org/10.1080/10408398.2021.1980762
30. Al-Dhabi NA, Arasu MV, Park CH, Park SU (2015) An up-to-date review of rutin and its biological and pharmacological activities. EXCLI J 14:59–63. https://doi.org/10.17179/excli2014-663
31. Mechchate H, Es-Safi I, Mohamed Al Kamaly O, Bousta D (2021) Insight into gentisic acid antidiabetic potential using *in vitro* and *in silico* approaches. Molecules 26:1932. https://doi.org/10.3390/molecules26071932
32. Ijaz S, Iqbal J, Abbasi BA, Ullah Z, Yaseen T, Kanwal S, Mahmood T, Sydykbayeva S, Ydyrys A, Almarhoon ZM, Sharifi-Rad J, Hano C, Calina D, Cho WC (2023) Rosmarinic acid and its derivatives: current insights on anticancer potential and other biomedical applications. Biomed Pharmacother 162: 114687. https://doi.org/10.1016/j.biopha.2023.114687
33. Sridhar A, Ponnuchamy M, Kumar PS, Kapoor A, Vo DVN, Prabhakar S (2021) Techniques and modeling of polyphenol extraction from food: a review. Environ Chem Lett 19:

3409–3443. https://doi.org/10.1007/s10311-021-01217-8

34. Yeasmen N, Orsat V (2023) Green extraction and characterization of leaves phenolic compounds: a comprehensive review. Crit Rev Food Sci Nutr 63(21):5155–5193. https://doi.org/10.1080/10408398.2021.2013771
35. Mizzi L, Chatzitzika C, Gatt R, Valdramidis V (2020) HPLC analysis of phenolic compounds and flavonoids with overlapping peaks. Food Technol Biotechnol 58(1):12–19. https://doi.org/10.17113/ftb.58.01.20.6395
36. Zhong L, Yuan Z, Rong L, Zhang Y, Xiong G, Liu Y, Li C (2019) An optimized method for extraction and characterization of phenolic compounds in *Dendranthema indicum* var. *aromaticum* flower. Sci Rep 9:7745. https://doi.org/10.1038/s41598-019-44102-9
37. Dalluge JJ, Nelson BC, Thomas JB, Sander LC (1998) Selection of column and gradient elution system for the separation of catechins in green tea using high-performance liquid chromatography. J Chromatogr A 793:265–274. https://doi.org/10.1016/S0021-9673(97)00906-0
38. Demirtas N, Sengul GF, Dizeci N, Yildirim O (2024) Exploring the chemical composition and nutritional properties of six edible mushroom species from Turkey. J Food Compos Anal 133:106477. https://doi.org/10.1016/j.jfca.2024.106477
39. Tel-Çayan G, Fındık C (2023) HPLC-DAD analysis, and determination of antioxidant, anticholinesterase and antidiabetic activities of *Polyporus squamosus* (Huds.) Fr. Mugla. J Sci Technol 9(2):73–80. https://doi.org/10.22531/muglajsci.1336470
40. Tel-Çayan G, Deveci E, Çayan F (2023) Study on phenolic and organic acid compositions and antioxidant and enzyme inhibition activities of Agaricomycetes mushroom species from Turkey. Int J Med Mushrooms 25(11):11–25. https://doi.org/10.1615/IntJMedMushrooms.2023050127
41. Deveci E, Tel-Çayan G, Çayan F, Yılmaz Altınok B, Aktaş S (2023) Chemical characterization, *in vitro* antioxidant, anti-cancer and enzyme inhibition activities of three edible mushroom species. J Food Meas Charact 17:5518–5533. https://doi.org/10.1007/s11694-023-02066-5
42. Çayan F, Tel-Çayan G, Deveci E, Duru ME, Öztürk M (2022) Isolation and identification of compounds from truffle *Reddellomyces westraliensis* and their antioxidant, cytotoxic and enzyme inhibitory activities. Process Biochem 121:553–562. https://doi.org/10.1016/j.procbio.2022.08.001
43. Çayan F, Tel-Çayan G, Deveci E, Duru ME, Türk M (2022) A detailed study on multifaceted bioactivities of the extracts and isolated compounds from truffle *Reddellomyces parvulosporus*. Int J Food Sci Technol 57(3):1411–1419. https://doi.org/10.1111/ijfs.15190
44. Çayan F, Tel-Çayan G, Deveci E, Duru ME (2021) A comprehensive study on phenolic compounds and bioactive properties of five mushroom species via chemometric approach. J Food Process Preserv 45:e15695. https://doi.org/10.1111/jfpp.15695
45. Çayan F, Tel-Çayan G, Deveci E, Duru ME (2021) HPLC-DAD characterization of phenolic profile and *in vitro* antioxidant, anticholinesterase, and antidiabetic activities of five mushroom species from Turkey. 3 Biotech 11:273. https://doi.org/10.1007/s13205-021-02819-3
46. Tel-Çayan G, Çayan F, Deveci E, Duru ME (2021) Phenolic profile, antioxidant and cholinesterase inhibitory activities of four *Trametes* species: *T. bicolor, T. pubescens, T. suaveolens,* and *T. versicolor.* J Food Meas Charact 15:4608–4616. https://doi.org/10.1007/s11694-021-01034-1
47. Çayan F, Deveci E, Tel-Çayan G, Duru ME (2020) Identification and quantification of phenolic acid compounds of twenty-six mushrooms by HPLC-DAD. J Food Meas Charact 14:1690–1698. https://doi.org/10.1007/s11694-020-00417-0
48. Çayan F, Tel-Çayan G, Deveci E, Öztürk M, Duru ME (2019) Chemical profile, *in vitro* enzyme inhibitory, and antioxidant properties of *Stereum species* (Agaricomycetes) from Turkey. Int J Med Mushrooms 21(11):1075–1087. https://doi.org/10.1615/IntJMedMushrooms.2019032893
49. Tel-Çayan G (2019) Phenolic profiles, antioxidant, and anticholinesterase activities of three *Gloeophyllum* species with chemometric approach. J Food Biochem 43(4):e12790. https://doi.org/10.1111/jfbc.12790
50. Deveci E, Tel-Çayan G, Duru ME, Öztürk M (2019) Chemical constituents of *Porodaedalea pini* mushroom with cytotoxic, antioxidant and anticholinesterase activities. J Food Meas Charact 13:2686–2695. https://doi.org/10.1007/s11694-019-00189-2
51. Deveci E, Tel-Çayan G, Duru ME (2019) Evaluation of phenolic profile, antioxidant and anticholinesterase effects of *Fuscoporia torulosa*. Int

J Second Metab 6:79–89. https://doi.org/10.21448/ijsm.496327

52. Cayan F, Deveci E, Tel-Cayan G, Duru ME (2018) Phenolic acid profile of six wild mushroom species by HPLC-DAD. Chem Nat Compd 54:985–986. https://doi.org/10.1007/s10600-018-2529-2
53. Tel-Çayan G, Deveci E, Çayan F, Duru ME (2018) Comparative assessment of phytochemical composition, antioxidant and anticholinesterase activities of two Basidiomycota truffle fungi from Turkey. Marmara Pharm J 22(1): 59–65. https://doi.org/10.12991/mpj.2018.41
54. Tel-Çayan G, Muhammad A, Duru ME, Öztürk M, Adhikari A, Türkoğlu A (2016) A new fatty acid ester from an edible mushroom *Rhizopogon luteolus*. Nat Prod Res 30(20): 2258–2264. https://doi.org/10.1080/14786419.2016.1160237
55. Tel-Çayan G, Öztürk M, Duru ME, Rehman MU, Adhikari A, Türkoğlu A, Choudhary MI (2015) Phytochemical investigation, antioxidant and anticholinesterase activities of *Ganoderma adspersum*. Ind Crop Prod 76:749–754. https://doi.org/10.1016/j.indcrop.2015.07.042
56. Palacios I, Lozano M, Moro C, D'Arrigo M, Rostagno MA, Martinez JA, Garcia-Lafuente-A, Guillamon E, Villares A (2011) Antioxidant properties of phenolic compounds occurring in edible mushrooms. Food Chem 128:674–678. https://doi.org/10.1016/j.foodchem.2011.03.085
57. Kim MY, Seguin P, Ahn JK, Kim JJ, Chun SC, Kim EH, Seo SH, Kang EY, Kim SL, Park YJ, Ro HM, Chung IM (2008) Phenolic compound concentration and antioxidant activities of edible and medicinal mushrooms from Korea. J Agric Food Chem 56(16): 7265–7270. https://doi.org/10.1021/jf8008553

Chapter 6

Extraction and Measurement of Ergosterol in Mushroom and Mushroom Substrate

Paramjeet Saroj and Shaojun Xiong

Abstract

Several distinct lipids may serve as useful indicators for differentiating between species of fungus. As a subclass of lipids, ergosterol is a particular kind of sterol. It is mostly present in fungal cell membranes. However, fungi differ in their cell size, shape, cell wall structure, and other attributes, so it is necessary to investigate extraction techniques to recover these bioactives effectively. The most widely utilized of these methods is the quantification of ergosterol, a biomarker that is mostly exclusive to fungi. The ergosterol from fungal-colonized substrates is extracted using methanol and measured by high-performance liquid chromatography (HPLC) with UV detection at 282 nm. This chapter designs a method to quantify the mycelia in fully colonized mushroom substrate, such as spent mushroom substrate (SMS).

Key words Mushroom, Ergosterol, Indicator, Extraction, Quantification, HPLC

1 Introduction

Mushrooms are valued for their culinary, therapeutic, and nutritional qualities. Bioactive substances, such as proteins, lipids, phenolic acids, and polysaccharides, are abundant in them. Although mushrooms are known for having a high protein, vitamin (including B-complex and D), mineral, and antioxidant content, they have a comparatively low-fat content. Nonetheless, in some species, the lipid fraction is important because it contains healthy elements, including ergosterol (a precursor to vitamin D2), polyunsaturated fatty acids (PUFAs), and other bioactive lipids [1–3]. The pharmaceutical, functional food, and nutraceutical industries are very interested in these lipids because of their well-known anti-inflammatory, antioxidant, and antibacterial qualities. To fully comprehend and take advantage of these chemicals' potential health benefits, effective lipid extraction techniques are necessary [4].

As a by-product of mushroom growth, spent mushroom substrate (SMS) is composed of both fungal mycelia and partly decomposed

Pablo Navarro Simarro (ed.), *Fungi*, Methods and Protocols in Food Science, https://doi.org/10.1007/978-1-0716-4989-3_6,
© The Author(s), under exclusive license to Springer Science+Business Media, LLC, part of Springer Nature 2026

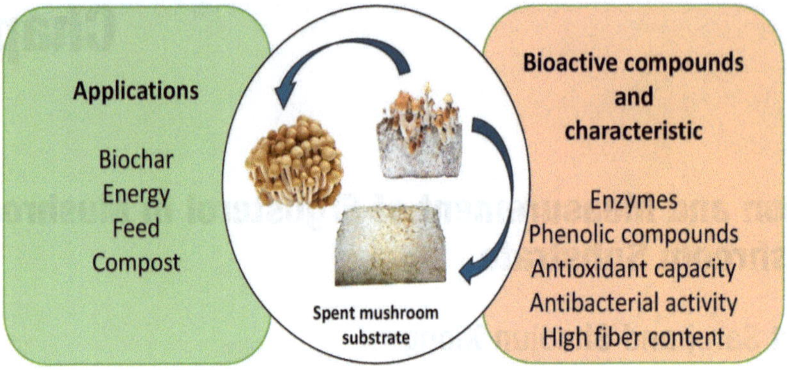

Fig. 1 Description of mushroom production, bioactive chemicals found in mushroom substrate, and their primary uses

1.1 Bioactive Compounds in Spent Mushroom Substrate

lignocellulose from initial substrate ingredients (such as wood or crop residues). SMS contains a variety of beneficial compounds (Fig. 1). These consist of vitamins, sterols, polysaccharides, proteins, and polyphenols that are mostly present in the fungal mycelium [5]. Furthermore, the process of mycelial growth releases bioactive substances into the environment, and fungi's breakdown of lignin and polysaccharides results in oligomeric molecules that may have bioactive properties. There are also useful chemicals in the lignocellulosic substrate's extractive fractions. While the bioactive compounds in edible fungal fruiting bodies have been extensively studied, there is still limited knowledge about the bioactive potential of SMS. Recovering these bioactive molecules from SMS represents an intriguing approach to its valorization [5–7].

Mushroom cultivation generates large amounts of SMS alongside the harvested mushroom fruitbodies. In 2022, global SMS production was estimated at approximately 242 million tons, with 5 kilograms of SMS produced for every kilogram of mushrooms grown [8]. Historically, SMS has been considered waste despite its high organic matter and micronutrient content, leading to environmental challenges. Over the past decade, the mushroom production industry has increasingly focused on addressing this issue, resulting in research aimed at finding economically and environmentally sustainable reuse methods [9].

Some effective strategies for SMS valorization [5, 9] include:

Agricultural applications: Using SMS as a natural fertilizer and soil amendment agent.

Mushroom cultivation: Supplementing SMS to enhance nutrient levels for subsequent cultivation cycles.

Bioremediation: Employing SMS as a soil bioremediation agent to address contamination.

Biofuel production: Utilizing SMS as a feedstock for biofuel generation.

These approaches not only mitigate environmental concerns but also contribute to a circular economy by transforming waste into valuable resources.

1.2 Ergosterol in Mushrooms

Ergosterol (ergosta-5,7,22-trien-3β-ol) is the most abundant sterol found in edible mushrooms, though its concentration can vary significantly. This variation is primarily influenced by factors such as growing conditions and post-harvest management. Most studies on ergosterol extraction focus on mushroom fruiting bodies that do not meet commercial standards or on fruitbody stipes [5, 10]. However, recent research has shown that *Lentinula edodes* (shiitake mushroom) spent mushroom substrate (SMS) can also serve as a potential source of ergosterol. By using ultrasound-assisted extraction, a novel method for extracting natural products from various biomaterials, ergosterol-rich extracts were obtained from *L. edodes* SMS [5]. In vitro tests demonstrated that the resulting extracts exhibited anticancer activity against three different cancer cell lines [11].

1.3 Lipid Extraction from Mushroom

Sterols, phospholipids, triglycerides, and other related substances are among the fat-soluble components that must be isolated in order to extract lipids from mushrooms. Lipid extraction can be done in a number of ways, such as pressing, solvent extraction, and supercritical fluid extraction. Solvent extraction is still the most widely used and economical of these techniques; it frequently makes use of organic solvents, including methanol, hexane, and chloroform.

1.4 Using Methanol as a Solvent to Extract Lipids

Methanol is a polar solvent that is frequently employed in lipid extraction because it can dissolve a variety of lipophilic compounds, such as fatty acids, sterols, and phospholipids. When it comes to extracting lipids from mushrooms, methanol has various benefits:

1. Methanol effectively removes lipids from biological materials, both polar and nonpolar. Mushrooms contain a mix of neutral lipids (like triglycerides) and polar lipids (like phospholipids and glycolipids), and methanol works especially well.
2. Methanol's polarity may cause it to extract lipid components and associated chemicals like phenolic compounds preferentially, which may affect the final extract's overall quality. To optimize lipid yield and purity, methanol is frequently mixed with other solvents, such as chloroform, in solvent mixes, such as the Bligh and Dyer method [12].
3. Methanol is a more appealing alternative for laboratory-scale extractions than harsher organic solvents like chloroform

because it is less hazardous, especially when taking into account the extracts' possible use in food or medicine.

Additionally, methanol's ability to break down complex cellular structures and disrupt cellular membranes makes it ideal for extracting lipids from mushrooms, which have intricated cell walls and membranes composed of polysaccharides and chitin.

2 Materials

1. Screw-cap extraction glass tubes (approx. 45 mL).
2. Vortex.
3. Magnetic stirrer and stirring bars.
4. Vacuum evaporator (BUCHI Syncore®, Germany) or N_2 gas.
5. Lyophilizer (Christ Alpha 1–4 LSCplus, Germany).
6. HPLC (Shimadzu, Canby, USA RP C18, UV detector set to 282 nm with reversed-phase C18 column (Ultra C18, 5 μm, 150 × 4.6 mm, Restek, USA)).
7. HPLC vials.
8. Pipettes and sterile syringes filters.
9. Incubator.
10. Centrifuge.
11. Methanol (analytical grade).
12. 2-propanol or isopropanol (analytical grade).
13. *n*-hexane (analytical grade).
14. Ergosterol standard ((CAS number: 57–87-4), purity >98%, Fisher Scientific, USA).
15. KOH (pellets, analytical grade).

3 Methods

3.1 Sample Preparation

To prepare the samples, first grind dried mushrooms or SMS into a < 0.2 mm in size using a grinder or homogenizer. Grinding helps to increase the surface area, thereby improving extraction efficiency. After grinding, store the powdered samples in airtight bags to preserve their quality. If fresh mushrooms or SMS are used, the water content may interfere with the extraction process. Therefore, it is necessary to freeze-dry the materials for at least 72 h before grinding.

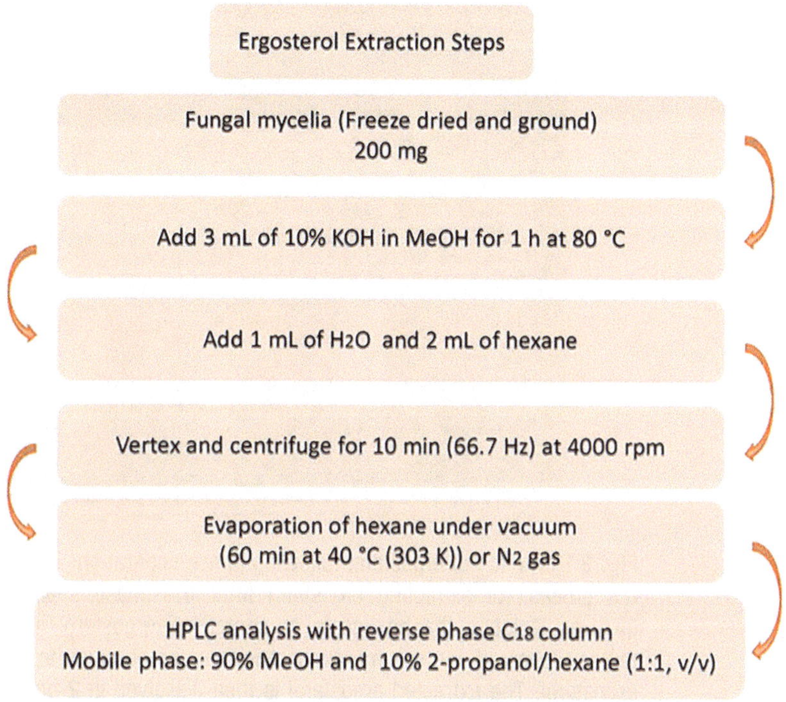

Fig. 2 Illustration of the steps involved in lipid extraction from fungal biomass [13, 14]

3.2 Solvent Extraction Steps

Figure 2 demonstrates the sequential process for lipid extraction, which includes the following steps:

1. Sample Preparation: Weigh freeze-dried 100 mg of mushroom or 200 mg of SMS material with a particle size less than 0.2 mm.
2. Saponification: Add 3 mL of 10% KOH in methanol to the ground material and incubate the mixture for 1 h at 80 °C.
3. Extraction: add 1 mL of water and 2 mL of hexane to the mixture. Vortex the mixture for 5–10 mins and centrifuge at 4000 rpm for 10 min and collect the hexane phase (upper transparent phase) (*see* Fig. 3). Repeat the extraction and pool the hexane phases (*see* **Note 1**).
4. Solvent Removal and Resuspension: Evaporate the hexane under vacuum at 40 °C for 60 mins or a gentle stream of nitrogen gas for 7–10 min (*see* **Note 2**). Dissolve the residue in 2-propanol/hexane (1:1) (*see* **Note 3**).
5. HPLC Analysis: Use a reversed-phase C18 column (Ultra C18, 5 μm, 150 × 4.6 mm, Restek, USA). Employ a mobile phase of 90% methanol and 10% (1:1) 2-propanol/hexane.

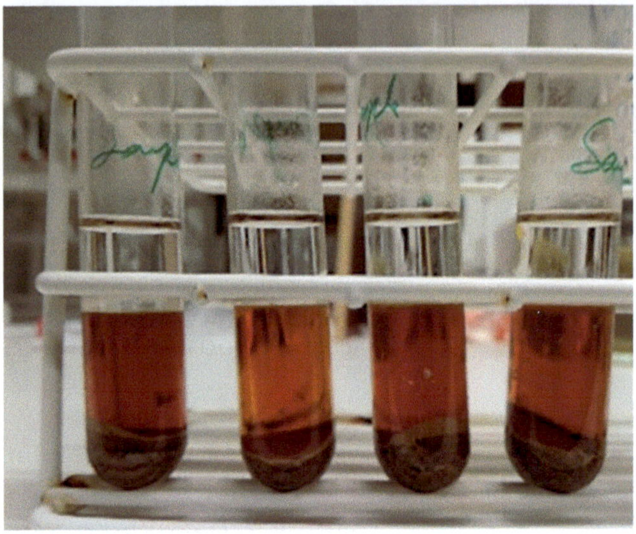

Fig. 3 Extraction of ergosterol following phase separation. The figure illustrates the process of extracting ergosterol after the phase separation step. This involves isolating the hexane layer containing ergosterol from the aqueous phase, followed by solvent evaporation under vacuum to concentrate the ergosterol. The extracted ergosterol is then dissolved in 2-propanol/hexane (1:1) for further analysis

3.3 Ergosterol Estimation by HPLC

1. HPLC Setup: Set the retention time for ergosterol to approximately 10 min. Mobile phases A and B, consisting of 90% methanol and 10% 2-propanol/hexane (1:1), respectively. Adjust the flow rate to 1 mL/min. Maintain the column temperature at 40 °C. Set the detection wavelength to 282 nm. Use an injection volume of 10 μL for both standards and samples (*see* **Note 4**).
2. Standard Curve Preparation: Prepare a standard curve using ergosterol stock solution in 2-propanol/hexane (1:1) at concentrations of 0, 20, 40, 60, 80, and 100 μg/mL.
3. Sample Analysis: Run the prepared standards on the HPLC system. Inject 10 μL of each sample and record the chromatographic data.
4. Peak Identification: Verify the identity of putative ergosterol peaks by co-injection of the ergosterol standard with the sample extract. Analyze of the UV spectrum for the characteristic absorption pattern.
5. Collect the data of measured peak areas for all samples and standards after the retention time is completed.
6. Calculate the ergosterol concentration in the samples by comparing the peak areas with the standard curve.

4 Notes

1. Repeating the extraction procedure on the same sample can enhance the overall extraction efficiency by recovering any residual ergosterol remaining in the sample after the initial extraction.
2. Evaporation under nitrogen involves using a gentle stream of nitrogen gas to remove hexane. The evaporation of hexane under nitrogen gas usually takes 7–10 min for each sample. This process prevents oxidation by displacing oxygen around the sample, making it suitable for compounds that are sensitive to air.
3. The solubility of the ergosterol standard is better in 2-propanol/hexane (1:1) than in methanol due to the chemical nature of ergosterol and the solvents. Ergosterol has a large hydrophobic region due to its fused ring system and hydrocarbon tail, along with a hydroxyl (-OH) group that provides some polarity. The combination of polar and nonpolar solvents in 2-propanol/hexane accommodates the amphiphilic nature of ergosterol. Hexane reduces the overall polarity of 2-propanol, improving the solubility of hydrophobic regions.
4. The injection volumes for both standards and samples should be identical to ensure accurate comparison.

Acknowledgments

This work was supported by Kempestiftelserna grant JCK-2125.

References

1. Saad AM, Sitohy MZ, Omar BA, El-Saadony MT, Sitohy B (2022) Mushroom; chemistry, bioactive components, and application. In: Current topics in functional food. IntechOpen. https://doi.org/10.5772/intechopen.104182
2. Cateni F, Gargano ML, Procida G, Venturella G, Cirlincione F, Ferraro V (2022) Mycochemicals in wild and cultivated mushrooms: nutrition and health. Phytochem Rev 21(2):339–383. https://doi.org/10.1007/s11101-021-09748-2
3. Sánchez C (2017) Bioactives from mushroom and their application. In: Food bioactives: Extraction and biotechnology applications, pp 23–57. https://doi.org/10.1007/978-3-319-51639-4_2
4. Das S, Prakash B (2022) Edible mushrooms: nutritional composition and medicinal benefits for improvement in quality life. In: Research and technological advances in food science. Academic Press, pp 269–300. https://doi.org/10.1016/B978-0-12-824369-5.00013-0
5. Martín C, Zervakis GI, Xiong S, Koutrotsios G, Strætkvern KO (2023) Spent substrate from mushroom cultivation: exploitation potential toward various applications and value-added products. Bioengineered 14(1):2252138. https://doi.org/10.1080/21655979.2023.2252138
6. Leong YK, Ma TW, Chang JS, Yang FC (2022) Recent advances and future directions on the valorization of spent mushroom substrate (SMS): a review. Bioresour Technol 344:

126157. https://doi.org/10.1016/j.biortech.2021.126157

7. Papoutsis K, Grasso S, Menon A, Brunton NP, Lyng JG, Jacquier JC, Bhuyan DJ (2020) Recovery of ergosterol and vitamin D2 from mushroom waste-potential valorization by food and pharmaceutical industries. Trends Food Sci Technol 99:351–366. https://doi.org/10.1016/j.tifs.2020.03.005
8. Ravlikovsky A, Pinheiro MNC, Dinca L, Crisan V, Symochko L (2024) Valorization of spent mushroom substrate: establishing the foundation for waste-free production. Recycling 9(3):44. https://doi.org/10.3390/recycling9030044
9. Kousar A, Khan HA, Farid S, Zhao Q, Zeb I (2024) Recent advances on environmentally sustainable valorization of spent mushroom substrate: a review. Biofuels Bioprod Biorefin 18(2):639–651. https://doi.org/10.1002/bbb.2559
10. Hoa HT, Wang CL, Wang CH (2015) The effects of different substrates on the growth, yield, and nutritional composition of two oyster mushrooms (*Pleurotus ostreatus* and *Pleurotus cystidiosus*). Mycobiol 43(4):423–434. https://doi.org/10.5941/MYCO.2015.43.4.423
11. Zou G, Li B, Wang Y, Yin X, Gong M, Shang J, Wei Y, Li X, Bao D (2021) Efficient conversion of spent mushroom substrate into a high value-added anticancer drug pentostatin with engineered *Cordyceps militaris*. Green Chem 23(24):10030–10038. https://doi.org/10.1039/D1GC03594K
12. Bligh EG, Dyer WJ (1959) A rapid method of total lipid extraction and purification. Can J Biochem Physiol 37(8):911–917. https://doi.org/10.1139/o59-099. PMID: 13671378
13. Kuijk V, Sandra JA, Sonnenberg ASM, Baars JJP, Wouter HH, José CR, Jorge Rencoret AG, Norbert CR, John WC (2017) Chemical changes and increased degradability of wheat straw and oak wood chips treated with the white rot fungi Ceriporiopsis subvermispora and *Lentinula edodes*. Biomass Bioenergy 105: 381–391. https://doi.org/10.1016/j.biombioe.2017.07.003
14. Niemenmaa O, Galkin S, Hatakka A (2008) Ergosterol contents of some wood-rotting basidiomycete fungi grown in liquid and solid culture conditions. Int Biodeterior Biodegrad 62(2):125–134. https://doi.org/10.1016/j.ibiod.2007.12.009

Chapter 7

Identification of Aromas and VOC in Solid-State Fermented Foods

Òscar Boronat Nielsen, Montserrat Riu-Aumatell, and Helena Martin-Gómez

Abstract

Aromas play a crucial role in shaping consumer acceptance of food products. Investigating the volatile organic compounds (VOCs) produced during fermentation provides insight into which microorganisms or substrates generate specific VOCs and how these compounds influence sensory perception and consumer preference.

This chapter describes a reproducible protocol for identifying and evaluating the VOCs in solid-state fermented foods using headspace solid-phase microextraction coupled with gas chromatography–mass spectrometry (HS-SPME-GC–MS) and sensory evaluation via check-all-that-apply (CATA). Using tempeh as a model, combining analytical analysis with sensory analysis allows us to better characterize the fermented product.

Key words Volatile organic compounds, Aroma, GC, MS, HS-SPME, CATA, Fermentation

1 Introduction

Aromas play a dominant role in how we perceive the flavor of foods. The sense of smell is estimated to account for up to ~80% of what is commonly perceived as flavor, underscoring its high importance in consumer perception [1, 2]. Consequently, volatile organic compounds (VOCs), which are responsible for a food's aroma, are essential in determining its overall sensory appeal. Thus, understanding which aroma compounds are present in a food and how they contribute to flavor and liking can provide insight into consumer preferences and guide the development of new appealing foods. Fermented foods are a diverse and abundant source of VOCs, making them interesting for studies on flavor development. During fermentation processes (e.g., solid-state fermentation), microorganisms metabolize components of the raw substrate into secondary metabolites, including many aromatic compounds [3, 4]. Profiling the VOCs in fermented products can reveal

Pablo Navarro Simarro (ed.), *Fungi*, Methods and Protocols in Food Science, https://doi.org/10.1007/978-1-0716-4989-3_7,
© The Author(s), under exclusive license to Springer Science+Business Media, LLC, part of Springer Nature 2026

which aromas might attract or repel consumers—knowledge essential for developing novel fermented foods with broad consumer acceptance [5].

VOCs can be analyzed using various analytical techniques, depending on the food matrix and the goals of the analysis. Gas chromatography–mass spectrometry (GC–MS) is one of the most widely used methods for detecting and identifying VOCs in foods [6]. This technique combines gas chromatography (GC), which separates volatile molecules, with mass spectrometry (MS), which detects and identifies them. In GC, the VOCs in a sample are first vaporized and then carried by an inert carrier gas (commonly helium or nitrogen) through a capillary column coated with a stationary phase. As the VOCs travel through the column, they interact with the stationary phase to varying degrees, resulting in different retention times based on their physicochemical properties. Once separated, the compounds enter the mass spectrometer, where they are ionized and subsequently fragmented. The resulting ions are analyzed based on their mass-to-charge (m/z) ratios, producing characteristic spectra that present a molecular fingerprint for each compound. These characteristic fragmentation patterns enable precise identification, while the relative intensities of the ion signals allow for qualitative and semiquantitative analysis. When used in combination with reference standards or calibration curves, GC–MS provides robust compound identification and quantification. By integrating the chromatographic peaks and comparing the mass spectra to reference libraries or authentic standards, GC–MS offers high specificity and sensitivity in the identification and quantification of VOCs [7].

For solid or complex samples (e.g., tempeh or koji), an initial solvent extraction step is often required to release VOCs into a liquid form for GC analysis [8–10]. Alternatively, headspace solid-phase microextraction (HS-SPME) provides a solvent-free sample preparation technique for VOC analysis. In this method, a fiber coated with an extracting phase is exposed to the headspace of a sealed vial containing the sample. To enhance VOC extraction, the sample is often preheated and incubated at a controlled temperature for a defined period, promoting the release and adsorption of volatiles onto the fiber [11]. After extraction, the VOC-laden fiber is inserted into the GC inlet, where the compounds are thermally desorbed for analysis. In this protocol, HS-SPME is employed to simplify the procedure and eliminate the need for solvent-based extraction.

However, instrumental analysis alone does not always align with human sensory perception. Some compounds may produce strong signals in GC–MS but have little to no impact on aroma, while others can be perceived by the human nose at concentrations below the detection limits of the instrument [12, 13].

Therefore, sensory analysis is essential to identify which aromas are perceptible and relevant to consumers. Sensory evaluation can be conducted using either a trained panel or untrained panelists. A trained panel, composed of a smaller group of participants, undergoes specialized training to recognize and describe specific aromas. In contrast, consumer tests require a larger number of participants, often exceeding 60, to collect statistically robust acceptance and preference data of the general population.

Each method has specific advantages and limitations. This study adopts a general consumer approach to assess product perception and identify the sensory attributes driving acceptance. To determine perceived aromas, check-all-that-apply (CATA) and rate-all-that-apply (RATA) methods are used. These techniques require participants to select or rate, respectively, the aroma descriptors they detect in each sample. Subsequently, statistical analysis reveals which aroma attributes are linked to overall liking, helping identify VOCs that enhance or reduce consumer acceptance [5].

By integrating instrumental volatilomic analysis with sensory evaluation, the study provides a comprehensive understanding of the samples' aromatic profiles. This combined approach enables both the identification of VOCs through GC–MS and the assessment of their sensory relevance, supporting the development of products with optimized flavor and enhanced consumer appeal. In some cases, it may be possible to directly associate specific VOCs with corresponding aroma descriptors. Based on this comparison, hypotheses can be made regarding the origin of these metabolites—whether they result from specific microbial metabolic pathways, originate from the raw substrate, or arise from interactions between microorganisms or between microbes and the substrate.

The goal of this chapter is to provide detailed protocols for the identification and sensory evaluation of volatile compounds in solid-state fermented foods, combining instrumental analysis (GC–MS) with consumer-based sensory methods.

2 Materials

1. Fresh tempeh sample.
2. Sodium chloride (optional, to enhance VOC release).
3. Analytical scale.
4. Gas chromatography–mass spectrometry (GC-MS) system with SPME attachment.
5. 20 mL headspace crimp vials and caps.
6. Vial crimper.
7. Magnetic hotplate stirrer.

8. HS-SPME fiber (50/30um DVB/CAR/PDMS, Stableflex (2 cm)).
9. 50 mL beaker.
10. Ultrasonic bath.
11. Sensory analysis room with individual cabinets.
12. RedJade Software.
13. XLSTAT Software.
14. RStudio Software.

3 Methods

3.1 Headspace Solid-Phase Microextraction (HS-SPME)

HS-SPME enables the extraction of VOCs from solid or liquid samples without the use of solvents. The choice of fiber can be adjusted according to the nature of the sample and the polarity of the target VOCs.

3.1.1 Sample Preparation and Pretreatment

1. Mix 3.5 g of fresh tempeh with 5.8 g of 20% (*w/v*) NaCl solution (*see* **Notes 1** and **2**) and grind with a mortar until a smooth paste is obtained (*see* **Note 3**).
2. Add 5 μL of a 15% hydroalcoholic solution of *2-pentanol, 4-methyl* (internal standard) (*see* **Note 4**).
3. Place 9.3 g of the previous mixture into a 20 mL HS-SPME vial and seal.
4. Add the vial to an ultrasonic water bath at 120 W and 40 KHz at 30 °C for 15 min (*see* **Note 5**).

3.1.2 Headspace Solid-Phase Microextraction (HS-SPME)

1. Place 30 mL of water in a 50 mL beaker and set it on a hot stirring plate. Adjust the temperature to 45 °C and allow it to equilibrate.
2. Place the vial into the beaker so that the water level reaches the vial cap, and let it equilibrate for 10 min (*see* **Note 6**).
3. Clean the SPME fiber by inserting it into the GC–MS inlet for 2 min before use (*see* **Note** 7).
4. After the equilibration at 45 °C for 10 min, insert the fiber into the vial and expose it to the sample headspace for 45 min.

3.2 Gas Chromatography–Mass Spectrometry (GC–MS) and VOC Analysis

3.2.1 GC–MS Run

1. Run the operation in gradient mode: hold initial temperature at 40 °C for 5 min, and increase it from 40 °C to 190 °C at 3 °C/min and from 190 °C to 220 °C at 10 °C/min and hold for 5 min using splitless injection mode (*see* **Note 8**).
2. Insert fiber into the GC–MS injector for thermal desorption and analysis. To prevent fiber damage, remove the fiber 5–10 min after the run begins.

3. Integrate the chromatographic peaks, either manually or automatically, and identify the corresponding compounds for each peak using spectral libraries (*see* **Note 9**).

3.2.2 GC–MS Analysis

1. Estimate the concentration of each VOC normalized to the internal standard:

$$\mathrm{VOC}\left(\frac{\mu\mathrm{g}}{\mathrm{g}}\mathrm{sample}\right)=\frac{\mathrm{Area\ VOC}}{\mathrm{Area\ IS}}\times\frac{750\ \mu\mathrm{g\ IS}}{9.3\mathrm{g\ Sample}}$$

2. Generate a heatmap to visualize the abundance of VOCs across the different samples.
3. Consult flavor and aroma databases (e.g., FlavorDB) to determine the sensory descriptors associated with each identified VOC.

3.3 Sensory Analysis (Check-All-That-Apply)

Check-all-that-apply (CATA) is a sensory evaluation method used to identify the attributes perceived by consumers in a sample. It is commonly combined with a hedonic (liking) question to explore relationships between specific sensory attributes and overall consumer preferences. The following protocol describes the application of the CATA method to evaluate three raw tempeh samples using a panel of 60 consumers.

3.3.1 Questionnaire Preparation

1. Prepare a questionnaire using RedJade or other suitable software. It should include:
   (a) A sociodemographic question to collect information such as gender, age, and nationality (*see* Fig. 1).
   (b) A liking + CATA question: for each sample, include a question asking consumers to rate their liking on a 1-to-9-point hedonic scale, or another odd-numbered

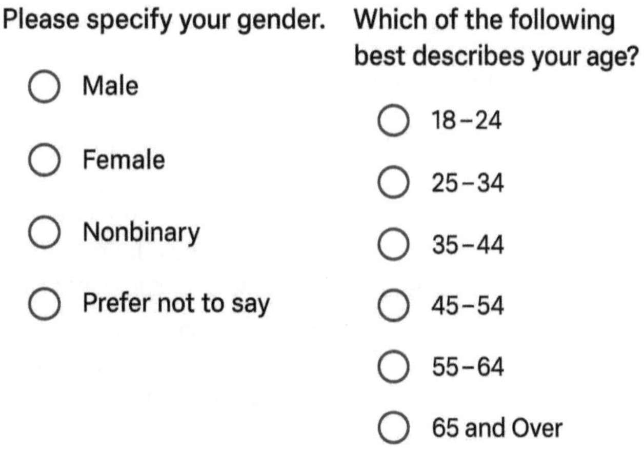

Fig. 1 Examples of sociodemographic questions

Please select the phrase that best describes how much you like or dislike this product.

Dislike Extremely	Dislike Very Much	Dislike Moderately	Dislike Slightly	Neither Like Nor Dislike	Like Slightly	Like Moderately	Like Very Much	Like Extremely
○	○	○	○	○	○	○	○	○

From the list of attributes below, please select those you believe characterize the product.

☐ Mild	☐ Salty	☐ Bitter
☐ Mushroom aroma	☐ Legume aroma	☐ Fishy aroma
☐ Grainy texture	☐ Floral aroma	☐ Umami
☐ Sour	☐ Ammonia aroma	☐ Cheesy aroma
☐ Soft texture	☐ Fruity aroma	☐ Sweet
☐ Fermented aroma	☐ Nutty aroma	☐ Roasted aroma
☐ Crunchy texture		

Fig. 2 Example of liking + CATA questions

Please rank the products listed below, in the order that you prefer them from most to least.

Start by clicking the product code which you like MOST, followed by clicking the product code which you like SECOND, continuing until the final product code you click is the one you like LEAST.

After you make your selections the product codes will appear in the space below with the product you liked MOST on the left to the product you liked LEAST on right.

If you'd like to change the order of your responses, click the "Reset" button in the lower right-hand corner.

Fig. 3 Example of a ranking question

scale. Then, add a CATA question with a list of attributes. These attributes can be selected based on prior literature or GC–MS results (*see* Fig. 2). Including between 15 and 28 attributes helps ensure reliable data without overwhelming participants (*see* **Note 10**).

(c) An optional preference or ranking question can be added if further information on consumer preference between samples is relevant for the study (*see* Fig. 3).

3.3.2 Sensory Analysis Implementation

1. Harvest and cut the fresh tempeh samples into 2 cm cubes and place them on plates, labeling each sample with a randomized three-digit code to ensure unbiased evaluation.
2. In a sensory analysis room equipped with individual booths, present the coded samples to consumers. Provide a glass of water and plain breadsticks to cleanse the palate between tastings.
3. Instruct the consumers to complete the questionnaire independently and without interaction with other participants.

3.3.3 Statistical Analysis

Analyze the data using sensory analysis software such as XLSTAT, applying Cochran's Q test followed by a post hoc test (e.g., Bonferroni).

3.4 GC–MS and Sensory Analysis Integrated Analysis

After analyzing the results obtained from both GC–MS and sensory analysis, an integrated approach should be conducted to comprehensively characterize the samples. Pearson correlation analysis is a straightforward analysis to explore linear relationships between VOC concentrations and sensory attributes, providing a visualization of their associations. In this protocol, R software was used to perform correlation analysis using the following approach:

1. Organize data by creating two separate worksheets: one for GC–MS data and one for sensory data. For GC–MS data, create a table with samples listed in rows and identified VOCs, with their respective abundances, as columns. For the sensory data, use the same sample codes as in the GC–MS data. Organize the table with columns for consumer ID, sample code, liking score, and sensory attributes. Each attribute should be coded using binary coding: enter "1" if the attribute was selected for the sample, and "0" if not (*see* **Note 11**).
2. Follow the specified script using R to generate a heatmap displaying the correlation between the VOCs and the sensory attributes:

```
### UPLOAD DATA AND LIBRARIES ###
# Upload both data files

#Install and upload the following packages
library(corrplot)
library(dplyr)
library(readxl)
library(caret)
library(pROC)

# View the first rows of each data set
head(Data_R_gc_ms)
```

```
head(Data_R_Sensorial)

### CLEAN AND PREPARE DATA ###
# Remove non-numeric columns like Sample from sensory data and
GC-MS
Data_R_Sensorial_clean <- Data_R_Sensorial[, sapply(Data_R_-
Sensorial, is.numeric)]
Data_R_gc_ms_clean <- Data_R_gc_ms[, sapply(Data_R_gc_ms, is.
numeric)]

# Remove consumer column
Data_R_Sensorial_clean <- Data_R_Sensorial %>% select(-Consu-
mer)

# Check the structure of the modified data
str(Data_R_Sensorial_clean)

# Scale GC-MS data
Data_R_gc_ms_scaled <- scale(Data_R_gc_ms_clean)

### APPLY PROPORTION AGGREGATION FOR SENSORY DATA ###
# Aggregate the sensory data by 'Sample' and calculate the
proportion (mean)
Data_R_Sensorial_aggregated <- Data_R_Sensorial_clean %>%
group_by(SAMPLE) %>%summarise(across(everything(), ~mean(.,
na.rm = TRUE)))
head(Data_R_Sensorial_aggregated)
### PREPARE THE DATA TO FURTHER COMBINATION ###
# Check they have the same rows
nrow(Data_R_Sensorial_aggregated)
nrow(Data_R_gc_ms_scaled)

# Change column names if there are any spaces or symbols
colnames(Data_R_Sensorial_aggregated) <- c("Consumer", "SAM-
PLE", "Umami", "Bitter", "Salty", "Sweet", "Acidic", "Legume",
"Floral", "Fruity", "Mushroom_Shiitake", "Truffle", "Ammonia",
"Fishy", "Nuts", "Toasted", "Butter", "Grass", "Soft", "Gran-
ular", "Crunchy", "Meaty", "Fried", "Smoky", "Fermented",
"Yeast")

# Tell to consider only attributes columns in sensory data
sensory_data <- Data_R_Sensorial_aggregated[, c("Umami", "Bit-
ter", "Salty", "Sweet", "Acidic", "Legume", "Floral", "Fru-
ity", "Mushroom_Shiitake", "Truffle", "Ammonia", "Fishy",
"Nuts", "Toasted", "Butter", "Grass", "Soft", "Granular",
"Crunchy", "Meaty", "Fried", "Smoky", "Fermented", "Yeast")]

# Combine data GC-MS and sensory data
```

```
combined_data <- cbind(Data_R_gc_ms_scaled, sensory_data)

# Remove zero variance results (ex: if fruity is not perceived
in any sample)
zero_variance_gcms <- apply(Data_R_gc_ms_scaled, 2, function
(x) var(x) == 0)
zero_variance_sensory <- apply(sensory_data, 2, function
(x) var(x) == 0)

# Clean GC-MS data from zero variance results
Data_R_gc_ms_scaled_clean <- Data_R_gc_ms_scaled[, !zero_var-
iance_gcms]

# Clean sensory data from zero variance results
sensory_data_clean <- sensory_data[, !zero_variance_sensory]

###PREPARE THE CORRELATION PLOT###
# Establish correlation matrix
cor_matrix_clean <- cor(Data_R_gc_ms_scaled_clean, sensory_-
data_clean)

# Subset the correlation matrix for VOCs (34 rows) and sensory
attributes (20 columns)
cor_matrix_voc_sensory <- cor_matrix_clean[1:34, 1:20]

# Create the plot from the correlation matrix
corrplot(cor_matrix_voc_sensory, method = "color", type =
"full", tl.col = "black", tl.srt = 60)

# Corroborate correlations for specific VOC and aromas
cor(Data_R_gc_ms_scaled[, "1-octanol"], sensory_data[,
"Fishy"])
```

3. Identify which VOCs are most abundant in samples where specific sensory attributes, previously shown to correlate positively or negatively with consumer liking, are more or less intense. This analysis can help to identify which VOCs may influence preference. By hypothesizing the microbial origin of these compounds, you can explore process optimization or strain selection strategies to enhance or reduce targeted VOCs, thereby improving the sensory quality and consumer appeal of the product.

4 Notes

1. This protocol is designed for tempeh samples, but adjustments may be necessary for other sample types depending on their specific properties. The goal is to obtain a chromatogram with well-resolved and high-quality peaks. To achieve it, parameters such as sample preparation, equilibration time, extraction time, and temperature should be optimized based on the nature of the sample.
2. The physical characteristics of the sample should be carefully considered. Depending on its water content, the volume of added water should be adjusted to achieve a uniform paste-like or liquid consistency. The addition of NaCl can enhance the extraction efficiency of certain VOCs. Concentrations up to 30% NaCl are commonly used to promote VOC extraction.
3. Tempeh samples can be frozen after harvesting while maintaining their organoleptic characteristics, such as VOCs. Whenever possible, it is recommended to flash-freeze samples in liquid nitrogen and store them at −80 °C until analysis.
4. The use of an internal standard (IS) enables semiquantitative analysis of VOCs. When selecting an IS, select a compound that is chemically similar to the target VOCs (e.g., an aldehyde or alcohol). Avoid IS with retention times that closely overlap with key VOCs in your sample to prevent chromatographic interferences. Additionally, the IS concentration should be adjusted according to the sample's VOC abundance, aiming for a signal intensity that falls below the highest analyte peak, ideally at a midpoint between the highest peak and baseline noise.
5. In this protocol, ultrasonication was used to improve the release of VOCs from the food matrix. However, this step may not be necessary for all sample types.
6. Extraction and equilibrium temperatures must be optimized according to sample characteristics. For heat-sensitive VOCs, temperatures should not exceed 50 °C, while for heavier VOCs, temperatures up to 60 °C may be convenient. Extraction temperature and time are closely related; increasing the extraction temperature typically accelerates the volatilization of VOCs, allowing for a shorter extraction time.
7. Similarly, equilibrium and extraction times should be carefully adjusted. Generally, an equilibrium time of 10 to 15 min and an extraction time of 30 to 60 min are applied. However, depending on the VOC profile of the sample, these times may require modification, as prolonged extraction may lead to VOC degradation.

8. Different SPME fibers exhibit varying affinities for VOCs depending on their polarity. This protocol uses a DVB/CAR/PDMS fiber due to its versatility. Nonetheless, alternative fibers may be more suitable depending on the sample matrix or the target analytes that you want to determine.
9. Chromatographic time and temperature settings should be fine-tuned according to the sample. If many VOCs elute within a short time window, extending this interval can improve separation and peak resolution.
10. A VOC library is essential for compound identification in GC–MS analysis, enabling comparison of mass spectra to known references. This study used the Wiley library, but other libraries are available and can be selected based on the sample type.
11. To evaluate the perceived intensity of different aromas in your samples, check-all-that-apply (CATA) can be replaced by rate-all-that-apply (RATA), which provides an intensity score for each attribute. Using RATA generally requires at least a minimally trained panel.
12. For further analysis, CATA data responses (binary yes/no) should be aggregated to align with GC–MS data by calculating the proportion of consumers selecting each attribute per sample. Alternatively, when using RATA, aggregation involves calculating the mean intensity rating for each attribute.

References

1. Hu Y, Badar IH, Zhang L, Yang L, Xu B (2024) Odor and taste characteristics, transduction mechanism, and perceptual interaction in fermented foods: a review. Crit Rev Food Sci Nutr. https://doi.org/10.1080/10408398.2024.2377292
2. Wrobel BB, Leopold DA (2004) Clinical assessment of patients with smell and taste disorders. Otolaryngol Clin N Am 37:1127–1142
3. Devanthi PVP, Gkatzionis K (2019) Soy sauce fermentation: microorganisms, aroma formation, and process modification. Food Res Int 120:364–374
4. Lindsay MA, Granucci N, Greenwood DR, Villas-Boas SG (2022) Identification of new natural sources of flavour and aroma metabolites from solid-state fermentation of agro-industrial by-products. Meta. https://doi.org/10.3390/metabo12020157
5. Mahmud MMC, Shellie RA, Keast R (2020) Unravelling the relationship between aroma compounds and consumer acceptance: coffee as an example. Compr Rev Food Sci Food Saf 19:2380–2420
6. Perez-Hurtado P, Palmer E, Owen T, Aldcroft C, Allen MH, Jones J, Creaser CS, Lindley MR, Turner MA, Reynolds JC (2017) Direct analysis of volatile organic compounds in foods by headspace extraction atmospheric pressure chemical ionisation mass spectrometry. Rapid Commun Mass Spectrom 31:1947–1956
7. Papadimitropoulos M-EP, Vasilopoulou CG, Maga-Nteve C, Klapa MI (2018) Untargeted GC-MS metabolomics metabolic profling. Methods Mol Biol 1738:133–147
8. Kim AJ, Choi JN, Kim J, Park SB, Yeo SH, Choi JH, Lee CH (2010) GC-MS based metabolite profiling of rice Koji fermentation by various fungi. Biosci Biotechnol Biochem 74:2267–2272
9. Hübschmann H-J (2025) Handbook of GC-MS: fundamentals and applications, 4th edn. Wiley-VCH, Weinheim
10. Jeleń H, Majcher M, Ginja A, Kuligowski M (2013) Determination of compounds responsible for tempeh aroma. Food Chem 141:459–465

11. Zhang R, Tang C, Jiang B, Mo X, Wang Z (2021) Optimization of hs-spme for gc-ms analysis and its application in characterization of volatile compounds in sweet potato. Molecules. https://doi.org/10.3390/molecules26195808
12. Brattoli M, Cisternino E, Rosario Dambruoso P, de Gennaro G, Giungato P, Mazzone A, Palmisani J, Tutino M (2013) Gas chromatography analysis with olfactometric detection (GC-O) as a useful methodology for chemical characterization of odorous compounds. Sensors (Switzerland) 13:16759–16800
13. Trimmer C, Arroyave R, Vuilleumier C et al (2023) Allosteric modulation of a human odorant receptor. Curr Biol 33:1523–1534.e4

Chapter 8

Multimycotoxin Analysis in Grains by UHPLC-MS/MS

Diego Bogantes-Ledezma, Andrea Irías-Mata, and Maria Vinas

Abstract

Mycotoxins are toxic fungal metabolites that pose significant food safety risks and cause substantial economic losses. Due to their widespread occurrence, regulatory limits vary across countries. Multimycotoxin analysis in diverse food matrices remains challenging due to their distinct chemical properties. This document presents a validated, cost-effective protocol for extracting and analyzing nine major mycotoxins (aflatoxins B1, B2, G1 and G2, deoxynivalenol, fumonisin B1, ochratoxin A, T–2 toxin, and zearalenone) in common beans, groundnuts, maize, rice, and wheat by means of ultrahigh performance liquid chromatography (UHPLC) equipment coupled to tandem mass spectrometry (MS/MS) with triple quadrupole. The method reduces costs by minimizing organic solvent use, replacing acetonitrile with methanol, and eliminating immunoaffinity or solid-phase columns. Additionally, it is time-efficient and environmentally friendly. This protocol provides a practical and sustainable approach to multimycotoxin detection, improving food safety monitoring.

Key words Mycotoxin, Multimycotoxin analysis, Analytical methods, Grains, Food safety

1 Introduction

Mycotoxins are secondary metabolites produced by some fungal species that can contaminate food products during preharvest and postharvest conditions. More than 400 mycotoxins have been identified but the most concerning mycotoxins found in cereals include aflatoxins, deoxynivalenol, fumonisins, ochratoxins, patulin, T–2 toxin, and zearalenone [1]. The occurrence of these toxins in food crops worldwide is high [2], resulting in economic losses and food safety risks. Exposure to mycotoxins can lead to health problems in humans causing diseases such as liver, breast, cervical and esophageal cancer, child growth impairment, and immunotoxicity [3].

The analysis of mycotoxins in grains matrices typically begins with an extraction procedure designed to isolate the target compounds from complex biological materials. This step is crucial to minimize matrix effects and ensure accurate downstream detection and includes acidified acetonitrile or methanol/water extraction

Pablo Navarro Simarro (ed.), *Fungi*, Methods and Protocols in Food Science, https://doi.org/10.1007/978-1-0716-4989-3_8,
© The Author(s), under exclusive license to Springer Science+Business Media, LLC, part of Springer Nature 2026

followed by a cleanup procedure using immunoaffinity columns or solid-phase extraction (SPE) [4]. Modified QuEChERS (Quick, Easy, Cheap, Effective, Rugged, and Safe) approaches can also be employed [5]. Following extraction, the identification and quantification of individual mycotoxins are nowadays commonly performed using techniques such as the enzyme-linked immunosorbent assay (ELISA) or using high-performance liquid chromatography (HPLC) coupled with mass spectrometry (HPLC-MS), fluorescence (HPLC-FLD), UV-visible (HPLC-UV), or photodiode array (HPLC-PDA) [6].

In recent years, multimycotoxin analytical protocols have gained preference, as they enable the simultaneous detection of multiple mycotoxins within a single run [7]. These approaches significantly improve analytical throughput, reduce overall analysis time, and lower operational costs, making it a more efficient and economically viable option for routine monitoring and risk assessment in food safety applications.

The protocol described here was validated (*see* **Note 1**) and is routinely employed in our laboratory for the analysis of key mycotoxins in grain samples, including aflatoxins B1, B2, G1, and G2, deoxynivalenol, fumonisin B1, ochratoxin A, T–2 toxin, and zearalenone. It is based on a modified QuEChERS (Quick, Easy, Cheap, Effective, Rugged, and Safe) extraction method, utilizing methanol as a cost-effective solvent and avoiding the use of immunoaffinity columns or SPE, which can significantly increase the cost of analysis. The protocol is not only economical but also rapid and straightforward, making it suitable for implementation in any laboratory equipped with HPLC-MS/MS instrumentation.

2 Materials and Equipment

2.1 Materials

1. Grains of common beans, maize, peanuts, rice, or wheat, ground to a 0.84 mm particle size.
2. 50 mL centrifuge tubes.
3. 40 mL test tubes.
4. Glass microfiber filters.
5. 2 mL glass vials for HPLC.
6. C18 column for HPLC Accucore aQ 100 × 2.1 mm, 2.6 μm.
7. 0.2 μm syringe filters.

2.2 Reagents and Solvent Mixtures

1. Ultrapure water.
2. Solvent mixture for extraction: methanol/water/formic acid (79/20/1, v/v/v).
3. Anhydrous $MgSO_4$.

4. Mobile phase A: water/methanol/formic acid (97/2/1, v/v/v).
5. Mobile phase B: methanol/water/formic acid (97/2/1, v/v/v).
6. Standard mixture of the analyzed mycotoxins (*see* **Note 2**): aflatoxin B1, B2, G1 and G2 (AFB1, AFB2, AFG1, and AFG2), fumonisin B1 (FB1), ochratoxin A (OTA), T2 toxin (T–2), zearalenone (ZEN), and deoxynivalenol (DON).

2.3 Equipment

1. A common blender.
2. Refrigerated centrifuge.
3. Vortex.
4. Ultrahigh performance liquid chromatography (UHPLC) equipment coupled to tandem mass spectrometry (MS—MS) with triple quadrupole.

3 Methods

3.1 Mycotoxin Extraction

1. Weigh 25 g of ground grains in a blender jar.
2. Add 100 mL of the solvent mixture for extraction.
3. Blend for 5 min at high speed in a common blender.
4. Allow the sample to rest for 10 min until the liquid and solid phases are separated.
5. Transfer 25 mL of the upper phase to a 50 mL centrifuge tube.
6. Add 4 g of anhydrous $MgSO_4$ and hand-shacked vigorously for 30 s.
7. Centrifuge at 1810 rcf for 5 min at 10 °C.
8. Transfer 10 mL supernatant to the test tubes.
9. Dilute with 10 mL ultrapure water and vortex for 30 s.
10. Filter through a glass microfiber filter.
11. Take at least 2 mL of the sample, filter through 0.2 μm syringe filter, and collect in vials for HPLC.

3.2 UHPLC Conditions

1. Set the instrument conditions to 45 °C column oven, flow rate of 400 μL/min, and injection volume of 5 μL.
2. Set the gradient program based on eluent B as follows: 5% for 0 min, 50% for 3 min, 75% for 6 min, 100% for 8 min, and 5% for 11 min. Percentage of eluent A is complementary. Total run time = 11 min.

3.3 MS-MS Conditions

1. Consider the mass spectrometric parameters for each mycotoxin (*see* **Note 3**). Note that some parameters (collision energy and RF lens) could be changed, depending on the characteristics of the UHPLC-MS/MS system.
2. Set the heated electrospray ionization (H-ESI) interface to positive-ion mode.
3. Set the static spray voltage to positive ion (3500 V)–negative ion (2500 V).
4. Set ion transfer tube and vaporizer temperatures at 342 °C and 358 °C, respectively.
5. Use argon as a collision gas at 199.98 Pa with a cycle time of 0.5 s.
6. Use 99% pure nitrogen as sheath, sweep, and auxiliary gas. Setting conditions will vary depending on the characteristics of the UHPLC-MS/MS system.
7. Measure using the selected reaction monitoring (SRM) scan mode.

3.4 Identification and Quantification

1. Prepare calibration points for each mycotoxin in the suitable ranges of concentrations (*see* **Note 2**), filter each solution through a 0.2 μm syringe filter, and collect in vials for HPLC.
2. Measure calibration points and mycotoxin extracts in the UHPLC-MS/MS system.
3. Identify mycotoxins in the extracts based on the mass spectrometry parameters and quantify based on the calibration curves.

4 Notes

1. Parameters of the validated protocol.

Grain	Mycotoxin	Linear range	R^{2a}	LOD (μg/kg)[a]	LOQ (μg/kg)[a]	R_E (%)[a]	RSD_R (%)[a]	RSD_r (%)[a]	Matrix effect (%)
Maize	AFB1	2.5–80	0.9973	0.9	2.9	68	15	13	35
	AFB2	5.0–80	0.9963	0.8	2.6	71	20	15	31
	AFG1	5.0–80	0.9941	2.6	8.8	98	19	11	−25
	AFG2	10.0–80	0.9694	3.4	11.5	99	30	13	18
	FB1	20.0–2000	0.9990	21.5	71.7	120	4	3	−32
	OTA	2.5–80	0.9835	0.6	1.9	84	12	11	20
	T-2	2.5–80	0.9940	0.9	3.1	117	25	12	−6
	ZEN	80–1000	0.9947	15.6	52.1	88	8	9	32
	DON	1000–2000	0.9625	63.4	211.3	85	n.d	8	23

(continued)

Grain	Mycotoxin	Linear range	R^{2a}	LOD (μg/kg)[a]	LOQ (μg/kg)[a]	R_E (%)[a]	RSD_R (%)[a]	RSD_r (%)[a]	Matrix effect (%)
Common beans	AFB1	5.0–80	0.9931	1.1	3.6	83	13	15	7
	AFB2	10.0–80	0.9915	2.7	9.0	97	18	13	0
	AFG1	5.0–80	0.9892	2.5	8.2	105	35	14	−47
	AFG2	10.0–80	0.9634	2.8	9.4	100	33	24	7
	FB1	20.0–2000	0.9976	4.9	16.5	122	10	5	−33
	OTA	2.5–80	0.9852	0.6	2.1	98	21	11	−3
	T-2	2.5–80	0.9850	0.7	2.3	107	34	8	−47
	ZEN	80–1000	0.9780	14.4	48	94	11	11	−18
	DON	1000–2000	0.9693	20.4	67.9	66	n.d	6	35
Rice	AFB1	2.5–80	0.9909	1.0	3.4	93	18	14	−3
	AFB2	2.5–80	0.9893	0.5	1.7	100	27	12	−18
	AFG1	5.0–80	0.9854	1.2	3.9	117	11	8	−73
	AFG2	2.5–80	0.9931	1.8	5.9	87	26	8	−15
	FB1	100–2000	0.9988	8.0	26.7	120	10	2	−42
	OTA	2.5–80	0.9943	0.5	1.6	97	18	9	−8
	T-2	2.5–80	0.9913	0.9	1.9	122	26	12	−16
	ZEN	40–300	0.9597	6.1	20.3	117	18	9	−56
	DON	1000–2000	0.9436	104.8	349.3	100	n.d	11	12
Groundnuts	AFB1	2.5–80	0.9945	0.5	1.6	77	22	9	17
	AFB2	2.5–80	0.9953	0.5	1.6	85	14	11	9
	AFG1	2.5–80	0.9949	1.2	3.9	87	18	9	−32
	AFG2	5.0–80	0.9626	2.3	7.8	87	79	14	27
	FB1	20.0–2000	0.9910	8.6	28.7	102	10	4	−31
	OTA	2.5–80	0.9560	0.8	2.5	86	37	16	9
	T-2	2.5–80	0.9940	0.7	2.3	131	16	11	−41
	ZEN	40–1000	0.9860	9.4	31.4	92	16	14	12
	DON	500–2000	0.9691	74.5	248.3	66	n.d	7	−26
Wheat	AFB1	2.5–80	0.9568	0.2	0.8	76	20	10	48
	AFB2	2.5–80	0.9974	0.3	1.1	72	17	12	25
	AFG1	2.5–80	0.9965	2.1	7.0	85	14	7	−9
	AFG2	2.5–80	0.9631	3.6	12.0	90	34	12	31
	FB1	100–2000	0.9988	45.7	152.2	117	n.d	3	−9
	OTA	2.5–80	0.9971	0.9	3.1	71	34	12	17
	T-2	2.5–80	0.9911	0.4	1.4	65	36	23	6
	ZEN	80–1000	0.9194	39.1	130.5	69	10	17	43
	DON	1000–2000	0.9674	28.6	95.3	74	n.d	7	28

[a]Coefficient of determination (R^2), limit of detection (LOD), limit of quantification (LOQ), extraction recovery (R_E), relative standard deviation of reproducibility (RSD_R), Relative Standard Deviation of repeatability (RSD_r)

2. Standard mixture of the analyzed mycotoxins.

   The working standard solutions are prepared in 100% MeOH (HPLC grade). Calibration points are prepared in the following ranges of concentrations: For AFB1, AFB2, AFG1, AFG2, T-2, ZEN, and OTA at 1, 5, 10, 25, 50, 75, and 100 ng/mL, for FB1 and DON at 500, 1000, 2500, 5000, and 10,000 ng/mL.

3. Mass spectrometric parameters.

Compound	Retention time (min)	Precursor (m/z)	Product (m/z)	Collision energy (V)	RF lens (V)
AFB1	5.54	313.03 $[M + H]^+$	285.00	21.93	139.82
			241.00	35.79	139.82
			269.00	29.52	139.82
AFB2	5.39	315.00 $[M + H]^+$	287.06	25.07	148.01
			259.00	28.66	148.01
			242.99	38.01	148.01
AFG1	5.13	329.00 $[M + H]^+$	243.07	25.17	151.65
			215.00	30.89	151.65
			283.00	24.01	151.65
AFG2	5.03	331.06 $[M + H]^+$	245.00	29.11	100.99
			257.00	29.62	100.99
			285.00	27.10	100.99
FB1	6.42	722.37 $[M + H]^+$	334.27	38.62	144.07
			352.33	34.17	144.07
			370.35	31.44	144.07
T-2	6.78	489.18 $[M + Na]^+$	245.10	25.98	166.52
			327.10	21.78	166.52
			387.11	19.96	166.52
ZEN	7.05	319.18 $[M + H]^+$	187.00	20.11	97.65
			283.04	12.02	97.65
			301.04	10.25	97.65
OTA	7.16	404.03 $[M + H]^+$	238.93	22.99	111.91
			358.00	13.69	111.91
			221.00	33.51	111.91
DON	1.08	297.152 $[M + H]^+$	249.06	10.253	110.697
			203.11	15.056	110.697
			231.06	12.629	110.697

References

1. Khodaei D, Javanmardi F, Khaneghah AM (2021) The global overview of the occurrence of mycotoxins in cereals: a three-year survey. Curr Opin Food Sci 39:36–42. https://doi.org/10.1016/j.cofs.2020.12.012
2. Eskola M, Kos, Gregor, Elliott, Christopher T, et al (2020) Worldwide contamination of food-crops with mycotoxins: validity of the widely cited 'FAO estimate' of 25%. Crit Rev Food Sci Nutr 60:2773–2789. doi:https://doi.org/10.1080/10408398.2019.1658570
3. Awuchi CG, Ondari EN, Nwozo S et al (2022) Mycotoxins' toxicological mechanisms involving humans, livestock and their associated health concerns: a review. Toxins 14:167. https://doi.org/10.3390/toxins14030167
4. Iqbal SZ (2021) Mycotoxins in food, recent development in food analysis and future challenges; a review. Curr Opin Food Sci 42:237–247. https://doi.org/10.1016/j.cofs.2021.07.003
5. Koesukwiwat U, Sanguankaew K, Leepipatpiboon N (2014) Evaluation of a modified QuEChERS method for analysis of mycotoxins in rice. Food Chem 153:44–51. https://doi.org/10.1016/j.foodchem.2013.12.029

6. Agriopoulou S, Stamatelopoulou E, Varzakas T (2020) Advances in analysis and detection of major mycotoxins in foods. Foods 9:518. https://doi.org/10.3390/foods9040518
7. Guan X, Feng Y, Suo D et al (2022) Simultaneous determination of 11 mycotoxins in maize via multiple-impurity adsorption combined with liquid chromatography–tandem mass spectrometry. Foods 11:3624. https://doi.org/10.3390/foods11223624

Chapter 9

Analysis of Mineral Element Contents in Fungi

Marin Senila and Eniko Kovacs

Abstract

Mushrooms are known for their ability to effectively uptake mineral elements from their substrates. For this reason, the content of macroelements (K, Na, Ca, Mg, S, P), microelements (Fe, Zn, Mn, Cu, Se) and potentially toxic elements (Cd, As, Pb, Ni, Cr, Co, Hg, etc.) in different species of fungi was widely studied worldwide. This chapter provides comprehensive procedures on sample pretreatment, digestion methods, and instrumental analysis options that can be employed to obtain reliable results. Although there are several analytical instruments that allow the direct determination of element contents in solid mushroom samples, in many cases, the analytical techniques used for metal determination require that the analyte be dissolved in a liquid aqueous solution. Thus, wet digestion procedures must be applied to the solid sample to bring the analytes into solution. Accurate data on mineral elements determination is crucial for evaluating mineral intake in human diets and for assessing the health risks posed by the ingestion of edible mushrooms due to their content of potentially toxic elements.

Key words Mushroom, Microelement, Macroelement, Toxic element, Spectrometry

1 Introduction

The content of macroelements (K, Na, Ca, Mg, S, P), microelements (Fe, Zn, Mn, Cu, Se), and potentially toxic elements (Cd, As, Pb, Ni, Cr, Co, Hg, etc.) in fungi is one of the most extensively studied topics related to edible mushrooms, since it is well-known that many species of mushrooms are accumulators of mineral elements. There are two main factors that influence the uptake of mineral elements in mushrooms. One factor is associated with the characteristics of the substrate (e.g., soil pH, organic matter, and mineral element contents in substrate), while the other is linked to the characteristics of the fungi (e.g., species of mushrooms, growth phases, part of the fruiting body) [1]. It was reported that fungi, through their mycelium, have the ability to release substances into the soil substrate that act as leaching agents, dissolving minerals, and leading to their transfer to the fruiting body [2].

For edible mushrooms, the uptake of minerals may be an advantage if we consider essential elements, such as Ca, K, P, Na,

Pablo Navarro Simarro (ed.), *Fungi*, Methods and Protocols in Food Science, https://doi.org/10.1007/978-1-0716-4989-3_9,
© The Author(s), under exclusive license to Springer Science+Business Media, LLC, part of Springer Nature 2026

Mg, Fe, Zn, Mn, Cu, or Se, since deficiencies of these elements in the human body may pose health problems. However, their content above certain thresholds, or the presence of nonessential elements, such as Cd, As, Pb, or Hg, may negatively affect human health [3, 4]. Considering these aspects, many studies have been completed all over the world on the content of potentially toxic elements (PTEs) in edible mushrooms, with significant attention on heavy metals along with the evaluation of potential health risks [5–7]. In certain studies, the content of essential elements was used to evaluate the potential contribution of mushroom consumption to the mineral intake in the human diet [8, 9].

The extensive advancement of analytical instrumentation enabled the development of different analytical protocols for mineral element determination in fungi. There are two possibilities to analyze mineral elements in fungi: analysis of mineral element contents in aqueous liquid samples after digestion and direct analysis of solid samples.

In analytical laboratories, three main groups of spectrometric techniques are the most used to measure elements in liquid solutions [10]. These are distinguished mainly according to the type of analyte detection or to the type of atomization source. In the first group of techniques, a mass spectrometer (MS) is used to measure ions of the analyte produced with the aid of inductively coupled plasma (ICP), resulting in the ICP-MS technique. The second group of spectrometric techniques uses an inductively coupled plasma as the radiation source for optical emission inductively coupled plasma (ICP-OES). In the third group of spectrometric techniques, the analytical signal is obtained by measuring the absorption of radiation emitted by a radiation source (e.g., a hollow cathode lamp) at a specific wavelength by the atoms of analyte. In atomic adsorption techniques, the atomization occurs either in a flame (F-AAS) or in a graphite furnace (GF-AAS). Because the sample introduction systems for these spectrometric techniques are designed for liquids, with small exceptions, the solid samples require digestion before instrumental analysis [11]. In the case of Hg, the direct measurement of a solid sample is possible using the thermal desorption atomic absorption spectrometry (TD-AAS) technique. Among spectrometric techniques, ICP-MS and ICP-OES have the benefits of high sample throughput due to the multi-elemental determination and wider working ranges [12].

Sample pretreatment and digestion represent a critical issue for obtaining reliable results. To obtain the dry mass of mushroom samples, these need to be dried (air drying, drying in an oven, or freeze-drying). The dried samples are then powdered before digestion. Different approaches can be used to extract mineral elements from solid mushroom samples into liquid aqueous solutions. Usually, the methodologies applicable to samples with a high organic load, such as mushrooms, are direct wet acid digestion or combustion, followed by acid digestion of the resulting ash [13, 14]. Depending on the scope of the study, other methods of extracting mineral elements into a liquid solution include extraction with complexing reagents, extraction using simulated body fluids (SBF), or microextraction. Nitric acid is typically used for acid wet digestion because many elements form soluble nitrates. Mixtures of HNO_3 with H_2O_2, $HClO_4$, HCl, HF, or H_2SO_4 are also used for the digestion of samples with organic matrices. Heating can also improve the digestion efficiency using hotplates or more advanced techniques, like microwave extraction, ultrasonic extraction, or infrared-assisted water extraction.

In the group of techniques that allows direct analysis of elements in a solid sample, Instrumental Neutron Activation Analysis (INAA) [15] and Energy Dispersive X-Ray Fluorescence [16] have been used to analyze mushrooms. However, the number of published papers reporting the use of these techniques for metals analysis in mushrooms is limited. For this reason, this chapter focuses on the pretreatment and digestion of mushroom samples, followed by instrumental determination using one of the atomic spectrometric techniques.

2 Materials

2.1 Equipment

1. Hot-plate, digestion system, or microwave-assisted digestion unit.
2. F-AAS, or GF-AAS, or ICP-OES, or ICP-MS spectrometer.
3. Analytical balance.
4. Freeze-drier or heating oven.
5. Grinder.
6. Fume hood.
7. Volumetric flasks, pipettes, Berzelius beakers.

2.2 Reagents

1. Distilled or ultrapure water.
2. Concentrated nitric acid (HNO_3) analytical grade.
3. Hydrogen peroxide (H_2O_2) 30% analytical grade, or concentrated perchloric acid ($HClO_4$) analytical grade, or concentrated sulfuric acid (H_2SO_4) analytical grade.
4. Multielement or monoelement stock solutions containing the analytes, for calibration of the spectrometer.
5. Certified Reference Materials (CRM) with appropriate matrices.
6. Mushrooms.

3 Methods

3.1 Pretreatment to Obtain Dry Powder

Powdered samples are required to obtain faster and quantitative digestion.

1. Wash freshly collected fruiting bodies of mushrooms with distilled water and chop them into small pieces with a plastic knife.
2. Dry in a heating oven at 105 °C until constant weight (about 16 h). If a freeze dryer is available, the mushroom samples can be lyophilized.
3. The dried samples are ground with a grinder to obtain a fine powder.
4. The mushroom powders can be stored at room temperature in closed recipients.

3.2 Moisture Content

1. Moisture content should be measured gravimetrically, on a separate subsample, by drying in an oven at 105 °C for 16 h, until constant weight.

3.3 Combustion and Acid Digestion of Ashes

1. Place 1 gram of the homogenized dried powdered mushrooms in a porcelain crucible and introduce in an oven at 550 °C for 6 h.
2. Add 5 mL concentrated HNO_3 to the resulting ash, cover with a glass watch, and heat to 120 °C for 2 h on a hotplate to dissolve the ash.
3. Pass the resulted solution through a 0.45 μm pore size filter paper in a pre-cleaned volumetric flask and dilute to 50 mL with distilled/ultrapure water. The resulted solution can be stored for further analysis.
4. Prepare a blank digest in the same way as the sample digestion.

3.4 Direct Acid Digestion of Powdered Mushrooms

3.4.1 Microwave-Assisted Acid Digestion

1. Weigh 0.5 grams of the homogenized dried powdered mushrooms into the pre-cleaned PTFE digestion vessels of the microwave digestion system.
2. Add 6 mL concentrated HNO_3 and 2 mL 30% H_2O_2 to the powdered mushrooms into the PTFE digestion vessels and let the mixture sit at room temperature for 16 h to react and decompose the organic tissues.
3. Introduce the PTFE digestion vessels into the digestion system, and apply the digestion program (with temperature and/or pressure control).
4. After cooling down at room temperature, open carefully the digestion vessels, and pass the solution in the pre-cleaned volumetric flasks through a 0.45 μm pore size filter paper.

5. Dilute to 50 mL with distilled/ultrapure water. The resulted solution can be stored for further analysis.
6. Prepare a blank digest using the same method as the sample digestion.
7. Prepare a Certified Reference Material with an appropriate matrix using the same procedure as for sample digestion.

3.4.2 Wet Acid Digestion on a Hot-Plate/Sand Bath

If a microwave digestion system is not available, the acid digestion can be carried out on a hot plate/sand bath, following the stages below.

1. Weigh 0.5 grams of the homogenized dried powdered mushrooms into the borosilicate glass Berzelius beakers (capacity 50 mL–150 mL).
2. Add 6 mL concentrated HNO_3 and 2 mL 30% H_2O_2 to the powdered mushrooms into the Berzelius beakers, then cover with watch glasses.
3. Let Berzelius beakers containing the mixture be covered in a fume hood for 16 h at room temperature to react.
4. Place the Berzelius beakers containing the mixture, covered with watch glasses, on the hot plate/sand bath, and increase the temperature to approximately 120 °C.
5. Maintain the temperature at approximately 120 °C for 2 h, to reach reflux conditions.
6. Monitor that the solution does not spill during heating and that the mixture does not run dry.
7. After cooling down to room temperature, pass the solution through a 0.45 μm pore size filter paper in pre-cleaned volumetric flasks.
8. Dilute to 50 mL with distilled/ultrapure water. The resulted solution can be stored for further analysis.
9. Prepare a blank digest using the same method as the sample digestion.
10. Prepare a Certified Reference Material with an appropriate matrix using the same procedure as for sample digestion.

3.5 Instrumental Determination of Elements Concentration Extracted in Solution by Extraction

3.5.1 Flame Atomic Absorption Spectrometry (F-AAS)

Flame atomic absorption spectrometry (F-AAS) is one of the earliest and simplest elemental analysis techniques. It is based on the principle that analyte atoms and ions produced into the flame absorb photons emitted by a source at a specific wavelength. The atomization source is typically air-acetylene or nitrous oxide-acetylene flame, depending on the analyte. A radiation source (usually a hollow cathode lamp - HCL) is used to emit photons at a wavelength that is characteristic of the analyte. The amount of radiation absorbed is proportional to the number of atoms and ions in the flame. A detector, typically a photomultiplier tube, measures the number of photons absorbed by the analyte atoms and ions. The main steps for the determination of elements by FAAS are as follows:

1. Preparation of calibration standards. Prepare at least five calibration solutions and a blank. The working range depends on the analyzed element and on the instrument configuration. The international standard ISO 8466-1:2021 [17] provides guidelines for the selection of the working range of calibration standards and evaluation of calibration curves. Typically, the working range in F-AAS is at sub-ppm to low-ppm levels.
2. Open the instrument according to the producer's instructions: open the HCL needed for analysis, start the flame.
3. Aspirate the standards into the flame in the increasing order of their concentrations and measure their absorbance signals.
4. Create a calibration curve by plotting the absorbances of the calibration standards against their known concentrations.
5. Aspirate the sample into the flame and measure the absorbance. Based on the absorbance of the sample and the calibration curve equation, the concentration of analyte in the solution is obtained.

3.5.2 Graphite Furnace Atomic Absorption Spectrometry (GF-AAS)

Graphite furnace atomic absorption spectrometry (GF-AAS) has similar principles to those of atomic absorption spectroscopy, as in the case of F-AAS, but uses a graphite furnace as the atomization source instead of a flame. Because of the use of higher temperatures for atomization, GFAAS is a highly sensitive analytical technique, with orders of magnitude lower detection limits compared to F-AAS. The main steps for the determination of elements by GF-AAS are as follows:

1. Preparation of calibration standards. Prepare at least five calibration solutions and a blank. The working range depends on the analyzed element and on the instrument configuration. Typically, the working range in GF-AAS is at sub-ppb to ppb level.
2. Open the instrument according to the producer's instructions: open the HCL needed for analysis, align the graphite furnace.
3. Set the instrument conditions and temperature program of the graphite furnace according to the instrument producer's recommendations.
4. Inject the standards into the instrument in the increasing order of their concentrations and measure their absorbance signals.
5. Create a calibration curve by plotting the absorbances of the calibration standards against their known concentrations.

6. Inject the sample into the instrument and measure its absorbance. Based on the absorbance of the sample and the calibration curve equation, the concentration of analyte in the solution is obtained.

3.5.3 Inductively Coupled Plasma Atomic Emission Spectrometry (ICP-OES)

In the inductively coupled plasma atomic emission spectrometry (ICP-OES) technique, plasma is used to emit light at characteristic wavelengths for each analyzed element, thereby ensuring the element's identification, while the number of emitted photons is proportionally related to the concentration of the analyte. The advantage of ICP-OES is its multielement determination capability. The main steps for the determination of elements by ICP-OES are as follows:

1. Preparation of calibration standards. Prepare at least five calibration solutions and a blank. Multielement calibration standards can be prepared for multielement determination. Working ranges depend on the analyzed elements and on the instrument configuration. Typically, the working range in ICP-AES is in the sub-ppm to hundred ppm range. A characteristic of ICP-OES is its large linear range of calibration curves spanning several orders of magnitude.
2. Open the instrument according to the producer's instructions: start the plasma, select the elements, and working wavelengths.
3. Introduce the standards into the instrument in increasing order of their concentrations and measure their emission signals.
4. Create calibration curves for each element by plotting the signal intensities of the calibration standards against their known concentrations.
5. Introduce the sample into the instrument and measure the emission signal. Based on the signal intensities of the sample and the calibration curve equations, the concentrations of analytes in the solution are obtained.

3.5.4 Inductively Coupled Plasma Mass Spectrometry (ICP-MS)

In the inductively coupled plasma mass spectrometry (ICP-MS) technique, plasma is used to produce ions. The resulted ions are then passed through a mass separator, which separates them according to their mass-to-charge (m/z) ratio. The abundance of ions is measured with a mass detector. In addition to its multielement determination capability, ICP-MS is more sensitive than ICP-OES. The main steps for the determination of elements by ICP-MS are as follows:

1. Preparation of calibration standards. Prepare at least five calibration solutions and a blank. Multielement calibration standards can be prepared for multielement determination. Working ranges depend on the analysed elements and on the

instrument configuration. Typically, the working range in ICP-MS is at sub-ppb to ppm level, with a large linear range of calibration curves of several orders of magnitude.

2. Open the instrument according to the producer's instructions: start the plasma, select the elements and m/z to be monitored.
3. Introduce the standards into the instrument in the increasing order of their concentrations and measure their signals.
4. Create calibration curves for each element by plotting the signal intensities of the calibration standards against their known concentrations.
5. Introduce the sample into the instrument and measure the signal. Based on this, the sample and the calibration curve equations, the concentrations of analytes in the solution are obtained.

3.6 Calculation of Element Content as the Dry and Wet Weights of Mushrooms

Once the concentration of the elements in the digestion solution has been determined, the content of the elements in the dry mass of the mushrooms can be calculated with the formula described below:

$$C_{M\ d.w.} = \frac{C_s * V * f}{m}\ [\text{mg/kg d.w.}] \qquad (1)$$

where $C_{M\ d.w.}$—ontent of element in mushroom (mg/kg, dry weight basis), C_S—concentration of element in the digested solution (mg/L) measured by one of the atomic spectrometry techniques, V—volume (mL) of the volumetric flask, f—dilution factor (if the sample was diluted), and m—mass (g) of the digested sample (dry weight basis).

Because the maximum allowable content of some elements in mushrooms is given on a wet weight basis, the conversion of mg/kg, dry weight to mg/kg, wet weight is necessary. This conversion can be performed using the formula:

$$C_{M\ w.w.} = \frac{(100 - \%\text{moisture})}{100} * C_{M\ d.w.}\ [\text{mg/kg w.w.}] \qquad (2)$$

where $C_{M\ w.w.}$—content of element in mushroom (mg/kg, wet weight basis); % moisture in water content measured by drying in an oven at 105 °C for 16 h, a mushroom sub-sample.

References

1. Dowlati M et al (2021) Heavy metals content in edible mushrooms: a systematic review, meta-analysis and health risk assessment. Trends Food Sci Technol 109:527–535
2. Smits MM, Wallander H (2017) Chapter 3 - Role of Mycorrhizal Symbiosis in mineral weathering and nutrient mining from soil parent material. In: Johnson NC, Gehring C,

Jansa J (eds) Mycorrhizal mediation of soil. Elsevier, pp 35–46
3. Briffa J, Sinagra E, Blundell R (2020) Heavy metal pollution in the environment and their toxicological effects on humans. Heliyon 6(9): e04691
4. Senila M et al (2024) Nutritional composition and health risk of toxic metals of some edible wild mushrooms growing in a mining area of Apuseni Mountains, Western Carpathians. J Food Compos Anal 128:106061
5. Kokkoris V et al (2019) Accumulation of heavy metals by wild edible mushrooms with respect to soil substrates in the Athens metropolitan area (Greece). Sci Total Environ 685:280–296
6. Barea-Sepúlveda M et al (2021) Metal concentrations in *Lactarius* mushroom species collected from southern Spain and northern Morocco: evaluation of health risks and benefits. J Food Compos Anal 99:103859
7. Falandysz J et al (2022) Total mercury and methylmercury (MeHg) in braised and crude *boletus edulis* carpophores during various developmental stages. Environ Sci Pollut Res 29(2): 3107–3115
8. Haro A et al (2020) Mineral elements content of wild growing edible mushrooms from the southeast of Spain. J Food Compos Anal 91: 103504
9. Senila M, Senila L, Resz M-A (2024) Chemical composition and nutritional characteristics of popular wild edible mushroom species collected from North-Western Romania. J Food Compos Anal 134:106504
10. Soylak M et al (2005) Determination of trace metals in mushroom samples from Kayseri. Tur Food Chem 92(4):649–652
11. Senila M (2024) Recent advances in the determination of major and trace elements in plants using inductively coupled plasma optical emission spectrometry. Molecules 29(13):3169
12. Douvris C et al (2023) How ICP-OES changed the face of trace element analysis: review of the global application landscape. Sci Total Environ 905:167242
13. Heleno SA et al (2015) Nutritional value, bioactive compounds, antimicrobial activity and bioaccessibility studies with wild edible mushrooms. LWT Food Sci Technol 63(2):799–806
14. Bankaji I et al (2023) Comparison of digestion methods using atomic absorption spectrometry for the determination of metal levels in plants. Separations 10(1):40
15. Řanda Z, Kučera J (2004) Trace elements in higher fungi (mushrooms) determined by activation analysis. J Radioanal Nucl Chem 259(1):99–107
16. Radulescu C et al (2010) Studies concerning heavy metals bioaccumulation of wild edible mushrooms from industrial area by using spectrometric techniques. Bull Environ Contam Toxicol 84(5):641–646
17. ISO 8466-1:2021 (2021) Water quality — Calibration and evaluation of analytical methods Part 1: Linear calibration function

Chapter 10

Fungal Characterization and Identification: Genomic and Transcriptomic Approaches

Pablo Navarro Simarro and Alberto José López-Jiménez

Abstract

Fungal diversity plays an essential role in the food industry by diversifying gastronomy and enabling the development of new flavors and products through fermentation. Genomic studies can be used to characterize new fungal strains and map their genetic variations. Moreover, fungi express different genes when confronted with different substrates, which has a direct influence the organoleptic properties of fermented products. This chapter discusses methods for extracting DNA from fungi, sequencing, and analyzing its genetic variations. In addition, it presents a detailed transcriptomic workflow, including RNA extraction, sequencing, and subsequent analysis with bioinformatics methods is described. Platforms such as Galaxy and R-based environments like Bioconductor will be explored to identify differentially expressed genes and their biological function.

Key words Mushrooms, RNA-seq, Genome, Transcriptome, Bioinformatics

1 Introduction

Identifying and characterizing fungal species is essential in both research and food industry settings to detect contamination or to unravel the biological diversity in fermented foods. One of the most important methods to identify fungal species is the amplification of the nuclear ribosomal internal transcribed spacer (ITS) region [1]. This sequence has been described as the universal barcode for fungi and can be amplified by PCR, sequenced, and taxonomically assigned using public databases. In addition, advances in sequencing methods have made it possible to reduce costs and time for whole-genome sequencing of organisms such as fungi. These techniques allow us to delve into fungal diversity, establish evolutionary relationships, and identify species with valuable applications in the food industry.

In the field of fermented food production, transcriptomic analysis provides key insights into the dynamic gene expression profiles that influence metabolic activity and product characteristics. Gene

Pablo Navarro Simarro (ed.), *Fungi*, Methods and Protocols in Food Science, https://doi.org/10.1007/978-1-0716-4989-3_10,
© The Author(s), under exclusive license to Springer Science+Business Media, LLC, part of Springer Nature 2026

expression during fermentation determines the synthesis of metabolites that shape the flavor, aroma, and chemical composition of the end product [2, 3]. The most popular technique nowadays to study gene expression is RNA sequencing (RNA-seq). This high-throughput sequencing method allows cDNA transcripts to be analyzed with high sensitivity and resolution, providing a comprehensive view of the transcriptome. To perform RNA-seq, first, total RNA is extracted from the biological sample, which is then retrotranscribed into cDNA to generate a sequencing library [4]. This cDNA is amplified, fragmented, and sequenced using high-throughput platforms such as Illumina. The reads obtained are aligned to a reference genome or transcriptome, allowing them to be mapped to specific genes. Once the reads are mapped, they are quantified to determine the relative abundance of each transcript, and the resulting data are normalized to correct for technical biases inherent to the protocol, such as variations in library size or gene length. Finally, statistical and bioinformatics tools are used to identify differentially expressed genes between experimental conditions, providing a detailed understanding of changes in gene expression [5].

Several bioinformatics tools are available for analyzing RNA-seq data, each with its specific approaches for differential gene expression mapping, quantification, and analysis. For mapping reads, programs such as Bowtie2 [6], HISAT2, BWA [7], and STAR [8] can be used. There are also alternative approaches, such as pseudo-alignment, in which tools such as Kallisto and Salmon stand out, offering advantages in terms of speed and accuracy when mapping transcripts without the need for full alignment. The selection of the appropriate tool depends on factors such as the nature of the reference (genome or transcriptome) and whether splicing events must be accounted for, a particularly important consideration in RNA-seq analyses. For differential expression analysis, a variety of software is used, including DESeq2, edgeR, and limma+voom [9].

From the perspective of functional analysis of differentially expressed gene groups, various tools are available to assess their biological significance. Some approaches rely on ranked gene lists, incorporating quantitative criteria such as the magnitude of change or statistical metrics (e.g., fgsea, clusterProfiler, ReactomePA) [10–12]. Others are based on non-ranked gene lists, considering only the presence of genes regardless of additional values, as in tools like DAVID, g:Profiler, or Panther, among others [13–15].

In addition, there are other tools, such as FastQC, for quality control of reads. Additionally, other software tools are required for the preprocessing of raw sequencing reads, with the aim of removing low-quality sequences, trimming adapter contamination from the sequencing process, and correcting other technical issues. Some of the most commonly used tools include Trimmomatic, Cutadapt, fastp, BBDuk, and Trim Galore [16–18]. This type of analysis may require substantial computational power, as some of the applications involved—particularly those for alignment, quantification, or large-scale enrichment analysis—can be resource-intensive. Fortunately, the Galaxy platform bundles many of its tools for remote use, enabling complex analyses to be performed on large databases by leveraging substantial computational power [19].

The RNA-seq method has enabled the evaluation of the differences in gene expression of filamentous fungi metabolizing various substrates, leading to significant advances, such as the identification of genes involved in the degradation of food by-products. Numerous studies have investigated the metabolic pathways of fungi like *Neurospora crassa* to metabolize plant carbohydrates, such as pectin, starch, or xylose [5, 20, 21]. In the field of fermented foods, it is worth highlighting the influence of the transcriptome in *Saccharomyces* to provide unique flavors to the great diversity of alcoholic beverages in the world, from wine to mezcal [22]. In addition, both genomic and transcriptomic methods allow us to analyze the diversity of traditional Indonesian foods such as tempeh, exploring which genes influence its characteristic flavor and health-promoting properties [23]. Such analyses will undoubtedly contribute to understanding the biological basis of the organoleptic properties of fungal fermented foods and may also facilitate the identification of novel fungal species.

This chapter will explore the methods of DNA and RNA extraction to obtain the highest quality in subsequent sequencing. In addition, in the case of genomic analysis, guidance will be provided on how to align sequencing results against reference genomes and the subsequent identification of single-nucleotide polymorphisms (SNPs). For transcriptome analysis, a pipeline for RNA-seq data with an integrated approach in both Galaxy and R will be described.

2 Materials

2.1 Fungal Identification by ITS Amplification

Material:

1. Thermocycler.
2. Automatic pipettes (1–1000 μL volumes).
3. PCR tubes.
4. 1.5 mL tubes.
5. Transilluminator.

Reagents:

1. DNA sample.
2. Primers (290 pb size amplification) [24] (*see* **Note 1**):

ITS1 (Forward) TCCGTAGGTGAACCTGCGG.

ITS2 (Reverse): GCTGCGTTCTTCATCGATGC.

3. dNTPs.
4. Taq DNA polymerase and buffer (with $MgCl_2$).
5. Nuclease-free water.
6. Agarose.
7. Ethidium bromide or other intercalating agents.

2.2 ARN Extraction

Equipment:

1. Mortar and pestle (for crushing the sample).
2. Cryocontainer with liquid nitrogen.
3. Refrigerated microcentrifuge (capable of 13,000 rpm and 7500 g).
4. Ice bath.
5. Lyophilizer (optional, for sample drying).
6. Automatic pipettes (1–1000 μL volumes).
7. Vortex.

Reagents:

1. TRIzol® Reagent.
2. Chloroform.
3. Isopropanol.
4. 75% ethanol prepared with RNase-free water.
5. RNase-free water.
6. 1.5 mL and 2.0 mL Eppendorf tubes.

2.3 Genomic and Transcriptomic Analyses

1. Access to Galaxy (usegalaxy.org), g:Profiler (biit.cs.ut.ee/gprofiler/) and STRING (string-db.org/).
2. Installed R and Rstudio.
3. The BiocManager, ggplot2, and pheatmap packages from R. From BiocManager install the DESeq2 and TopGO packages.

3 Methods

3.1 Genomic DNA Extraction and Fungal Identification

3.1.1 Genomic DNA Extraction

1. In the case of filamentous fungi, it is recommended to grow in potato dextrose broth medium (or specific medium for that strain) for at least 24 h to obtain sufficient biomass for DNA extraction. If the fungus does not grow in liquid medium, it can be grown on agar or, if necessary, on another type of solid

medium, although the extraction yield may be lower (*see* **Note 2**).

2. There are various approaches to extracting high-quality DNA, ranging from traditional manual methods to commercially available kits [25]. Manual protocols allow for customization depending on the sample type, but they can be time-consuming. On the other hand, commercial kits, such as spin column-based systems, offer a faster and more standardized approach, minimizing contamination and ensuring reproducibility. However, these kits often yield lower DNA quantities and can be costly, making them less suitable for large-scale studies. The choice between manual methods and kits depends on the specific needs of the study, balancing factors such as DNA yield, purity, time, and cost.
3. DNA purity and concentration can be assessed using spectrophotometric methods, such as measuring absorbance at 260 nm and 280 nm, where an A260/280 ratio of approximately 1.8–2.0 indicates high-purity DNA suitable for downstream applications. Also, the 230/260 ratio should be between 2.0 and 2.2 to avoid the presence of contaminants. In addition to conventional spectrophotometers, devices such as the NanoDrop allow for rapid and small-volume quantification, making them especially convenient for routine checks. Furthermore, instruments like the Bioanalyzer or TapeStation can provide more detailed assessments of DNA integrity and fragment size distribution, which are particularly valuable when high-quality DNA is required for sensitive applications such as next-generation sequencing.

3.1.2 Fungal Identification by ITS Amplification

Accurate fungal identification is a critical step prior to downstream analyses, as undetected contamination can compromise the reliability of subsequent results. For this reason, amplification and sequencing of the internal transcribed spacer (ITS) region is commonly employed as a molecular barcode for taxonomic identification.

The protocol for PCR amplification of the ITS region begins with the preparation of the reaction mixture in a final volume of 25 μL, which includes the following: 1–2 μL of template DNA, 0.5 μL of each primer (ITS1 and ITS2, 10 μM), 12.5 μL of master mix with Taq polymerase, dNTPs and buffer, and nuclease-free water to complete the volume. PCR is performed in a thermocycler with the following program: initial denaturation at 95 °C for 3 min, followed by 30–35 cycles of denaturation at 95 °C for 30 s, alignment at 55–60 °C for 30 s, extension at 72 °C for 1 min, and a final extension at 72 °C for 5–10 min. Subsequently, the amplified products are verified by electrophoresis in 1% agarose gel stained with ethidium bromide or another DNA chelator and visualized in a UV transilluminator. The purified PCR products are then sequenced, and the resulting sequences are compared against reference databases—such as the comprehensive NCBI nt database—using tools like BLASTn [26].

3.1.3 Genome Sequencing

In recent years, there has been an explosion in sequencing capacity, driven by advances in next-generation sequencing (NGS) and other high-throughput technologies [27]. This chapter will not focus on explaining these methods step by step, as laboratories without the necessary infrastructure can outsource sequencing to specialized companies, ensuring high-quality results without the need for in-house sequencing capabilities.

3.1.4 De Novo Assembling and Identification of SNPs

This section will show a protocol to identify genetic variants of fungal individuals using Galaxy (https://usegalaxy.org/) and UCSC browser (https://genome.ucsc.edu/):

1. Import the sequencing data into Galaxy (Fig. 1) and perform quality control according to points 1, 2, and 3 of Subheading 3.2.3 (*see* **Note 3**).
2. Align the reads against the reference genome with Bowtie2 through Galaxy. Note that if data from the fungal sequencing are paired-end, it is necessary to enter two files. The indexed reference genome of several species can be selected in Bowtie2. If no reference is available for the species of interest, it will need to be manually uploaded from an external database (e.g., NCBI or Ensembl) and indexed it accordingly. After mapping with Bowtie, it is advisable to assess the percentage of reads that align concordantly to the genome, as well as the proportion of reads that map to multiple locations. These metrics are indicative of mapping quality; deviations from expected values may suggest contamination with other species, sequencing issues, or problems during sample preparation.
3. The Galaxy FreeBayes tool can be used for the detection of genetic variants. The input will be a Bowtie2 BAM file format, and the output will be a list in Variant Call Format ".vcf," in which all genetic variants of the individual to be tested are located. It is recommended to apply post-calling filters to the

```
@LH00714:15:22HHTNLT4:4:1101:24726:2177 2:N:0:TATGCCTTAC+TGATGTGTCT
GGGGGGGGGGGGGGGGGGGGGGGGGGGGGGGGGGGGGGGGGGGGGGGGGGGGGGGGGGGGGGGGGGGGGGGGGGGGGGGGGGGGGGGGGGGGGGGGGGGAGGGGGGGGGGATTCTGTGGTAAGGGGGACCCGGGCTACGTGGTTGCCGCT
+
-I--99II9-I9I9II99-III9I9-9999-9--99-9-9-9-9---99---9999---99-99----999--9-9-99-9999-99--99-99--9--9999-------------------9-9-----------9----9-9-9--
@LH00714:15:22HHTNLT4:4:1101:24743:2177 2:N:0:TATGCCTTAC+TAATGTGTCT
GAAATTAAGGATTCTATCTAGCTTCTAACGGTCGTGAGTTTAAAGATAGTCGCAAAGTGTGCTCAACAAAAACAAGTCTTCATTCAATAGATGTGAAAATGCTCTATCGTAGCTGCAGGATCACAGCTTTCATCCTCATCATCTTCCT
+
IIIIIIIIIIIIIIIIIIIIIIIIIIIIIIIIIIIIIII-IIII9IIIIIIIIIIIIIIIIII9IIIIIIIIIIIIIIIIIII9I-IIIIIIIIII9IIIIIIIIII-IIIIIIII-III9-I9IIIIII-I9-III9IIIIIII
```

Fig. 1 Example of the Illumina sequencing product in FASTQ format. Each read consists of a top row with the sequence identifier, another with the nucleotide chain of the sequence, and finally, the Q scores (Quality) based on ASCII characters

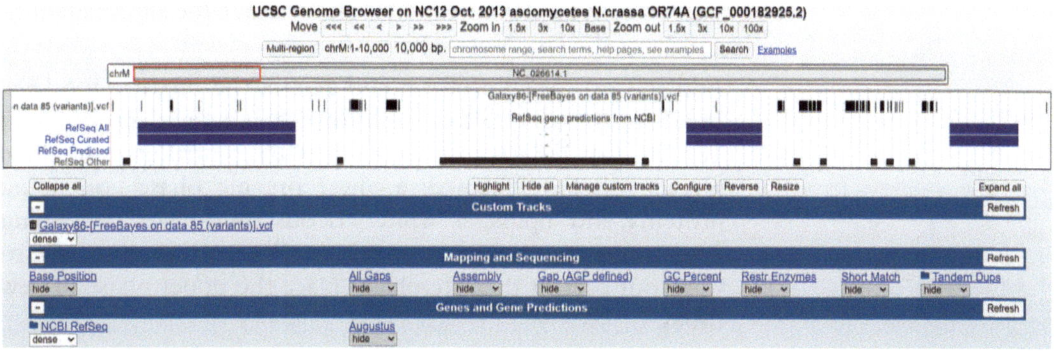

Fig. 2 View from the UCSC Genome Browser (https:/genome.ucsc.edu/). The reference genome used is *Neurospora crassa* OR74A, and the NCBI RefSeq annotation for the first 10,000 bp of the mitochondrial genome is displayed. As a Custom Track, the freeBayes results for a strain of *N. crassa* are shown, with the black lines in this track representing DNA sequence variations

variant set–for example, using tools such as VCFfilter or bcftools filter–in order to remove low-quality or poorly supported variant calls and improve the reliability of the results.

4. One way to visualize the set of genetic variants in the genome is by using the UCSC Genome Browser. To do this, select the reference genome to work with and display only the NCBI RefSeq section under Genes and Gene Predictions. The file from FreeBayes can then be uploaded as a custom track and displayed against the reference genome (Fig. 2). Alternatively, if the reference genome is not available through UCSC Genome Browser, alternative approaches include the local installation of genome browsers such as Integrated Genome Browser (IGB) or Integrative Genomics Viewer (IGV). These tools allow manual integration and indexing of custom genomes, as well as the loading of annotation files and other relevant data for visualization and analysis.

3.2 RNA Sequencing

3.2.1 RNA Extraction

In this case, the TRIzol-based extraction method [28] will be described:

1. First, the biological sample of interest will be collected and immediately immersed in liquid nitrogen to avoid RNA degradation. This sample can be handled directly while frozen, or it can be lyophilized. In both cases, the sample should be crushed with a mortar or pestle until its cell structure is disrupted.
2. Take a sample of 50–100 mg of fresh sample or 10–20 mg of lyophilized sample (*see* **Note 2**) and place it in an Eppendorf tube.
3. Add 1 mL TRIzol®, shake the mixture, and incubate at room temperature for 10 min. Subsequently, centrifuge at

13,000 rpm for 10 min at 4 °C and transfer the supernatant to another tube (*see* **Note 4**).

4. Add 0.2 mL of chloroform and agitate the mixture for 15 s. Incubate the sample for 3 min on ice and then centrifuge at 13,000 rpm for 10 min at 4 °C. After centrifugation, three phases will have formed: a lower organic phase containing proteins and lipids, a white viscous interphase containing DNA, and an upper phase rich in RNA. With care not to wash out part of the interphase, place the upper phase in new tubes.
5. Add 0.5 mL isopropanol to the upper phase, shake vigorously and incubate on ice for 10 min. Then, centrifuge at 13,000 rpm for 10 min at 4 °C and discard the supernatant. The resulting pellet is the RNA.
6. To wash the RNA from impurities, add 1 mL of 75% ethanol with RNAse-free water. Shake to release the pellet and centrifuge at 7500 g for 5 min at 4 °C. Once centrifuged, discard the supernatant and allow the remaining ethanol to evaporate for 5 min by leaving the Eppendorf lid open.
7. Finally, dissolve the pellet in 50–100 μL of RNAse-free water. This solution can be stored at −80 °C or sent for sequencing. In the latter case, it is recommended to lyophilize the sample to avoid degradation events.

3.2.2 RNA Sequencing

The previous section explained how to extract RNA from biological samples. These samples can be sent to specialized companies or sequenced in the laboratory if the appropriate equipment is available. The focus of this chapter is not on giving a detailed protocol for sequencing, although the basis for the preparation of RNA libraries using the Illumina TruSeq Stranded mRNA kit is provided:

1. Purification of mRNA: mRNA is isolated from total RNA using magnetic beads with poly-T oligos, which specifically capture transcripts with poly-A tails. The sample is then repeatedly washed with the appropriate buffer to remove ribosomal RNA and other impurities.
2. RNA fragmentation: The purified mRNA is fragmented using divalent cations and high temperature, generating fragments of a suitable size for sequencing.
3. cDNA synthesis: Reverse transcriptase is used together with random primers for the conversion of mRNA into first-strand cDNA. Actinomycin D is incorporated to prevent DNA-dependent synthesis and improve strand specificity. Next, the second cDNA strand is synthesized by replacing dTTP with dUTP in the tagging mixture, allowing selective inactivation of the complementary strand at later stages.

4. Preparation for adapter ligation: An adenosine is added to the 3′ ends of the fragments to prevent unwanted ligation. Adapters with unique indices are ligated to the double-stranded cDNA fragments for identification of individual samples in multiplex sequencing.
5. Library enrichment: For fragment selection Magnetic beads are used to select fragments with an optimal insert size for the sequencing platform. Libraries are then amplified using a minimum number of PCR cycles to avoid amplification bias. For library validation, fragment size distribution is analyzed using agarose gel electrophoresis or a bioanalyzer. Finally, the final concentration of the libraries is determined using fluorometry or qPCR before sequencing.
6. Sequencing: The prepared libraries are loaded onto an Illumina system for high-throughput sequencing. Reads from the first end (Read 1) will map to the antisense strand, while reads from the second end (Read 2) will map to the complementary strand.

3.2.3 Quality Control, Reference Genome Alignment of Reads, and Counting

This section will explain the workflow used to upload the RNA sequencing files to Galaxy and perform a quality control with the FASTQC tool and an alignment against the reference genome using RNA STAR (*see* **Note 5**). In addition, a count of the different reads associated with a gene will be performed using the featureCounts tool. A comprehensive view of the RNA-seq data analysis process can be found in Fig. 3.

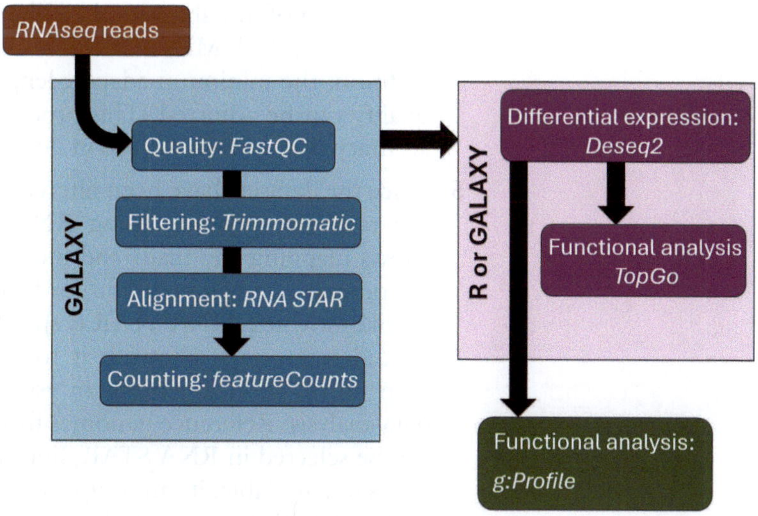

Fig. 3 Tool pipeline used to analyze RNA-seq data, from read acquisition to biological significance analysis. The tools are available on Galaxy, R, or online platforms

1. RNA-seq data can be imported into Galaxy directly from the upload section on the left-hand side of the screen. Galaxy supports compressed files (.gz) that can be uploaded directly from the host files. Once imported, the files will appear in green on the right side of the screen (History) if they have been uploaded successfully.
2. The Galaxy tools can be accessed by clicking on the Tools tab on the left side of the screen. In this space, it is possible to type the name of the desired tool, in this case, FASTQC, to evaluate the quality of the reads. This tool allows for analyzing one dataset (select "single dataset") or several datasets (select "multiple datasets"). In this window, it is possible to optionally select a list of contaminants or to limit the minimum size of the reads to be considered. The report generated by FastQC includes a detailed analysis with the following key statistics and metrics (*see* **Note 6**): total number of sequences, sequence length, quality per base, quality per tile, quality scores per sequence, base content per position, GC content per sequence, N content per base, sequence length distribution, sequence duplication levels, over-represented sequences and adapter content (Fig. 4).
3. The MultiQC tool can then be used to generate a combined report of the set of sequences. This tool can be used to obtain combined reports from other programs; just select the source program in the "Which tool was used to generate logs?" section.
4. Readings can be filtered using the Trimmomatic tool in Galaxy. Single-end or pair-end reads in FASTQ format are supported. The option of initial removal of Illumina-specific adapters and sequences (ILLUMINACLIP) is available for selection. In addition, the minimum adapter length and minimum allowed quality can be adjusted. This process ensured the quality and accuracy of the reads for use in downstream analysis.
5. Once the datasets have been filtered, they must be aligned with the reference genome using RNA STAR [8]. This process allows mapping the reads and indicating their position in the genome, which will be useful for further analysis (*see* **Note 7**). Unlike Bowtie2, RNA STAR is specifically designed to handle spliced transcripts, allowing it to accurately align reads that span intron-exon junctions–an essential feature for RNA-seq data analysis. Reference genomes from a wide variety of species can be selected in RNA STAR, but in case the required organism is not available, it can be uploaded to Galaxy from databases such as NCBI. Finally, run the program to obtain a file in BAM format containing the different reads mapped and classified by gene. As in the mapping described in Subheading 3.1.4, it is

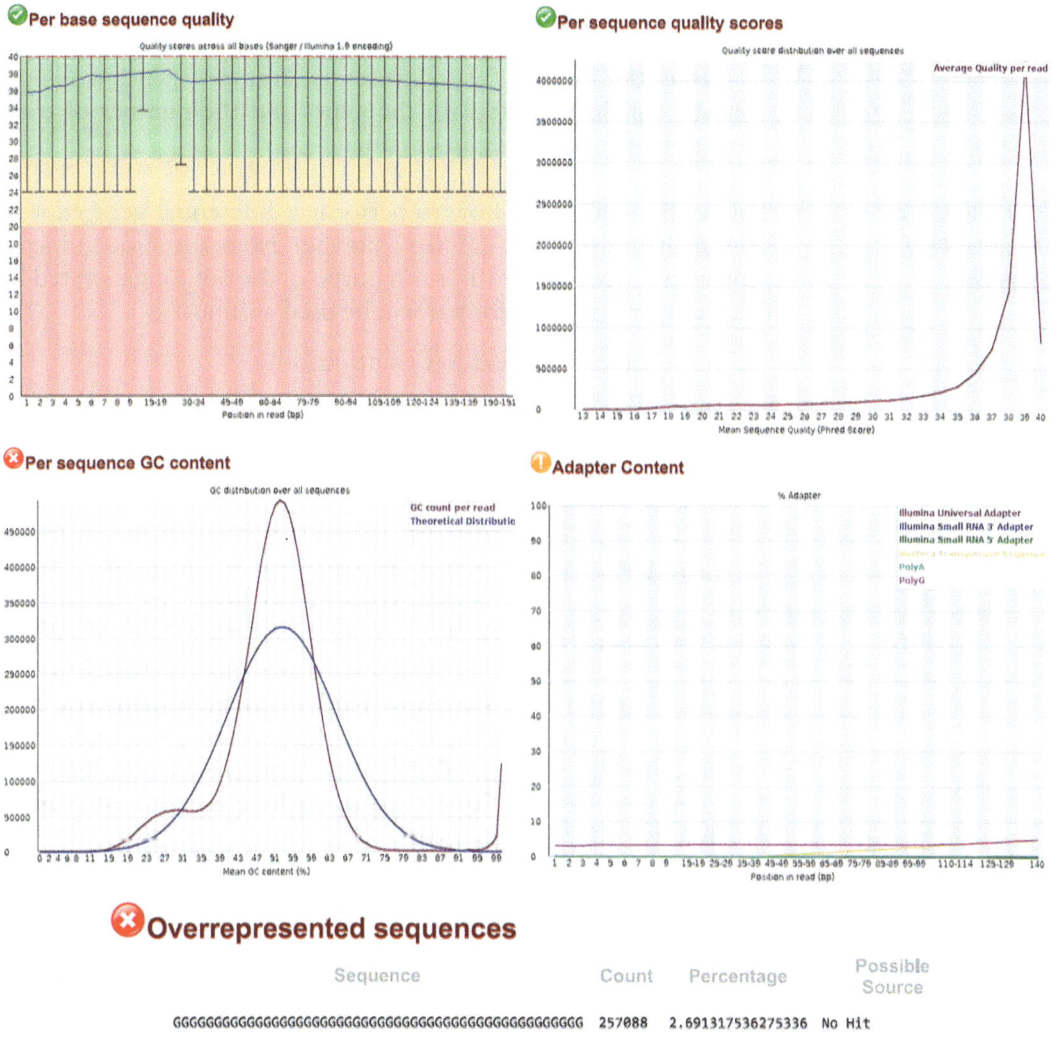

Fig. 4 Example of the quality control output of a set of reads using the FASTQC tool in Galaxy. Green ticks indicate high quality for the tested parameter, an exclamation mark signifies a moderate issue, and a red cross indicates that the quality is very low for that parameter

important to examine the percentage of reads that align concordantly, as well as the proportion of multimapping reads, since significant deviations from expected values may indicate contamination or issues in previous sample processing steps.

6. BAM files from RNA STAR are now loaded into featureCounts. This tool counts the number of reads assigned to each gene, which allows for subsequent analyses to compare the expression levels of different transcripts. For this, the strand orientation type is selected, and a gff file with the genome annotation is added as support. In the case of fungi, eukaryotes, the feature type "exon" should be selected to include only reads matching this category within the GFF file. In

addition, "gene_id" (or any other ID indicating the feature "gene" in the annotation file) should be set as the attribute to group the features (exons) into metafeatures (genes). As output, a plain text file with the genes and the number of transcripts identified will be obtained.

3.2.4 Analysis of Differential Gene Expression

This section will show how to perform a differential gene expression analysis using the DESeq2 package from the Bioconductor project in R (*see* **Note 8**). For this purpose, the output generated by featureCounts in CSV format will be used as input.

1. Import files from featureCounts into R.
2. Install BiocManager and DESeq2 if not already installed:

```
install.packages("BiocManager")
BiocManager::install("DESeq2")
```

3. Load DESeq2:

```
library("DESeq2")
```

4. Then, combine the samples excluding the gene name column, define the names of the columns, and assign the groups. The order of the groups will define the reference group. For this example code, the X and Y groups will be used as if they were different RNA-seq samples:

```
# Combine the samples with the command cbind (excluding the
Geneid column)
combined <- cbind(X1[, 2], X2[, 2], X3[, 2], Y1[, 2], Y2[, 2],
Y3[, 2])
# Define column names
colnames(combined) <- c("X1", "X2", "X3", "Y1", "Y2", "Y3")
# Define the grups (Y as baseline)
group <- factor(c("X", "X", "X", "Y", "Y", "Y"))
```

5. The next step is to create an object of class DESeqDataSet, which stores the count data and experimental design information:

```
dds <- DESeqDataSetFromMatrix(countData = combined,
colData = DataFrame(group),
design = ~ group)
```

6. Then, the data is normalized and the results are obtained. Normalization is necessary as it is highly unlikely that two RNA-seq samples will have the same number of reads. DESeq2 uses normalization based on library size and gene

variability to adjust expression values and make them comparable between samples. Once the data has been normalized, the comparison can be performed to generate the results.

```
# Normalize the data and fit the model
dds <- DESeq(dds)
# Obtain results of the X vs Y comparison
res <- results(dds, contrast = c("group", "X", "Y"))
summary(res)
```

7. The results are organized with each row corresponding to a different gene. Each column corresponds to one of the following parameters:
   - logFC (Log Fold Change): Represents the change in gene expression between the groups compared, expressed on a base 2 logarithmic scale. Positive values indicate higher expression in the experimental group (X in this case), while negative values indicate lower expression in group X compared to the reference group (Y) (*see* **Note 9**).
   - logCPM (Log Counts Per Million): Indicates the average abundance of the gene in all samples, on a base 2 logarithmic scale and normalized per million mapped reads. Provides an idea of the basal expression of the gene.
   - LR (Likelihood Ratio Test Statistic): The likelihood ratio test statistic used to assess the significance of differential expression. The higher the value, the greater the evidence that the gene is differentially expressed.
   - PValue: This is the raw p-value associated with the hypothesis test on differential expression. Low values indicate greater confidence that the gene is differentially expressed (*see* **Note 10**).
8. To understand the differences in differential gene expression, several graphs can be plotted (Fig. 5):
   - PCA (Principal Component Analysis), which allows visualizing the global variability of gene expression among samples, grouping those with similar expression profiles.
   - MA-plot, showing the relationship between mean expression (X-axis) and change in expression (logFC, Y-axis), highlighting differentially expressed genes based on their statistical significance.
   - Volcano plot, which combines log-fold change with statistical significance (p-value or FDR), allowing the identification of highly regulated genes with significant differences in their expression.

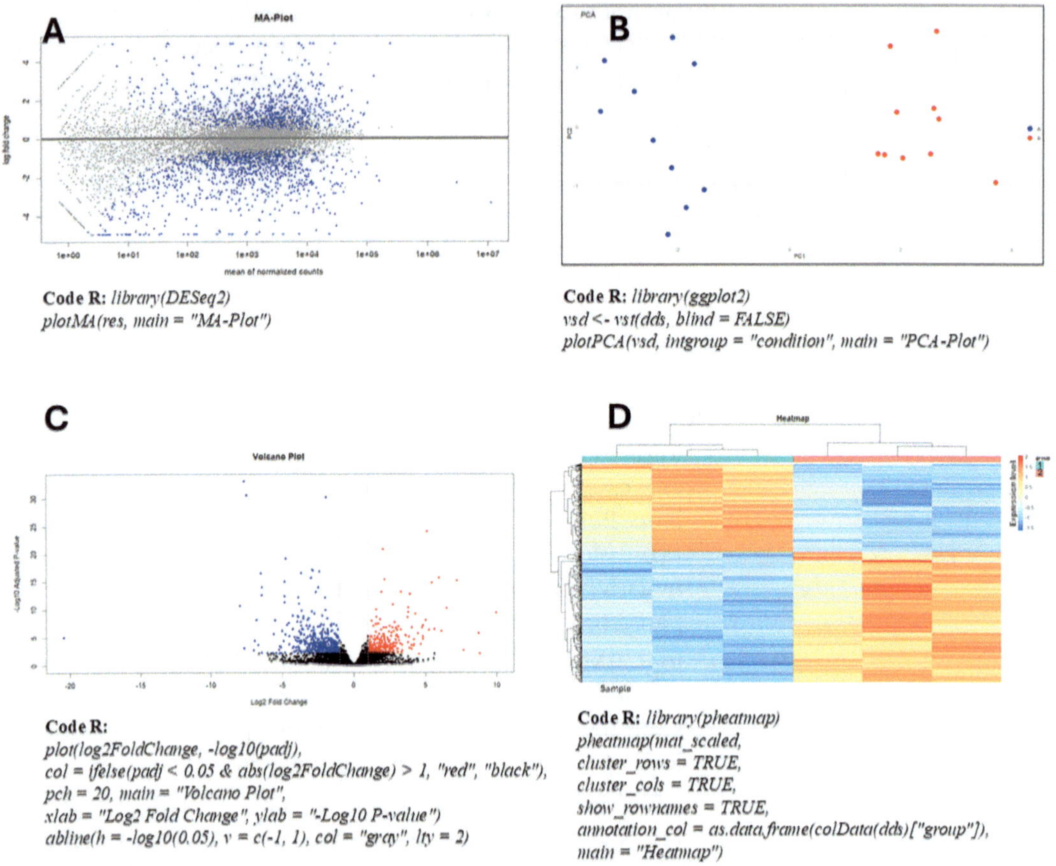

Fig. 5 A set of useful plots for differential gene expression analysis, accompanied by generic R code snippets to assist with their visualization. The figure includes an MA plot (**a**), PCA plot (**b**), Volcano plot (**c**), and Heatmap (**d**)

– Heatmaps, which help to visualize gene expression patterns through a map of colors scored according to logFC, showing clusters of genes with similar profiles between experimental conditions. For this plot, it is recommended to use only genes with a *p*-value less than 0.05 or 0.01.

3.2.5 Biological Significance and Routes

A key aspect of transcriptomic data analysis is to elucidate the biological role of differentially expressed genes. These data can help us to understand whether the fungal cellular machinery is focusing on carbohydrate degradation, cell wall remodeling, gene regulation, and protein expression, among other functions. In this section, we will show how biological significance analysis can be obtained with the R package TopGo or with more intuitive online tools such as g:Profiler. These two tools classify genes into three functional categories:

- *Biological Process* (BP): This category covers biological processes involving differentially expressed genes. It includes metabolic pathways, responses to external stimuli, developmental processes, and cell cycle-related functions.
- *Molecular Function* (MF): Describes the biochemical activities carried out by genes, such as protein binding, catalytic activity and enzymatic functions.
- *Cellular Component* (CC): Identifies the subcellular locations or cellular structures where gene products exert their functions, such as membranes, organelles, and protein complexes.

TopGo

1. Install the TopGo package in R if it is not already installed:

```
if (!requireNamespace("BiocManager", quietly = TRUE))
install.packages("BiocManager")
BiocManager::install("topGO")
```

2. Load TopGo:

```
library(topGO)
```

3. Genes with significant differential expression should be selected. In this case, we select those with a log2FoldChange greater than 1 or less than −1 and with a padj less than 0.01. In addition, in this chapter, the code to separate them into upregulated and downregulated with respect to the reference group will be displayed:

```
# Load differentially expressed genes
upregulated_genes <- rownames(res[res$log2FoldChange > 1 & res
$padj < 0.01, ])
downregulated_genes <- rownames(res[res$log2FoldChange < -1 &
res$padj < 0.01, ])
# Create a vector with all analyzed genes
all_genes <- unique(c(upregulated_genes, downregulat-
ed_genes))
# Assign values: 1 for overexpressed, -1 for underexpressed,
0 for others
gene_list <- rep(0, length(all_genes))
names(gene_list) <- all_genes
gene_list[upregulated_genes] <- 1
gene_list[downregulated_genes] <- -1
```

4. Performing the gene ontology analysis: To perform the GO term enrichment analysis, it is necessary to have a gene2GO annotation dictionary, which relates each gene to its

corresponding GO terms. Generally, this file has two columns: one with the gene identifiers and another with the corresponding GO terms, separated by commas if a gene is associated with multiple terms. These dictionaries can be obtained from public databases such as Ensembl (biomaRt), Gene Ontology Consortium, NCBI, or organism-specific resources such as fungalgenomes.org for fungi. They can also be generated manually from GFF files or genomic database annotations.

```
# Load annotation dictionary
gene2GO <- read.table("gene2go.txt", header = TRUE, sep =
"\t")
# Create data object for topGO
GOdata <- new("topGOdata",
ontology = "BP", # "BP" for Biological Process, "MF" for
Molecular Function, "CC" for cellular component
allGenes = gene_list,
annot = annFUN.gene2GO,
gene2GO = gene2GO,
geneSelectionFun = function(x) x != 0)
```

5. Fisher's test is used to determine the significance of the enriched GO terms:

```
# Perform Fisher's statistical test
resultFisher <- runTest(GOdata, algorithm = "classic", sta-
tistic = "fisher")
# Obtain the most significant GO terms
GO_results <- GenTable(GOdata, classicFisher = resultFisher,
topNodes = 10)
print(GO_results)
```

6. For easier interpretation, the most significant GO terms can be plotted on a bar chart.

```
# Filter the 10 most enriched GO terms
GO_filtered <- GenTable(GOdata, classicFisher = resultFisher,
topNodes = 10)
# Create a bar chart to visualize the most significant GO terms
barplot(as.numeric(GO_filtered$Significant),
names.arg = GO_filtered$Term,
las = 2,
col = "skyblue",
horiz = TRUE,
main = "Enriched GO terms",
xlab = "Number of genes",
ylab = "GO term")
```

g:Profiler

g:Profiler (biit.cs.ut.ee/gprofiler/) is an online tool to perform functional gene enrichment analysis, including Gene Ontology (GO) terms, KEGG/Reactome pathways, and annotations of databases such as Ensembl and UniProt [13]. It is widely used to identify biological functions associated with sets of differentially expressed genes in various organisms. Unlike TopGo, this tool does not require programming skills.

To use it, you need to create a text file with the names of differentially expressed genes that meet the desired statistical significance criterion (e.g., padj <0.05). Then, access the g:Profiler web page, paste the list of genes in the corresponding field and select the study organism. The results are presented in a table with the enriched terms, their p-adj (FDR) values, the number of genes involved in each category, and links to further information. You can also generate interactive graphs to visualize the functional distribution of genes and download the data in formats such as CSV or PNG.

3.2.6 Protein Interactions

The codes of differentially expressed genes can be entered into tools such as STRING (https://string-db.org/) [29] to analyze their interactions at the protein level. This platform allows the identification of protein-protein interaction networks, revealing whether genes belong to the same metabolic pathways, protein complexes, or share biological functions. This type of analysis is essential to contextualize changes in gene expression within cellular and functional processes, providing a more integrated view of the underlying biological mechanisms.

The output of STRING is presented in the form of a protein interaction network, where each node represents a protein encoded by a gene and the edges indicate functional associations based on experimental evidence, databases, co-expression, co-occurrence, and phylogenetic similarity. In addition to the interaction graph, STRING provides a table with details about each connection, including the source of evidence and a confidence score. Functional enrichments, such as GO terms and KEGG pathways, can also be visualized, allowing interpretation of whether differentially expressed genes are involved in specific biological processes or are part of the same functional complex.

3.3 Reverse Transcription-Quantitative PCR (rt-qPCR)

RT-qPCR is a molecular technique that allows real-time quantification of gene expression from messenger RNA (mRNA) previously retrotranscribed into complementary DNA (cDNA). In the field of fermentations with filamentous fungi, this technique is used to monitor the expression of different genes of interest, as is the case with industrially important enzymes such as laccase expressed by *Pleurotus ostreatus* [30]. In addition, this technique is necessary to validate the results of RNA-seq, in order to verify that the differentially expressed genes are really doing so [31].

For a detailed RT-qPCR protocol, see Chapter 11.

4 Notes

1. Other primers may be chosen from the literature or may be specifically designed by the researchers. It is possible that for certain taxonomic groups, it is preferable to use other primers.
2. In all cases, it is recommended to freeze-dry the sample to obtain a higher extraction yield and to facilitate the grinding of the biomass.
3. To check for contamination with genetic material from other organisms, NCBI BLAST + blastn or Kraken2 can be used in Galaxy.
4. A lower sample weight will result in a poor amount of RNA. A higher sample weight could result in saturation, preventing the reagents from correctly extracting the RNA and thus resulting in a low yield.
5. From now on, the Eppendorff tube should always remain on ice.
6. To align RNA-seq reads, RNA STAR is used instead of Bowtie2. This is because RNA STAR is specifically designed to handle RNA-seq data, which includes the detection of spliced alignments in eukaryotic genes, something that Bowtie2 does not do directly.
7. In the case of working with pair-end reads, it must be indicated, and the data must be processed in pairs. Also, it is essential to consider whether the sequencing data is stranded or unstranded, as this affects how reads are assigned to genes, particularly in regions with overlapping transcripts. Additionally, one must determine whether the data is forward-stranded or reverse-stranded, depending on the library preparation protocol used. If this information is not readily available, tools such as "Infer Experiment" in the Galaxy platform can help infer the strand specificity of the dataset by analyzing the orientation of aligned reads relative to annotated transcripts.
8. The use of DESeq2 in R allows more flexibility in working with the data. If you lack sufficient programming skills, you can work with DESeq2 within Galaxy.
9. It is recommended to prepare a dataset with genes whose logFC is less than −1 or greater than 1, as these represent at least twofold changes in expression and ensure that only

differentially expressed genes with biological relevance are analyzed in some tests.

10. It is recommended to elaborate a dataset considering only genes with p-value less than 0.05 or 0.01, which indicates a significant difference in gene expression with high statistical confidence.

Acknowledgements

This work was supported by grant 230025CONV from the Junta de Comunidades de Castilla-La Mancha (co-financed European Union FEDER funds).

References

1. Schoch CL, Seifert KA, Huhndorf S, Robert V, Spouge JL, Levesque CA, Chen W, Fungal Barcoding Consortium et al (2012) Nuclear ribosomal internal transcribed spacer (ITS) region as a universal DNA barcode marker for fungi. Proc Natl Acad Sci 109:6241–6246. https://doi.org/10.1073/pnas.1117018109
2. Mendes I, Sanchez I, Franco-Duarte R, Camarasa C, Schuller D, Dequin S, Sousa MJ (2017) Integrating transcriptomics and metabolomics for the analysis of the aroma profiles of *Saccharomyces cerevisiae* strains from diverse origins. BMC Genomics 18:455. https://doi.org/10.1186/s12864-017-3816-1
3. Smid EJ, Hugenholtz J (2010) Functional genomics for food fermentation processes. Annu Rev Food Sci Technol 1:497–519. https://doi.org/10.1146/annurev.food.102308.124143
4. Finotello F, Di Camillo B (2015) Measuring differential gene expression with RNA-seq: challenges and strategies for data analysis. Brief Funct Genomics 14:130–142. https://doi.org/10.1093/bfgp/elu035
5. Samal A, Craig JP, Coradetti ST, Benz JP, Eddy JA, Price ND, Glass NL (2017) Network reconstruction and systems analysis of plant cell wall deconstruction by *Neurospora crassa*. Biotechnol Biofuels 10:225. https://doi.org/10.1186/s13068-017-0901-2
6. Langmead B, Salzberg SL (2012) Fast gapped-read alignment with Bowtie 2. Nat Methods 9:357–359. https://doi.org/10.1038/nmeth.1923
7. Li H, Durbin R (2009) Fast and accurate short read alignment with Burrows–Wheeler transform. Bioinformatics 25:1754–1760. https://doi.org/10.1093/bioinformatics/btp324
8. Dobin A, Davis CA, Schlesinger F, Drenkow J, Zaleski C, Jha S, Batut P, Chaisson M, Gingeras TR (2013) STAR: ultrafast universal RNA-seq aligner. Bioinforma Oxf Engl 29:15–21. https://doi.org/10.1093/bioinformatics/bts635
9. Liu S, Wang Z, Zhu R, Wang F, Cheng Y, Liu Y (2021) Three differential expression analysis methods for RNA sequencing: limma, EdgeR, DESeq2. J Vis Exp:e62528. https://doi.org/10.3791/62528
10. Yu G (2024) Thirteen years of clusterProfiler. Innovation 5. https://doi.org/10.1016/j.xinn.2024.100722
11. Yu G, He Q-Y (2016) ReactomePA: an R/Bioconductor package for reactome pathway analysis and visualization. Mol BioSyst 12:477–479. https://doi.org/10.1039/C5MB00663E
12. Korotkevich G, Sukhov V, Budin N, Shpak B, Artyomov MN, Sergushichev A (2021) Fast gene set enrichment analysis. bioRxiv 060012. https://doi.org/10.1101/060012
13. Kolberg L, Raudvere U, Kuzmin I, Adler P, Vilo J, Peterson H (2023) g:Profiler—interoperable web service for functional enrichment analysis and gene identifier mapping (2023 update). Nucleic Acids Res 51:W207–W212. https://doi.org/10.1093/nar/gkad347
14. Sherman BT, Hao M, Qiu J, Jiao X, Baseler MW, Lane HC, Imamichi T, Chang W (2022) DAVID: a web server for functional enrichment analysis and functional annotation of gene lists (2021 update). Nucleic Acids Res 50:W216–W221. https://doi.org/10.1093/nar/gkac194

15. Thomas PD, Ebert D, Muruganujan A, Mushayahama T, Albou L-P, Mi H (2022) PANTHER: making genome-scale phylogenetics accessible to all. Protein Sci 31:8–22. https://doi.org/10.1002/pro.4218
16. Bolger AM, Lohse M, Usadel B (2014) Trimmomatic: a flexible trimmer for Illumina sequence data. Bioinforma Oxf Engl 30: 2114–2120. https://doi.org/10.1093/bioinformatics/btu170
17. Martin M (2011) Cutadapt removes adapter sequences from high-throughput sequencing reads. EMBnetjournal 17:10–12. https://doi.org/10.14806/ej.17.1.200
18. Chen S, Zhou Y, Chen Y, Gu J (2018) fastp: an ultra-fast all-in-one FASTQ preprocessor. Bioinformatics 34:i884–i890. https://doi.org/10.1093/bioinformatics/bty560
19. Batut B, van den Beek M, Doyle MA, Soranzo N (2021) RNA-seq data analysis in galaxy. In: Picardi E (ed) RNA bioinformatics. Springer, New York, pp 367–392
20. Xue F, Zhao Z, Gu S, Chen M, Xu J, Luo X, Li J, Tian C (2023) The transcriptional factor Clr-5 is involved in cellulose degradation through regulation of amino acid metabolism in *Neurospora crassa*. BMC Biotechnol 23:50. https://doi.org/10.1186/s12896-023-00823-4
21. Li J, Lin L, Li H, Tian C, Ma Y (2014) Transcriptional comparison of the filamentous fungus *Neurospora crassa*growing on three major monosaccharides D-glucose, D-xylose and L-arabinose. Biotechnol Biofuels 7:31. https://doi.org/10.1186/1754-6834-7-31
22. Ibáñez C, Pérez-Torrado R, Morard M, Toft C, Barrio E, Querol A (2017) RNAseq-based transcriptome comparison of *Saccharomyces cerevisiae* strains isolated from diverse fermentative environments. Int J Food Microbiol 257:262–270. https://doi.org/10.1016/j.ijfoodmicro.2017.07.001
23. Tamang JP, Das S, Lama S, Utama GL, Safitri R, Balia RL, Thapa N (2025) Analysis of meta-transcriptomics and identification of genes linked to bioactive peptides and vitamins in Indonesian *tempe*. Food Res Int 202: 115757. https://doi.org/10.1016/j.foodres.2025.115757
24. White TJ, Bruns T, Lee S, Taylor J (1990) Amplification and direct sequencing of fungal ribosomal RNA genes for phylogenetics. In: PCR protocols: a guide to methods and applications. Academic Press, Amsterdam, pp 315–322
25. Abdel-Latif A, Osman G (2017) Comparison of three genomic DNA extraction methods to obtain high DNA quality from maize. Plant Methods 13:1. https://doi.org/10.1186/s13007-016-0152-4
26. Altschul SF, Gish W, Miller W, Myers EW, Lipman DJ (1990) Basic local alignment search tool. J Mol Biol 215:403–410. https://doi.org/10.1016/S0022-2836(05)80360-2
27. Shendure J, Balasubramanian S, Church GM, Gilbert W, Rogers J, Schloss JA, Waterston RH (2017) DNA sequencing at 40: past, present and future. Nature 550:345–353. https://doi.org/10.1038/nature24286
28. Rio DC, Ares M, Hannon GJ, Nilsen TW (2010) Purification of RNA using TRIzol (TRI reagent). Cold Spring Harb Protoc 2010:pdb.prot5439. https://doi.org/10.1101/pdb.prot5439
29. Szklarczyk D, Kirsch R, Koutrouli M, Nastou K, Mehryary F, Hachilif R, Gable AL, Fang T et al (2023) The STRING database in 2023: protein-protein association networks and functional enrichment analyses for any sequenced genome of interest. Nucleic Acids Res 51:D638–D646. https://doi.org/10.1093/nar/gkac1000
30. Castanera R, Pérez G, Omarini A, Alfaro M, Pisabarro AG, Faraco V, Amore A, Ramírez L (2012) Transcriptional and enzymatic profiling of *Pleurotus ostreatus* laccase genes in submerged and solid-state fermentation cultures. Appl Environ Microbiol 78:4037–4045. https://doi.org/10.1128/AEM.07880-11
31. Everaert C, Luypaert M, Maag JLV, Cheng QX, Dinger ME, Hellemans J, Mestdagh P (2017) Benchmarking of RNA-sequencing analysis workflows using whole-transcriptome RT-qPCR expression data. Sci Rep 7:1559. https://doi.org/10.1038/s41598-017-01617-3

Part II

Assessment of Health-Promoting Effects

Chapter 11

Study of the Anti-inflammatory Potential of Mushrooms

Cristian Martínez Fajardo and Susana López-López

Abstract

Pathologies associated with chronic inflammatory processes are currently a public health problem. The dysregulation of the immune system can produce certain pathologies that end up causing diseases such as psoriasis, rheumatoid arthritis, and Crohn's disease, among others. There are methods to control and alleviate this symptomatology associated with poor regulation of the inflammatory response, ranging from synthetic drugs and medicines to bioactive compounds of natural origin. Since ancient times, fungi have aroused great interest in traditional medicine. The great potential of mushrooms to treat various foods, such as different types of infections, anxiety, hypertension, or depression, is well known. All this therapeutic potential lies mainly in the amount of bioactive molecules present in mushrooms, which makes them a very important source of pharmacological resources. For these reasons, it is of particular interest to study how the different compounds can have an effect on the control of the inflammatory response produced by the immune system.

Key words Mushroom, Bioactive compounds, Immunomodulators, Inflammation, Nitrites, RT-qPCR

1 Introduction

The inflammatory process has been extensively studied, as many diseases are currently the result of chronic inflammatory processes, including psoriasis, rheumatoid arthritis and Crohn's disease, among many others. Immunomodulation is a vital strategy to raise the body's defense and recently has gained a broad interest. Using immunomodulators to enhance host defenses has proved to be an effective strategy for combating inflammatory diseases [1].

Diet plays a major role in the development of inflammatory diseases, which is why it is interesting to study the different bioactive compounds in foods and to establish the therapeutic potential they may have against inflammatory processes. Edible mushrooms play a very small role in our diet, and these products have a wide variety of pharmacological properties.

Among the numerous immunomodulatory compounds of fungal origin are those attracting attention as potential candidates for

Pablo Navarro Simarro (ed.), *Fungi*, Methods and Protocols in Food Science, https://doi.org/10.1007/978-1-0716-4989-3_11,
© The Author(s), under exclusive license to Springer Science+Business Media, LLC, part of Springer Nature 2026

biomedical and pharmaceutical applications. Understanding the different categories of immunomodulatory compounds found in medicinal mushrooms can provide valuable insights into their potential applications in promoting tissue regeneration and maintaining immunological balance [2, 3].

In medicinal mushrooms, the primary categories of chemicals that possess immunomodulatory effects include terpenes and terpenoids, polysaccharides (specifically β-d-glucans, as well as polysaccharide peptides and polysaccharide proteins), lectins, and immunomodulatory proteins [4]. Terpenes and terpenoids have been found to exhibit immunomodulatory activity in terms of their involvement in the regulation of genes in nuclear factor-kappa B (NF-κB) pathway [5].

Polysaccharides also have the potential to stimulate different components of the immune system [6]. For example, the polysaccharide was able to promote immune responses through different pathways, such as affecting the cell cycle and promoting B-lymphocyte proliferation. Other polysaccharides such as fucose, mannose, glucose, and galactose also showed immunomodulatory activity in RAW 264.7 cells by inducing NO, IL-6, and TNF-α secretion through up-regulation of their mRNA expression [7].

Lectins, among other effects, have been shown to induce the formation of nitrites by immune system cells, as well as increasing TNF-α expression and the secretion of pro-inflammatory interleukins. All this enhances lymphocyte activation and increases the production of macrophage-activating factors. In addition, many fungal lectins have been shown to have significant antiviral, mitogenic, antibacterial, and antioxidant properties [8, 9].

Regarding the proteins present in fungi, it has been observed that they can enhance the synthesis of different cytokines, including IL-2, IL-4, and TNF-α [10]. In addition, studies with different proteins have shown very interesting results in terms of anticancer properties, such as inhibition of cell proliferation or induction of apoptosis [11].

Bioactive compound	**Mushroom species**	**Mechanism of action**	**References**
Polysaccharides	*Flammulina velutipes*	Decreased levels of CD4+ CD8+, with increased level in IL-10 in medium	[12]
	Lentinus polychrous	Decreased expression of iNOS, IL-1β, IL-6, TNF-α, and COX-2	[13]
	Caripia montagnei	Reduction of NO level, IL-1ra, IL-10, and IFN-γ	[14]

(continued)

Bioactive compound	Mushroom species	Mechanism of action	References
Terpenoids	*Inonotus obliquus*	Inhibition of mRNA expression of iNOS and COX-2	[15]
	Ganoderma lucidum	Downregulation of iNOS and COX-2	[16]
	Cyathus africanus	Inhibition of NO production	[17]
Proteins	*Phallus indusiatus*	Reduction of NO production and downregulation NF-kappaB activation	[18]
	Cordyceps sinensis	Decreased level of TNF-α, IL-1β	[19]
Phenolics compounds	*Albatrellus caeruleoporus*	Inhibition of NO production	[20]
	Lactarius deliciosus	Inhibition of NO production and expression of iNOS, IL-1β, and IL6 mRNAs	[21]
Lectins	*Agaricus bisporus*	Inhibition of NO production	[22]
	Pleurotus eryngii	Reduction of pro-inflammatory cytokines, e.g., TNF-α, IL-6, and IL-1β	[23]

In this chapter, we highlight some of the studies that can be performed to assess the immunostimulatory potential of different fungal compounds, including the evaluation of nitrite production and RNA expression of different proteins involved in the inflammatory process.

2 Materials

1. RAW264.7 macrophage cells from ATCC (ATCC no. TIB-71).
2. DMEM (GIBCO BRL, Gaithersburg, MD) with 1% penicillin–streptomycin, 10 or 5% FBS, and 4 mM L-glutamine.
3. LPS or cytokines.
4. Fresh, dried or lyophilized mushrooms extract.
5. TRIzol reagent.
6. RevertAid First Strand cDNA Synthesis Kit (Thermo Fisher).
7. Primer mixture.

8. SYBR Green.
9. Sulfanilic acid 10 mM.
10. N-1-(naphthyl)ethylenediamine (NEDA) 10 mM.

3 Methods

3.1 Cell Culture Model

1. To optimize, the process, cells should be seeded at a density of 6–8 × 10^4 cells per cm^2 in DMEM with 10% FBS.
2. Once the cells are seeded, we must wait about four hours for the cells to adhere. Once the cells are adhered, the medium should be changed to DMEM with 5% FBS (*see* **Note 1**).
3. After changing the medium, treat with the mushroom extracts at concentrations of your choice. Incubate overnight.
4. The next day, activate the cultures with LPS or cytokines. The activated cultures are incubated for 8 hours.
5. After the incubation period, the medium is aspirated and 1 mL of TRIzol is added to each culture plate.

3.2 RNA Isolation

1. Add 200 μL of chloroform. Shake and incubate for 3 min (*see* **Note 2**).
2. Centrifuge at 12000 *g* at 4 °C for 15 min.
3. Collect in a new Eppendorf the first phase of the supernatant from the tube.
4. Add 500 μL of isopropanol, shake, and incubate for 10 min.
5. Centrifuge at 12000 g at 4 °C for 10 min. Discard supernatant.
6. Add 1 mL of 75% ethanol. Vortex and centrifuge at 7500 g at 4 °C for 5 min.
7. Discard the supernatant and allow the pellet to dry for 5–10 min.
8. Add 40–50 μL of free water.
9. Heat in Thermoblock at 55–60 °C for 10–15 min.
10. Store at −80 °C until use.

3.3 cDNA Synthesis and RT-qPCR

1. Add 1 ng of RNA and dilute with H_2O free to a volume of 11 μL.
2. Add 1 μL of Oligo(dT) Primer, and heat the samples at 65 °C for 5 min.
3. Add RT reaction mix according to Table 1.
4. After adding the reaction mix, incubate the samples at 42 °C for 60 min.

Table 1
Anti-inflammatory mechanisms of bioactive compounds from edible and medicinal mushrooms

Reagent	Volume per reaction (μL)
5X reaction buffer 250 mM Tris-HCl (pH 8.3), 250 mM KCl, 20 mM $MgCl_2$, 50 mM DTT	4
10 mM dNTP mix	2
RevertAid RT (200 U/μL)	1
RiboLock RNase inhibitor (20 U/μL)	1

5. Bring the cDNA to a volume of 200 μL and store at −20 °C until use.
6. Prepare the qPCR mix; for each reaction, use 5 μL of Syber green, 0.4 μL forward primer, and 0.4 μL reverse primer.
7. To load the PCR plate, use 4.2 μL of cDNA and 5.8 μL of qPCR mix.
8. Place in the thermal cycler and use the following program: 95 ° C for 5 min, [95 °C for 3 s, 60 °C for 30 s] × 40, 72 °C for 5 min.

3.4 Evaluation of Nitrite Production

1. Collect 1 mL of the culture medium in which the cell experiments have been carried out (*see* **Note 3**).
2. Add 100ul of 10 mM sulphanilic acid.
3. Add 100ul of 10 mM NEDA.
4. Shake to homogenize the solution and wait a few seconds to observe the color change (*see* **Note 4**).
5. Quantify in spectrophotometer at 540 nm. Lower absorbance values will indicate a lower amount of nitrite in the medium and therefore a higher anti-inflammatory capacity.

4 Notes

1. It is important to lower the level of FBS in the medium when treating the cells with the treatments of interest, as this way, the cells will be more susceptible to incorporate the different compounds inside them. When treating the cells, make sure that the compound of interest is homogeneously distributed in the culture medium.
2. For RNA isolation, it is very important to work at low temperature using ice. RNA is very sensitive to temperature degradation, and it is essential to take special care with the material used during this process to ensure that everything is free of RNases.

3. In order to carry out the nitrite production test, the culture medium used must be a medium without phenol red, as the presence of nitrites causes the medium to change to a pink color, which means that the medium with phenol red is not suitable for carrying out this test.
4. It is very important to add the reagents in this order for the color change to occur correctly.

References

1. Liu Y, Yang J, Guo Z, Li Q, Zhang L, Zhao L, Zhou X (2024) Immunomodulatory effect of *Cordyceps militaris* polysaccharide on RAW 264.7 macrophages by regulating MAPK signaling pathways. Molecules 29(14):3408. https://doi.org/10.3390/molecules29143408
2. Minato K (2010) Mushrooms: Immunomodulating activity and role in health promotion. In: Watson RR (ed) Dietary components and immune function. Humana Press, Totowa, pp 529–539
3. Zhao S, Gao Q, Rong C, Wang S, Zhao Z, Liu Y, Xu J (2020) Immunomodulatory effects of edible and medicinal mushrooms and their bioactive immunoregulatory products. J Fungi 6:269
4. Drzewiecka B, Wessely-Szponder J, Świeca M, Espinal P, Fusté E, Fernández-De La Cruz E (2024) Bioactive peptides and other immunomodulators of mushroom origin. Biomedicine 12(7):1483. https://doi.org/10.3390/biomedicines12071483
5. Gao Y, Zhou S, Chen G, Dai X, Ye J (2002) A phase I/II study of a *Ganoderma lucidum* (Curt.: Fr.) P. Karst. Extract (Ganopofy) in patients with advanced cancer. Int J Med Mushrooms 4:8
6. Maity P, Sen IK, Chakraborty I, Mondal S, Bar H, Bhanja SK, Mandal S, Maity GN (2021) Biologically active polysaccharide from edible mushrooms: a review. Int J Biol Macromol 172:408–417
7. Zhong RF, Yang JJ, Geng JH, Chen J (2021) Structural characteristics, anti-proliferative and immunomodulatory activities of a purified polysaccharide from *Lactarius volemus* Fr. Int J Biol Macromol. https://doi.org/10.1016/j.ijbiomac.2021.10.049
8. Zhang GQ, Sun J, Wang HX, Ng TB (2009) A novel lectin with antiproliferative activity from the medicinal mushroom *Pholiota adiposa*. Acta Biochim Pol 56:415–421
9. Ngai PHK, Ng TB (2004) A mushroom (*Ganoderma capense*) lectin with spectacular thermostability, potent mitogenic activity on splenocytes, and antiproliferative activity toward tumor cells. Biochem Biophys Res Commun 314:988–993
10. Li Q, Wang X, Chen Y, Lin J, Zhou X (2010) Cytokines expression induced by *Ganoderma sinensis* fungal immunomodulatory proteins (FIP-Gsi) in mouse spleen cells. Appl Biochem Biotechnol 162:1403–1413
11. Li S, Jiang Z, Sun L, Liu X, Huang Y, Wang F, Xin F (2017) Characterization of a new fungal immunomodulatory protein, FIP-Dsq2 from *Dichomitus squalens*. J Biotechnol 246:45–51
12. Wu DM, Duan WQ, Liu Y, Cen Y (2010) Anti-inflammatory effect of the polysaccharides of golden needle mushroom in burned rats. Int J Biol Macromol 46(1):100–103. https://doi.org/10.1016/j.ijbiomac.2009.10.013
13. Fangkrathok N, Junlatat J, Sripanidkulchai B (2013) *In vivo* and *in vitro* anti-inflammatory activity of *Lentinus polychrous* extract. J Ethnopharmacol 147(3):631–637. https://doi.org/10.1016/j.jep.2013.03.055
14. Queiroz LS, Nascimento MS, Cruz AKM, Castro AJG, Moura MDFV, Baseia IG, Araújo RM, Benevides NMB, Lima LFA, Leite EL (2010) Glucans from the *Caripia montagnei* mushroom present anti-inflammatory activity. Int Immunopharmacol 10(1):34–42. https://doi.org/10.1016/j.intimp.2009.09.015
15. Park YM, Won JH, Kim YH, Choi JW, Park HJ, Lee KT (2005) *In vivo* and *in vitro* anti-inflammatory and anti-nociceptive effects of the methanol extract of *Inonotus obliquus*. J Ethnopharmacol 101(1–3):120–128. https://doi.org/10.1016/j.jep.2005.04.003
16. Thyagarajan A, Sliva D (2009) Suppression of the inflammatory response by triterpenes isolated from the mushroom *Ganoderma lucidum*. Int Immunopharmacol 9(11):1272–1280. https://doi.org/10.1016/j.intimp.2009.07.011
17. Han J, Chen Y, Bao L, Yang X, Liu D, Li S, Zhao F, Liu H (2013) Anti-inflammatory and cytotoxic cyathane diterpenoids from the

medicinal fungus *Cyathus africanus*. Fitoterapia 84(1):22–31. https://doi.org/10.1016/j.fitote.2012.10.001

18. Sedtananun S, Sujiwattanarat P, Hoang Tra MH, Suphakun P, Seetaha S, Auetragul A, Choowongkomon K, Tabtimmai L (2025) *Phallus indusiatus* extracts promoted MCF-7 apoptosis under TNFα-induced tumor microenvironment by attenuating NF-kappaB and Akt activation. Asian Pac J Cancer Prev 26(2): 479–487. https://doi.org/10.31557/APJCP.2025.26.2.479
19. Qian GM, Pan GF, Guo JY (2012) Anti-inflammatory and antinociceptive effects of cordymin, a peptide purified from the medicinal mushroom *Cordyceps sinensis*. Nat Prod Res 26(24):2358–2362. https://doi.org/10.1080/14786419.2012.658800
20. Quang DN, Hashimoto T, Arakawa Y, Kohchi C, Nishizawa T, Soma GI, Asakawa Y (2006) Grifolin derivatives from *Albatrellus caeruleoporus*, new inhibitors of nitric oxide production in RAW 264.7 cells. Bioorg Med Chem 14(1):164–168. https://doi.org/10.1016/j.bmc.2005.08.005
21. Moro C, Palacios I, Lozano M, D'Arrigo M, Guillamón E, Villares A, Martínez JA, García-Lafuente A (2012) Anti-inflammatory activity of methanolic extracts from edible mushrooms in LPS activated RAW 264.7 macrophages. Food Chem 130(2):350–355. https://doi.org/10.1016/j.foodchem.2011.07.049
22. Ditamo Y, Rupil LL, Sendra VG, Nores GA, Roth GA, Irazoqui FJ (2016) *In vivo* immunomodulatory effect of the lectin from edible mushroom *Agaricus bisporus*. Food Funct 7(1):262–269
23. Kerezoudi EN, Zervakis GI, Pletsa V, Kyriacou A, Brummer RJ, Rangel I (2025) *Pleurotus eryngii* mushrooms fermented with human fecal microbiota protect intestinal barrier integrity: immune modulation and signaling pathways counter deoxycholic acid-induced disruption in healthy colonic tissue. Nutrients 17(4):694. https://doi.org/10.3390/nu17040694

Chapter 12

Evaluating Prebiotic Potential

Helena Araújo-Rodrigues, Ana Sofia Salsinha, João B. Relvas, Freni K. Tavaria, and Manuela Pintado

Abstract

Mushrooms represent a promising source of bioactive macromolecules with numerous biological properties, including prebiotic potential. Although most studies focus on polysaccharide extracts, emerging evidence suggests that diverse mushroom macromolecules, such as proteins, peptides, fatty acids, and polyphenolic compounds, may also exert prebiotic effects. This chapter comprehensively explores the methodological approaches for the evaluation of the prebiotic properties of edible fungi, emphasizing the gastrointestinal simulation and gut microbiota modulation approaches. The standardized INFOGEST protocol simulates human gastrointestinal conditions, helping to understand the bioaccessibility of mushroom-derived compounds. Moreover, in vitro intestinal permeability assays are used to determine the colon-available fraction. Human fecal fermentation assays are frequently employed to assess mushrooms' prebiotic potential and its impact on gut microbiota modulation. These studies typically monitor key parameters, including their impact on bacterial population dynamics, production of SCFAs, and other relevant metabolic variations (e.g., amino acid metabolism). By integrating in vitro digestion and fermentation models with advanced molecular biology, microbiological, and chemical techniques, this chapter provides a systematic outline for evaluating the prebiotic potential of mushroom compounds and understanding their interaction with gut microbiota composition and function.

Key words Gastrointestinal simulation, Prebiotics, Gut microbiota modulation, Bacterial dynamics, Ammonia production, Short-chain fatty acids, Amino acid production

1 Introduction

Mushrooms are a rich source of bioactive macromolecules with diverse nutritional and biological properties. Polysaccharides, such as β-glucans, α-glucans, and chitin, are the most prevalent group, which have been studied for their antitumor, anti-inflammatory, and immunomodulatory activities [1–3]. Beyond these bioactivities, there is growing interest in the prebiotic potential of mushrooms due to the relevant role of gut microbiota in numerous disorders. Gut microbiota dysbiosis has been directly connected

Authors Helena Araújo-Rodrigues and Ana Sofia Salsinha have equally contributed to this chapter.

Pablo Navarro Simarro (ed.), *Fungi*, Methods and Protocols in Food Science, https://doi.org/10.1007/978-1-0716-4989-3_12,
© The Author(s), under exclusive license to Springer Science+Business Media, LLC, part of Springer Nature 2026

to several diseases, including cardiovascular, metabolic, and neurological diseases, highlighting the importance of dietary interventions that promote microbial dynamics balance [1].

The gut microbiota is involved in the fermentation of non-digestible polysaccharides such as β-glucan, chitin, hemicellulose, galactans, mannans, and xylans, producing metabolites beneficial to host health [1, 3, 4]. With advancements in microbiome research, the definition of the prebiotics concept was expanded in 2017 by Gibson to include any substrate that is selectively utilized by host microorganisms, conferring health benefits [5]. In addition to polysaccharides, other bioactive macromolecules present in mushrooms–such as proteins, peptides, fatty acids, and phenolic compounds—may also influence gut microbiota composition and function [6].

The structural complexity of mushroom polysaccharides, for instance, characteristic β-glycosidic bonds of β-glucans, determines their high resistance to gastrointestinal tract conditions, exerting prebiotic effects in the colon [1, 2, 4]. However, a comprehensive evaluation of the gastrointestinal tract's impact on mushroom macromolecules is crucial, especially when assessing the prebiotic potential of the whole mushroom matrix. The INFOGEST protocol, a standardized in vitro digestion model, is widely used worldwide to simulate human gastrointestinal conditions. This protocol takes into account physiological and biological parameters such as enzyme ratios and activity, pH levels, and digestion times, providing a controlled system for evaluating food digestion and bioavailability. This protocol offers significant advantages over traditional animal and human intervention studies, addressing time, cost, and ethical issues [7–10], enabling the preliminary selection of the most promising formulations or ingredients for subsequent validation in animal and human trials.

When mushroom polysaccharides reach the colon, gut microbiota metabolize these compounds, potentially enhancing intestinal barriers through mechanical, chemical, immunological, and biological mechanisms [1, 11]. Through mushroom and mushroom-derived compounds consumption, the gut microbiota diversity is reshaped, increasing the relative abundance of beneficial bacteria genera (e.g., *Bacteroides*, *Lactobacillus*, *Bifidobacterium*, *Akkermansia*, and *Parabacteroides*) [12]. Gut microbiota modulation promotes the production of short-chain fatty acids (SCFAs) and amino acids, which provide multiple health benefits. SCFAs are an energy source for intestinal epithelium cells but also promote a beneficial gut environment and health [13]. The increase in these key metabolites decreases the luminal pH, inhibiting the growth of pathogenic microorganisms while promoting the growth of probiotic bacteria [14] and the absorption of essential minerals and nutrients [13]. In addition to biological barriers, SCFA impacts the normal function of the intestinal mechanical, chemical, and immunological barrier, for example, increasing the mucus thickness and promoting the expression of genes involved with colon tight junction protein and mucin production [11]. SCFAs also play key roles in metabolic homeostasis, cancer, and the gut-brain axis. Amino acids also play several important roles in human health, with amino acid dysregulation reported in numerous diseases [1].

This chapter provides a comprehensive description of the methodologies used to evaluate the prebiotic potential of mushrooms, focusing on gastrointestinal digestion and in in vitro human fecal fermentation approaches, as well as bacterial dynamics analysis, SCFA and amino acid quantification, and total concentration of ammoniacal nitrogen.

2 Materials

2.1 Raw Material

1. Fresh, dried, or lyophilized mushrooms (in extract form or whole mushrooms).

2.2 General Laboratory Equipment and Material

1. Ultrapure water (UP) generated by a Milli-Q system (Millipore, Bedford, MA, USA).
2. An incubator adjustable to 37 °C and 50–250 rpm.
3. A standard vortex and stirring plate.
4. A pH meter and electrode.
5. A standard laboratory centrifuge.
6. A Thermoblock adjustable until 100 °C.
7. Micropipettes and tips.
8. Eppendorf tubes (2 mL), plastic centrifuge tubes (15 and 50 mL), and glass tubes (15 mL).
9. Volumetric flasks and glass beakers.
10. 0.22 μm membrane filters.

2.3 In Vitro Gastrointestinal Simulation

1. Enzymes: α-amylase from human saliva (ref. A0521, Sigma-Aldrich, St. Louis, MO, USA—*see* **Note 1**), rabbit gastric extract (RGE, ref. RGE -15, Lipolytech, Marseille, France), bovine bile (ref. B3883, Sigma-Aldrich), porcine pancreatin (ref. P7545, Sigma-Aldrich) or individual intestinal porcine enzymes: porcine trypsin (ref. T0303), bovine chymotrypsin (ref. C7762), porcine pancreatic amylase (ref. A3176), porcine pancreatic lipase (ref. L3126), and porcine pancreatic colipase (ref. C3028) from Sigma-Aldrich.
2. Simulated fluids: $CaCl_2$ (ref. 2382, Merck, Damdstadt, Germany), NaOH (ref. 9141, Merck), HCl (ref. 320,331, Sigma-Aldrich), KCl (ref. 4936, Merck), KH_2PO_4 (ref. P0662, Sigma-Aldrich), $NaHCO_3$ (ref. 6329, Merck), NaCl (6404, Merck), $MgCl_2(H_2O)_6$ (5833, Merck), $(NH_4)_2CO_3$ (207861, Sigma-Aldrich).

3. Inhibitors (if applicable): For more information regarding inhibition strategies, please refer to the original INFOGEST 2.0 protocol [7].

2.4 In Vitro Intestinal Permeability Simulation

1. Dialysis membranes (3.5 kDa pre-wetted RC tubing, ref. 15380732, Spectrum, California, USA).
2. Transwells (Milicell hanging cell culture insert, PET 0.4 μm, 12-well, 48/pk, Merck).
3. HT29-MTX E12 (ref. ECACC 12040401) and Caco-2 (ref. ECACC 86010202) cell lines (ATCC, Manassas, Virginia, USA).

2.5 In Vitro Human Fecal Fermentation

1. Anaerobic chamber, Whitley A35 Workstation (Don Whitley Scientific, UK).
2. Feces collection materials: OxoidTM AnaeroGenTM 2.5 L sachet (ref. AN0025A, Thermo Fischer Scientific) and OxoidTM AnaeroJarTM 2.5 L (ref. AG0025A, Thermo Fischer Scientific).
3. Phosphate-buffer saline tablets (PBS tablets, ref. P4417-50TAB, Sigma-Aldrich).
4. Batch fermentation jar (maintaining sterility, anaerobic environment, and pH monitoring), FerMac 260 pH controller (Electrolab Biotech Ltd., Gloucestershire, UK), and water bath for temperature control.
5. Hydrochloric acid prepared at 0.5 mol/L and sodium hydroxide prepared at 1.0 mol/L.
6. Fermentation medium: cysteine hydrochloride monohydrate (ref. 1.02839.0100, Merck), tryptone soya broth without dextrose (ref. T3938, Merck), ammonium chloride (ref. 101,145, Merck), sodium chloride (ref. 746,398, Sigma-Aldrich), calcium chloride dihydrate (ref. C7902, Sigma-Aldrich), calcium chloride hexahydrate (ref. 442,909, Sigma-Aldrich), potassium phosphate dibasic trihydrate (ref. P5504, Sigma-Aldrich), potassium phosphate dibasic (ref. P3786, Sigma-Aldrich), resazurin sodium salt (ref. 199,303, Sigma-Aldrich), bactopeptone (peptone bacteriological, ref. RM001, Himedia Labs, Einhausen, Germany), peptone from animal tissue (ref. P5905, Sigma-Aldrich), yeast nitrogen base without amino acids (ref. Y0626-250G, Sigma-Aldrich), yeast extract (ref. Y1625, Sigma-Aldrich), sodium bicarbonate (ref S6014, Sigma-Aldrich), magnesium chloride hexahydrate (ref. 459,337, Carlo Erba Diagnostics, Milano, Italy), Tween® 80 (ref. P1754, Sigma-Aldrich), hemin (ref. H9039, Sigma-Aldrich), vitamin k1 (ref. 95,271, Sigma-Aldrich), bile salts (ref. 48,305, Sigma-Aldrich), trace mineral supplement solution (ATCC, Virginia, USA).

7. Positive control: fructooligosaccharides (FOS, ref. P-FOS28, Megazyme, Wicklow, Ireland) or inulin from chicory (ref. I2255, Sigma-Aldrich).

2.6 Bacterial Population Analysis

1. Tris-EDTA buffer (10× concentrate, ref. PPB010, Sigma-Aldrich).
2. Lysozyme (ref. P00698, Sigma-Aldrich).
3. iQ SYBR Green Supermix (ref. 1708882, Bio-rad Laboratories, Hercules, CA, USA).
4. NZY Tissue gDNA kit for DNA extraction (ref. MB13502, NZYTech, Munich, Germany).

2.7 Total Ammonia Production

1. $MgSO_4$ (ref. M7506, Sigma-Aldrich) prepared at 1 M.
2. NH_4Cl (ref. 213,330, Sigma-Aldrich) as standard.
3. An ion selective electrode.

2.8 Short-Chain Fatty Acids (SCFAs)

1. Analytical standards: acetic acid (C2; 99.7%; ref. 695,092), propionic acid (C3; 99.5%; ref. 94425), i-butyric acid (iC4; 99%; ref. I1754); n-butyric acid (C4; 99.5%; ref. 19,215), i-valeric acid (iC5; 99.8%; ref. 75,054), n-valeric acid (C5; 99.8%; ref. 75,054), i-caproic acid (iC6; 99%; ref. 277827), n-caproic acid (C6; 99%; ref. 21,529), and L(+)-lactic acid (98%; ref. 1750) from Sigma-Aldrich.
2. High-performance liquid chromatography (HPLC): a Beckman Coulter System Gold HPLC (Knauer, Berlin, Germany) coupled to IR detector using Aminex HPXa cation exchange column (Aminex HPX-87H 300 × 7.8 mm column; Bio-RAD Laboratories).
3. Sulfuric acid (95–98%; ref. 258,105; Sigma-Aldrich) prepared at 5 mM in UP water.
4. Gas Chromatography—Flame Ionization Detector (GC-FID): Agilent Technologies 6850 GC (Santa Clara, CA, USA); nitroterephthalic acid modified polyethyleneglycol (PEG) column (DB- FFAP, 25 m, 0.25 mm i.d., 0.25 μm film thickness; Agilent Technologies).
5. Sulfuric acid prepared at 50% w/v and ethyl ether.

2.9 Amino Acid Analysis

1. Sample preparation: Percloric acid (ref. 244252; Sigma-Aldrich) prepared at 0.6 N or C18 cartridges (SEP-PAK Vac 3 cc, 500 mg, Waters).
2. High-performance liquid chromatography (HPLC): a liquid chromatograph equipped with a high-resolution fluorescence detector (Agilent Technologies, Inc., California, USA) and a Chromolith® Performance RP18 column (4.6 × 100 mm; Merck, Darmstadt, Germany).

3. Mobile phase: disodium phosphate (ref. S0876, Sigma-Aldrich), sodium tetraborate (ref. 221732; Sigma-Aldrich), sodium azide (ref. S2002; Sigma-Aldrich), acetonitrile (ref. 34851; Sigma-Aldrich), and methanol (ref. AX1699M; Supelco, Bellefonte, Pennsylvania, USA).
4. Reagents for derivatization: boric acid (ref. B0394, Sigma-Aldrich), sodium tetraphenylborate (ref. T25402, Sigma-Aldrich), 2-mercaptoethanol (ref. M6250, Sigma-Aldrich).
5. Amino acid standards: gamma-amino butyric acid (GABA; ref. A5835), aspartic acid (Asp; ref. A9256), glutamic acid (Glu; ref. G-1251), cysteine (Cys; ref. 168149), asparagine (Asn; ref. A-0884), serine (Ser; ref. S-4500), histidine (His; ref. H-8125), glutamine (Gln; ref. G-7029), threonine (Thr; ref. T-8625), arginine (Arg; ref. A-8094), alanine (Ala; ref. A-7627), tirosine (Tyr; ref. 8371), valine (Val; ref. V-0500), methionine (Met; ref. M-9625), tryptophan (Trp; ref. T-0254), phenylalanine (Phe; ref. P-2126), isoleucine (Ile; ref. I2752), leucine (Leu; ref. L-5652), lysine (lys; ref. L-5626), norvaline (ref. N7502) and homoserine (ref. S25783 344) from Sigma-Aldrich. Glycine (ref. 33226H; Honeywell).

3 Methods

3.1 *In Vitro* Gastrointestinal Simulation

The following Gastrointestinal tract digestion protocol described here is based on the INFOGEST 2.0 protocol [7].

3.1.1 Protocol Preparation

1. Enzymes: The recommended enzymes are detailed in Subheading 2 of this chapter. Enzyme activity assays are recommended before digestion. For more information regarding the measurement of enzyme activity, please refer to the original INFOGEST 2.0 protocol [7].
2. pH adjustments (optional but recommended): run a time-lagged assay (before performing the final/main assay) with your samples to determine the volumes of NaOH and HCl required in each step to adjust pH. Register the volumes (NaOH and HCl) needed for pH adjustment throughout the protocol.
3. Stock solution preparation: Prepare the stock solutions in distilled or ultrapure water, as described in the mentioned INFOGEST 2.0 protocol [7]. Store the stock solutions at −20 °C for a maximum of 1 year.
4. Simulated digestion fluids preparation: Prepare the digestion fluids as recommended in the INFOGEST 2.0 original protocol [7]. For $CaCl_2(H_2O)_2$ addition, *see* **Note 2**. Store the digestion fluids at −20 °C for a maximum of 1 year.

3.1.2 Protocol Considerations

1. Enzymes preparation: Enzymes should be prepared/resuspended on the day of the assay and maintained on ice during the experiment. α-amylase and RGE should be resuspended in ultrapure water. Bovine bile and pancreatin should be resuspended in simulated intestinal fluid (SIF).
2. Sample containers: Regarding the sample containers used during the assay, please refer to **Note 3**.
3. Enzyme and simulated fluids content: concerning the amounts of enzymes, simulated fluids, water, and $CaCl_2(H_2O)_2$ (*see* **Note 4**).
4. Replicates and controls: It is recommended to use duplicates or triplicates of each sample. Use water as a negative control instead of a sample as an enzyme/bile salts blank.
5. Aliquot collection/sampling:

   The sampling in the oral, gastric, and intestinal endpoints should only be performed if it is essential to assess the impact of each gastrointestinal phase in the food matrix. The aliquot collection from requires the use of a higher sample and enzyme amounts. When collecting aliquots from different digestion phases, ensure sample homogeneity to minimize variability.

   Two sampling options are available (*see* Fig. 1):

   Option 1 consists of starting the experiment with individual containers (tubes, flasks, etc.) for each of the aliquots of the samples, allocating one tube for each aliquot.

   Option 2 consists of taking aliquots directly from a single sample, ensuring proper homogenization before collection to diminish heterogeneity.

3.1.3 Gastrointestinal Tract Digestion Protocol

Oral Phase

1. Pre-warm all the digestion-simulated fluids at 37 °C.
2. Mix food with simulated salivary fluid (SSF) at 1:1 (wt/wt) final ratio to achieve a swallowable bolus with a paste-like consistency comparable to that of tomato paste or mustard at the end of the oral phase. If the consistency of the bolus is thicker than such a paste, add water to achieve the proper consistency. Record the volume of the water used, as it will be needed for the gastric phase (**step 9**).
3. Add $CaCl_2(H_2O)_2$ to achieve a 1.5 mM concentration in SSF.
4. Add the salivary α-amylase.
5. Add the remaining water to achieve a 1× SSF concentration.
6. Incubate the samples at 37 °C and 200 rpm for 2 min.
7. If an aliquot is taken at this step, salivary amylase should be inactivated.

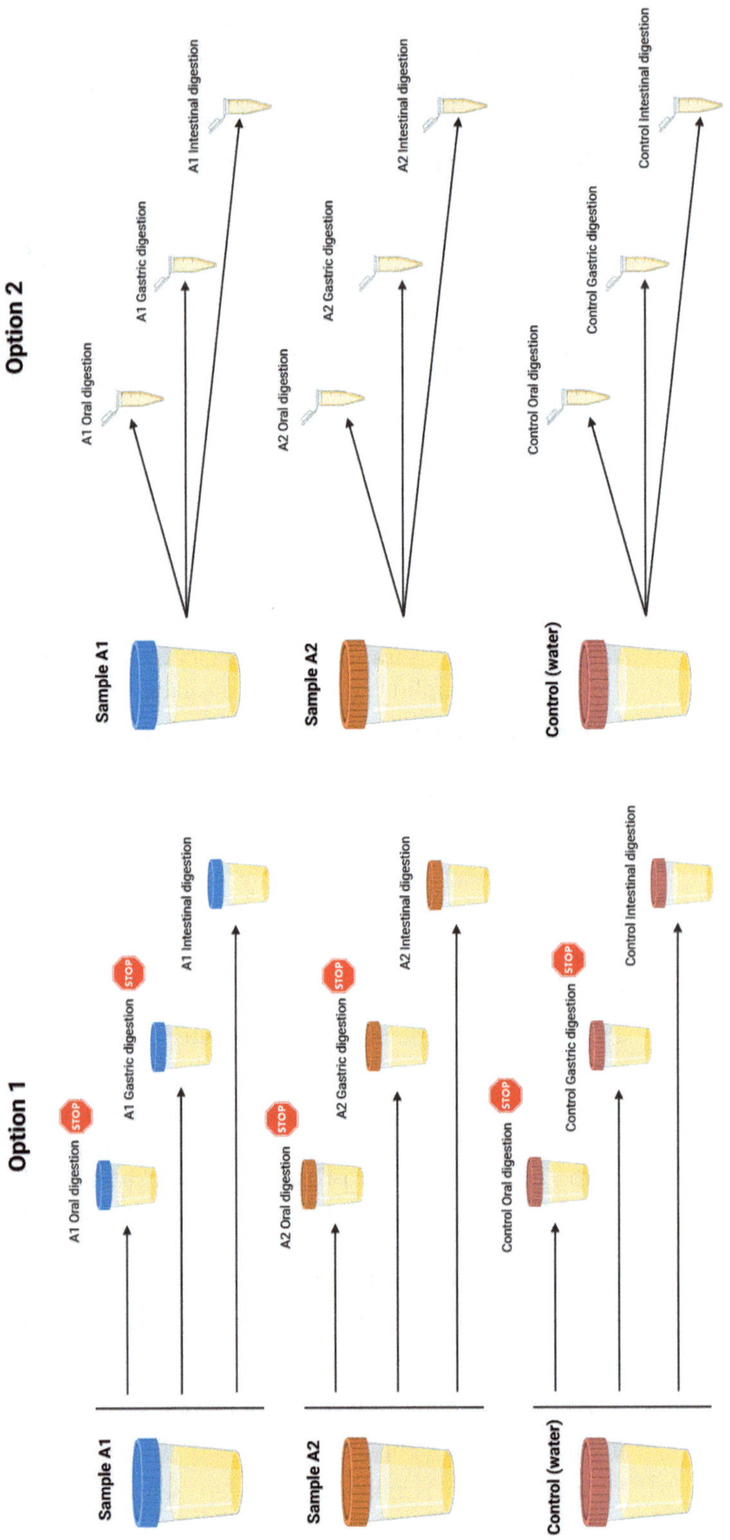

Fig. 1 Different methods for aliquot collection/sampling during digestion

Gastric Phase

8. If water was added to achieve the tomato paste consistency in **step 2**, please consider the added volume in the following steps.
9. Add the simulated gastric fluid (SGF) electrolyte stock solution to the oral bolus to achieve a final ratio of 1:1 (vol/vol).
10. Adjust the pH to 3.0 by adding a defined volume of HCl previously determined during a pH-test adjustment experiment.
11. Add $CaCl_2(H_2O)_2$ solution to achieve a final concentration of 0.15 mM in SGF.
12. If required, add the porcine pepsin solution prepared in water (*see* **Note 5**).
13. Add the gastric lipase solution (RGE) prepared in water to achieve an activity of 60 U/mL in the final digestion mixture.
14. Verify the pH and adjust it to 3.0 if necessary.
15. Add water to achieve a 1× concentration of SGF.
16. Incubate the samples at 37 °C and 130 rpm for 2 h.
17. If an aliquot is taken at this step, pepsin and lipase should be inactivated.

Intestinal Phase

18. Add the simulated intestinal fluid (SIF) electrolyte to the gastric chyme to achieve a final ratio of 1:1 (vol/vol).
19. Adjust the pH to 7.0 by adding a defined volume of NaOH previously determined during a pH-test adjustment experiment.
20. Add the bile solution to the SIF/gastric chime solution to reach a final concentration of 10 mM.
21. Place the solution in a rotating wheel mixer (orbital) at 37 °C for at least 30 min to achieve complete bile solubilization.
22. Add $CaCl_2(H_2O)_2$ solution to reach a concentration of 0.6 mM in SIF.
23. Perform the intestinal phase with pancreatin or with individual enzymes (*see* the original INFOGEST 2.0 protocol [7]).
24. Verify the pH and adjust it to 7.0 if necessary.
25. Add water to achieve a 1× concentration of the SIF.
26. Incubate the samples at 37 °C at 45 rpm for 2 h.
27. If an aliquot is taken, intestinal enzymes (amylase, lipase, trypsin, and chymotrypsin) should be inactivated.

3.1.4 Enzyme Inactivation

During gastrointestinal digestion, natural enzyme inactivation occurs due to pH variations. However, during aliquot collection, enzymes remain active, which may result in the continued hydrolysis of macronutrients. To ensure accuracy and reduce lab-to-lab variability, it is crucial to apply physical or chemical inactivation methods when testing critical samples. This is particularly important when multiple parameters are evaluated, making immediate analysis after digestion difficult. The INFOGEST 2.0 protocol [7] suggests enzyme inactivation methods, including temperature-based inactivation, pH-based inactivation, and enzymatic inhibitors.

A recent study suggests that different macromolecule groups present in *Pleurotus ostreatus* (carbohydrates, proteins, lipids and phenolic compounds) are affected differently by inactivation and storage methods during digestion [15]. Based on these findings, the authors recommended:

1. Aliquot freezing instead of freeze-drying, as freezing seems to minimize residual enzymatic activity and protect mushroom compounds from degradation.
2. Thermal inactivation as a simple, rapid, and cost-effective approach for carbohydrate-rich matrices. However, this method may have limitations when studying molecules with tertiary structure-dependent functionality.
3. Enzymatic inhibitors as a potential alternative, but their use should consider possible enzyme-inhibitor interactions.

The selection of the inactivation method should be based on the mushroom matrix composition and the specific research objective. Table 1 summarizes the tested inactivation strategies for *P. ostreatus* species, but it is important to consult all the methods recommended by the INFOGEST 2.0 protocol [7]. Concerning α-amylase inhibitor from *Triticum aestivum* tested in the oral phase, it was not described by INFOGEST 2.0 protocol but reviewed by Sales et al. [16].

3.2 In Vitro Intestinal Permeability Simulation

3.2.1 Centrifugation Method: INFOGEST Protocol Extension

Pérez-Burillo and collaborators [17] elaborated a protocol describing an in vitro batch fermentation procedure to be carried out after an in vitro gastrointestinal digestion, which is an extension of the previous protocol described by Brodkorb et al. [7] for studying in vitro gastrointestinal digestion. The sample preparation procedure for in vitro colonic fermentation considering a simulated intestinal permeability by centrifugation, is described next:

1. Centrifuge the in vitro digested samples (obtained as described in Subheading 3.1. In vitro gastrointestinal simulation) at 4000×*g* for 10 min at 4 °C. The solid residue corresponds to the undigested fraction, and the supernatant to the absorbable fraction.

Table 1
Tested enzyme inactivation methods for *Pleurotus ostreatus* species

Inactivation method	Oral phase (α-amylase inactivation)	Gastric phase (RGE inactivation)	Intestinal phase (pancreatic enzymes inactivation)
Temperature	100 °C for 5 min		
pH	pH < 4.5	pH > 8	pH < 4.5
Enzymatic (inactivators)	α-Amylase inhibitor from *Triticum aestivum* (wheat seed)—(1 U of inhibitor to reduce the activity of 2 U of α-amylase by 50%)	Pepstatin A (highly selective inhibitor for asparly proteases—Final concentration of 0.5 to 1 μM) and Orlistat (lipase inhibitor—10 μL/mL of a 100 mM solution in ethanol—1 mM final concentration)	Pefabloc SC (trypsin and chymotrypsin inhibitor; 5 mM)

2. Weigh the solid residue obtained after the in vitro digestion process.
3. It is recommended to ferment 0.5 g of solid residue to ensure that there is enough sample for bacteria to ferment.
4. It is known that (on average) 10% of the supposedly absorbable fraction in the large intestine is not absorbed. Add the corresponding volume (10%) of the in vitro digestion supernatant (absorbable fraction reaching the large intestine). The calculation of the 10% (v/v) volume is as follows:

Calculate 10% (v/v) of the supernatant that corresponds to the total amount of solid residue (undigested fraction) available from that digestion reaction. Since only 0.5 g of solid residue is going to be fermented, calculate the proportional volume of supernatant to add.

Example described in the original [17]: if the in vitro digested sample has 3 g of solid residue and 37 mL of supernatant, 10% of supernatant is 3.7 mL, which would correspond to the total 3 g of solid residue. So, to ferment 0.5 g of solid residue, we would add the following amount of the supernatant:

$$V\,(10\%) = \frac{0.5\ \text{g} \times 3.7\ \text{mL}}{3\ \text{g}} = 0.62\ \text{mL of supernatant}$$

3.2.2 Size Exclusion Method: Dialysis Membranes

To understand the possible intestinal absorption of the bioactive compounds in your sample, after digestion, a dialysis process with 3.5 kDa membranes (dialysis tubing cellulose membranes) can be performed.

1. As depicted in Fig. 2, place the final content of the digestion inside the dialysis tubing (*see* **Note 6**) and place it in an appropriate recipient.
2. Fill the recipient with a significant volume of water [18]/ $NaHCO_3$ (1 M) [19]/NaCl (10 mM) [20] on the outside (enough to cover all your sample that is inside the membrane).
3. Measure the volume of water added.
4. Incubate the dialysis samples overnight with enough agitation (≈50 rpm) at 37 °C.
5. After the incubation time is over, collect both the membrane content (retentate) and the water (permeate).
6. Freeze-dried both fractions for further use in in vitro colonic fermentation assays.

3.2.3 Transepithelial Diffusion Across Intestinal Caco-2/HT29-MTX Cell Layers

The intestinal permeability of the study samples can be evaluated in a manner that more closely mimics the in vivo process by using a Caco-2/HT29-MTX seeded in Transwell inserts, effectively replicating the absorptive epithelium of the human intestine (Fig. 3). Although this method does not yield a sufficient amount of colon-available fraction sample for use in colonic fermentation studies, it

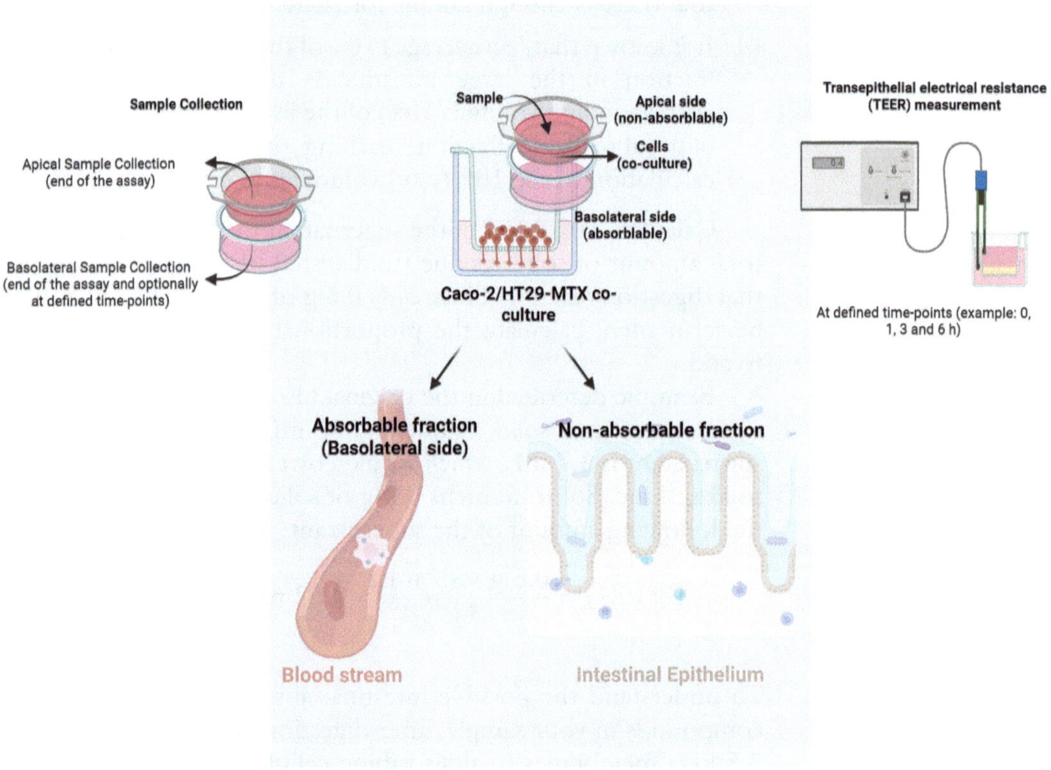

Fig. 2 Transepithelial diffusion across intestinal (Caco-2/HT29-MTX) cell layers

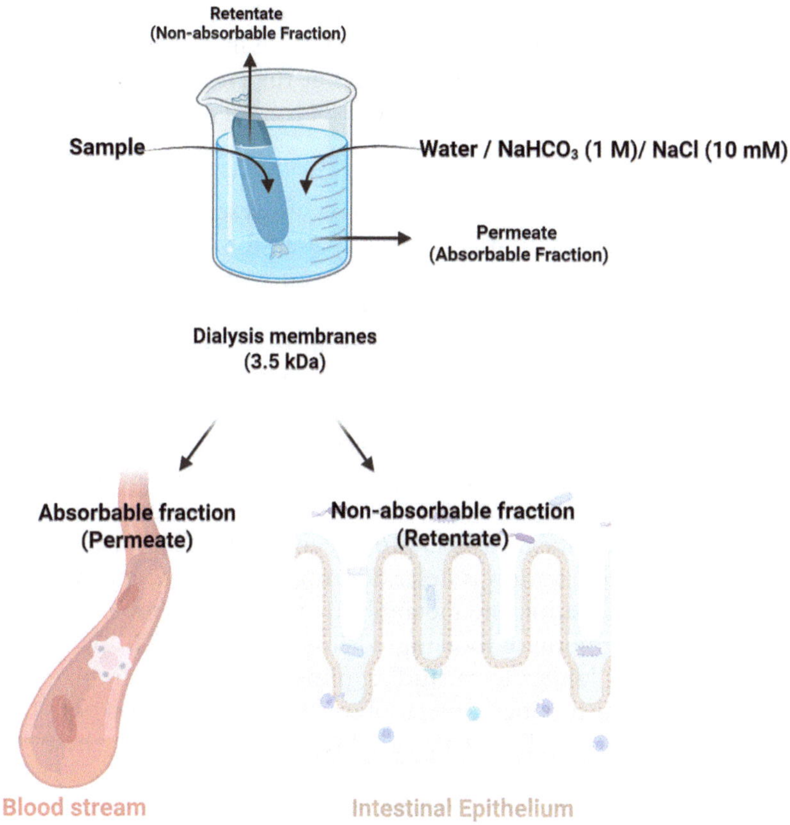

Fig. 3 In vitro Intestinal Permeability Simulation using a Size Exclusion Method: Dialysis Membranes

remains valuable for validating results obtained from previously described intestinal permeability simulation approaches, namely, the Centrifugation Method (INFOGEST Protocol Extension) and the Size Exclusion Method (Dialysis Membranes). For more information about the mentioned method, please refer to the cited bibliography. Nevertheless, the main considerations are described next.

1. For an adequate food digesta samples preparation (*see* **Note** 7) for downstream Caco-2 monoculture or co-culture studies, please refer to the Kondrashina and collaborators [21] study.
2. HT29-MTX E12 (ECACC 12040401) and Caco-2 (ECACC 86010202) cell lines are co-cultured in Transwells in a 9:1 ratio of Caco-2 and HT29 cells, respectively, to reach a monolayer with a final density of 3×10^5 cells/membrane [22].
3. Cell layer integrity transepithelial electrical resistance (TEER) is used to assess the cell layer integrity of the Caco-2/HT29 co-culture.
4. For the apparent permeability (Papp) assessment the medium on the apical side (donor side) is replaced with the sample [23].

5. The Papp can be calculated for a specific bioactive compound considering its concentration in both apical (donor) and basolateral (receptor) sides at the end of the experimental time, as previously described [24].

3.3 In Vitro Human Fecal Fermentation

In this section, the fecal inoculum preparation and different in vitro fecal fermentation protocols are described. The static in vitro batch fermentation is extensively used for human fecal fermentation due to its feasibility and cost-effectiveness, making it accessible for researchers with limited resources. Two versions of this batch fermentation system are described, namely, without agitation and pH control and with agitation and pH control for short-term studies. The first approach corresponds to more basic fermentation conditions, and the authors recommend the use of the first approach for preliminary screening studies, while the second provides a more physiologically relevant environment, maintaining optimal conditions for microbial growth and metabolism. Additionally, dynamic fermentation is described as an alternative approach, reflecting the adaptation and metabolism of the gut microbiota under continuous ingestion of food ingredients. This approach application is essential for longer-term studies, preventing the accumulation of microbial end-products metabolism by the addition of fresh medium [25]. The schematic representation of these fermentation approaches is present in Fig. 4.

3.3.1 Fecal Inoculum Pool Preparation

1. Fresh fecal samples were obtained voluntarily from healthy human donors (*see* **Note 8**) and promptly transferred into sterile plastic containers under anaerobic conditions, for example, to OxoidTM AnaeroJarTM 2.5 L with OxoidTM AnaeroGenTM 2.5 L sachet. Collection and processing of feces were completed within a maximum of 2 h under anaerobic conditions to preserve microbial viability.
2. Similar proportions of each donor's fecal sample were pooled to create a uniform and representative inoculum.
3. The fecal inoculum pool was diluted at 10% (w/w) in phosphate-buffer saline (PBS) solution (0.1 M, pH 7.4) and homogenized in a stomacher at 460 paddle-beats per min for 2 min.
4. For long-term storage at −80 °C, 5% (v/v) of DMSO should be added [26]. Before use in the in vitro colonic fermentation assay, the fecal inoculum must be washed as follows:
   (i) Thaw the inoculum under anaerobic conditions for 1 h.
   (ii) Vortex the sample and centrifuge at 4696×*g* for 5 min at 4 °C.
   (iii) Discard the supernatant and resuspend the pellet in PBS (0.1 M, pH 7.4).

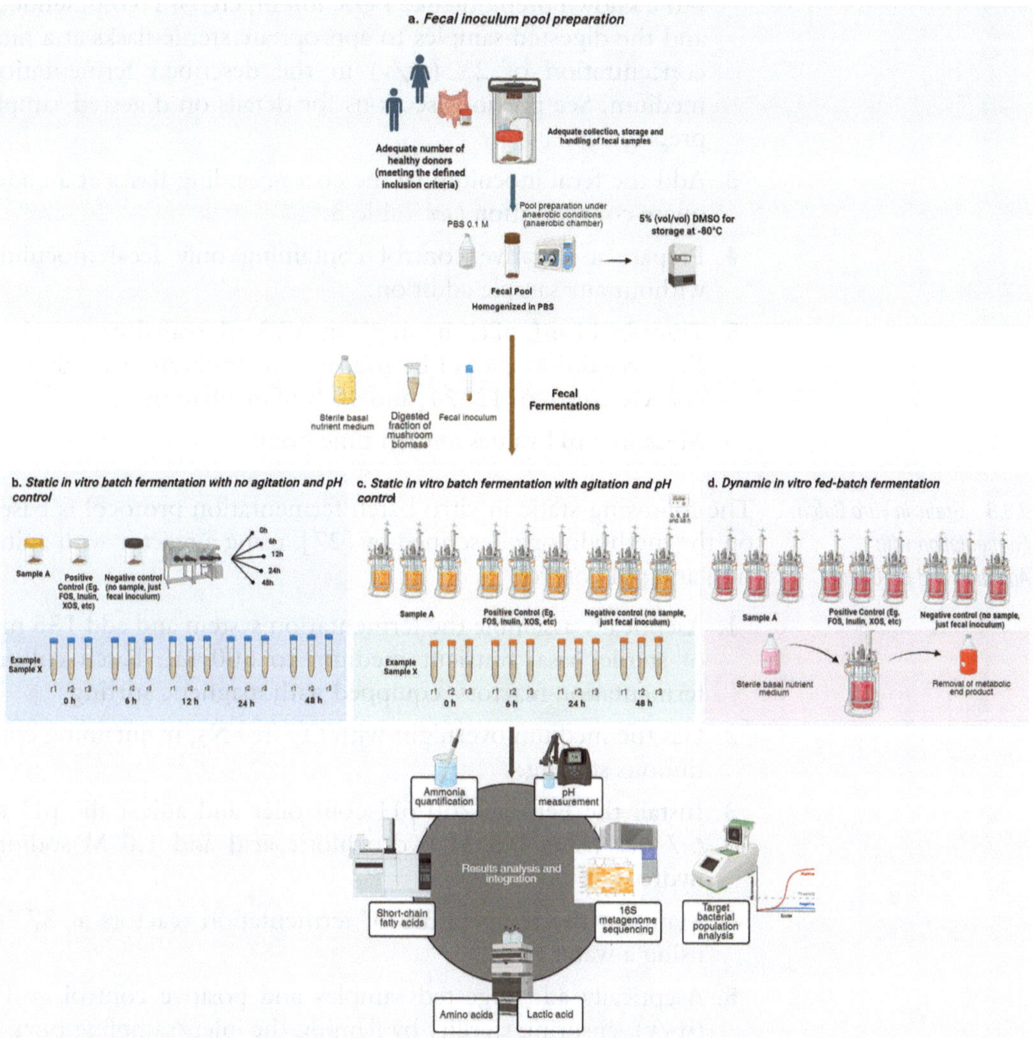

Fig. 4 Schematic representation of (**a**) fecal inoculum preparation, (**b**) static in vitro batch fermentation with no agitation and pH control; (**c**) static in vitro batch fermentation with agitation and pH control (reactor), and (**d**) in vitro dynamic fermentation

(iv) Repeat the same procedure two more times (three cycles in total).

(v) After the final centrifugation, resuspend the pellet in PBS (0.1 M, pH 7.4) to obtain the initial volume (10 mL or 15 mL).

3.3.2 Static In Vitro Batch Fermentation with No Agitation and pH Control

1. Various examples of fermentation media are described in Table 3.
2. Conduct the fecal fermentations in an anaerobic workstation by adding the positive control (adaptable for each experiment,

but a known prebiotic like FOS, inulin, etc., is recommended) and the digested samples to appropriate sterile flasks at a final concentration of 2% (w/v) in the described fermentation medium. See previous sections for details on digested sample preparation.

3. Add the fecal inoculum to the corresponding flasks at an adequate concentration (*see* Table 3.
4. Prepare a negative control containing only fecal inoculum without any sample addition.
5. Transfer 11 mL of each sample into 13 mL sealed sterile tubes. Each condition should be prepared in triplicate with aliquots collected at 0, 6, 12, 24, and 48 h of incubation.
6. Measured pH values at each time point.

3.3.3 Static In Vitro Batch Fermentation with Agitation and pH Control

The following static in vitro batch fermentation protocol is based on the methodology described by [27], using a reactor with agitation and pH control.

1. Aseptically assemble the fermentation system and add 135 mL of sterile basal nutrient medium to 300 mL batch culture fermentation reactors, equipped with magnetic stirring.
2. Gas the medium overnight with O_2-free N_2, maintaining continuous stirring.
3. Install the FerMac 260 pH controller and adjust the pH to 6.7–6.9 using 0.5 M hydrochloric acid and 1.0 M sodium hydroxide.
4. Maintain the temperature of fermentation reactors at 37 °C using a water bath.
5. Aseptically add digested samples and positive control at 1% (w/v), ensuring sterility by flaming the inlet/sampling port.
6. After complete homogenization of the substrates in the basal medium, inoculate each reactor with 15 mL of fecal inoculum.
7. Collect aliquots of 10 mL at 0, 6, 24, and 48 h of incubation.
8. During the fermentation period, replace the amount of acid and base and note the volume used.

3.3.4 Dynamic In Vitro Fed-Batch Fermentation

This method follows a similar procedure to Subheading 3.3.3 but incorporates a system that allows the introduction of fresh medium and removal of metabolic end product. For long-term fermentation studies, additional sampling points are typically applied, depending on the research objectives.

3.4 Bacterial Community Analysis

3.4.1 DNA Extraction (Fermentation Fluids)

1. Centrifuge 4 mL of the fecal fermentation sample for 10 min at 4 °C.
2. Dissolve the resultant pellet in 1 mL of TE buffer.
3. Perform a centrifugation at 4000×*g* for 10 min (4 °C).
4. Add 180 μL of a lysozyme solution (10 mg/mL in TE buffer) to the pellet.
5. The solution is incubated at 37 °C for 2 h and subsequently centrifuged at 4000×*g* for 10 min at 4 °C.
6. According to the manufacturer's instructions, use the resulting pellet for DNA extraction using the NZYTissue gDNA isolation kit (*see* **Note 9**).

3.4.2 Bacterial Community Analysis: 16S Amplicon Metagenomic Sequencing

The bacterial population can be analyzed using the16S amplicon metagenomic sequencing. In this approach, the 16S rRNA genes within the V3–V4 or V4–V5 regions are amplified using specific primers tagged with barcodes. α-Diversity metrics assesses the diversity, richness, and evenness of microbial communities, while β-diversity evaluates differences in community composition across sample groups, providing insights into their complexity and variation.

3.4.3 Target Bacterial Population Analysis: Quantitative Real-Time PCR

The 16S rRNA gene from the Firmicutes, *Lactobacillus* spp., *Enterococcus* spp., *Clostridium leptum*, Bacteroidetes, *Prevotella* spp., *Bifidobacterium* spp., *Akkermansia* spp., and *Roseburia* spp. groups has been successfully quantified in previous studies [28–31], using specific primers, according to a real-time polymerase chain reaction (RT-PCR) using iQ SYBR Green Supermix and a CFX96 Touch™ Real-Time PCR Detection System. The primer sequences and the RT-PCR conditions were adapted and optimized based on a previous study (Table 2) [32].

3.5 Quantification of Ammonia

When studying a mushroom matrix instead of a polysaccharide extract, it is important to determine the total concentration of ammoniacal nitrogen, as mushrooms are typically a good source of protein, with values between 19–35% on dry weight [1], and to guarantee that ammonia levels are within safe levels [27] (*see* **Note 10**).

The following protocol for total ammoniacal nitrogen quantification is based on the methodology described by Carvalho and collaborators [26, 27].

1. Centrifuge the collected samples at 4696×*g* for 5 min at 4 °C.
2. Add 300 μL of 1 M $MgSO_4$ to 3 mL of the collected supernatant to modify the ionic strength.

Table 2
Primer sequence information and genomic DNA standard used for gut microbiota analysis

Target group	Primers sequence (5′-3′; F—forward, R—reverse)	Annealing temperature (°C)	Genomic DNA standard	
Firmicutes Total	**F:** ATG TGG TTT AAT TCG AAG CA **R:** AGC TGA CGA CAA CCA TGC AC	55	*Lactobacillus gasseri* (ATCC 33323)	DSM20243
Bacteroidetes	**F:** CAT GTG GTT TAA TTC GAT GAT **R:** AGC TGA CGA CAA CCA TGC AG	55	*Bacteroides vulgatus* (ATCC 8482)	DSM 1447
Lactobacillus spp.	**F:** CAC CGC TAC ACA TGG AG **R:** AGC AGT AGG GAA TCT TCC A	59	*Lactobacillus gasseri* (ATCC 33323)	DSMZ 20243
Roseburia spp.	**F:** TAC TGC ATT GGA AAC TGT CG **R:** CGG CAC CGA AGA GCA AT	60	*Roseburia hominis*	DSM 16839
Bifidobacterium spp.	**F:** CGC GTC TGG TGT GAA AG **R:** CCC CAC ATC CAG CAT CCA	60	*Bifidobacterium longum* subsp. Infantis	DSM 20088
Clostridium leptum subgroup	**F:** GCA CAA GCA GTG GAG T **R:** CTT CCT CCG TTT TGT CAA	58	*Clostridium leptum*	DSM 753
Akkermansia spp.	**F:** CAG CAC GTG AAG GTG GGG AC **R:** CCT TGC GGT TGG CTT CAG AT	62	*Akkermansia muciniphila*	DSM 22959
Prevotella spp.	**F** CAC RGT AAA CGA TGG ATG CC **R** GGT CGG GTT GCA GAC C	45	*Prevotella nigrescens*	DSMZ 13386
Enterococcus spp.	**F** CCC TTA TTG TTA GTT GCC ATC ATT **R** ACT CGT TGT ACT TCC CT TGT	55	*Enterococcus gilvus*	DSMZ 15689

Table 3
Examples of fermentation media used in various gut microbiota simulation studies, along with the corresponding percentage of fecal inoculum

Medium components	Type of fermentation protocol	Studied matrices	Fecal inoculum	References
Trypticase soya broth (TSB) without dextrose (5.0 g/L), bactopeptone (5.0 g/L), yeast nitrogen base (5.0 g/L), cysteine-HCl (0.5 g/L), 1.0% (v/v) of salt solution A (100.0 g/L NH_4Cl, 10.0 g/L $MgCl_2{\cdot}6H_2O$, 10.0 g/L $CaCl_2{\cdot}2H_2O$), 0.2% (v/v) of salt solution B (200.0 g/L $K_2HPO_4{\cdot}3H_2O$), 0.2% (v/v) resazurin solution (0.5 g/L) and 1% (v/v) trace minerals (ATCC, 10 mL/L). The media components are dissolved in distilled water, and the pH is adjusted to 6.8 before sterilization by autoclave	In vitro batch-static (no agitation and no pH control)	Mushroom biomass Microbial-Fructo-oligosaccharides Enriched oils (Omega-3, conjugated fatty acids, fish oil and pomegranate oil) Brewer's spent grain Olive pomace *Fucus vesiculosus* Phlorotannins (crude and ethyl acetate extract) Melon peel (*Cucumis melo L. inodorus*) flour Tomato flours Pineapple by-product flours	**Fecal slurry** prepared at 10% feces (w/v) in phosphate buffer or reduced physiological salt solution (RPS), which consisted of 0.5 g/L cysteine hydrochloride and 8.5 g/L NaCl **Fecal inoculum** added to each fermentation tube at 2% (v/v)	[28, 30, 31, 37, 38]
Peptone solution: 15 g of peptone in 1 L of Milli-Q water, adjust the pH to 7.0 **Reductive solution:** 312 mg of cysteine and 312 mg of sodium sulfide in 2 mL of 1 M NaOH, make up the volume to 50 mL with Milli-Q water **Resazurin solution:** solution at 0.1% (wt/vol)—1 mg of resazurin in 1 mL of Milli-Q water **Final fermentation medium:** 1 L of peptone solution with 50 mL of reductive solution and 1.25 mL of resazurin for each liter of fermentation medium	In vitro batch fermentation protocol (with agitation and no pH control)	INFOGEST 2.0 protocol extension	**Fecal slurry** prepared at 32% feces (w/v) in phosphate buffer **Fecal inoculum** added to each fermentation tube at 20% (v/v)	[17]

(continued)

Table 3
(continued)

Medium components	Type of fermentation protocol	Studied matrices	Fecal inoculum	References
Basal nutrient media: peptone (2 g/L), yeast extract (2 g/L), NaCl (0.1 g/L), K_2HPO_4 (0.04 g/L), KH_2PO_4 (0.04 g/L), $MgSO_4{\cdot}7H_2O$ (0.01 g/L), $CaCl_2{\cdot}6H_2O$ (0.01 g/L), $NaHCO_3$ (2 g/L), Tween® 80 (2 mL/L), vitamin K (10 μL/L), and L-cysteine HCl (0.5 g/L). The media components are dissolved in distilled water and incubated at 90 °C under agitation until complete dissolution. Then, turn off the heat and add bile salts (0.5 g/L) and resazurin (4 mg/L). The medium is autoclaved for 20 min at 121 °C. After autoclaving, hemin (0.05 g/L) is added, and the pH is adjusted to 7.0 under aseptic conditions	In vitro batch fermentation—with agitation and pH control	Mushroom biomass and fruiting bodies Insect flour Plant-based functional beverage Ham formulations	**Fecal slurry** prepared at 10% feces (w/v) in phosphate buffer **Fecal inoculum** added to each fermentation tube at 10% (v/v)	[26, 39]

3. Measure the ammonia concentration using a 9663-ion selective electrode at 20 °C and pH (6.7–6.9), following the manufacturer's instructions. Each replicate should be measured in duplicate.
4. Prepare a calibration curve of NH_4Cl (2 mM to 55 mM).
5. Calculate the total ammonia nitrogen concentration using the following equation:

$$\frac{[NH_4^+]}{[NH_3 + NH_4^+]} = 1 - \frac{1}{1 + 10^{(pKa - pH)}}$$

Where:

[NH4+] = ammonium ion concentration

[NH3 + NH4+] = total ammonia nitrogen concentration

pKa = acid dissociation constant, which varies with temperature (T) and is calculated using the following equation:

$$pKa = 4 \times 10^{-8} T^3 + 9 \times 10^{-5} T^2 - 0.0356T + 10.072$$

3.6 Quantification of SCFAs

The following protocol for SCFA quantification by HPLC is based on the methodology described by Ribeiro et al. [28].

3.6.1 Quantification by High-performance Liquid Chromatography (HPLC)

1. Sample preparation:
   (i) Centrifuge the collected samples at 4000×*g* for 5 min at 4 °C.
   (ii) Collect the supernatant and filter it through a 0.22 μm membrane for HPLC analysis.
2. Chromatographic conditions:
   (i) Injection volume: 30 μL.
   (ii) Mobile phase: 5 mM of sulphuric acid.
   (iii) Flow rate: 0.5 mL/min.
   (iv) Column temperature: 55 °C.
3. Identification and quantification:
   (i) Prepare standard solutions of acetic acid (C2), propionic acid (C3), butyric acid (C4), and isobutyric acid (iC4) at different concentrations.
   (ii) For static in vitro batch fermentation with no agitation and pH control, it is important to measure lactic acid (*see* **Note 11**).
   (iii) Identify SCFAs by comparing sample peaks' relative retention times (RT) with those of pure standards.
   (iv) Construct a calibration curve (0.2–20 mg/mL) for SCFA quantification.

3.6.2 Quantification by Gas Chromatography-Flame Ionization Detector (GC-FID)

The following protocol for SCFA quantification by GC-FID is based on the methodology described by Scortichini and collaborators [33]:

1. Optimization of sample amount and selection of internal standard (IS):
   (i) Prepare standard curves for the most abundant SCFAs (C2, C3, and C4 at 7.00×10^3 μM) and the less abundant SCFAs (iC4, iC5, C5, and C6 at 1.00×10^3 μM) dissolved in ethyl ether.
   (ii) Determine the optimal sample amount for extraction by testing different amounts (*see* **Note 12**) at different time points. Run experiments to identify the SCFAs present in fermented samples (follow the "Extraction of SCFAs" section, but without the addition of IS).
   (iii) Select the most suitable internal standard (IS) that is absent in the samples (*see* **Note 13**).
2. Preparation of IS:
   (i) Prepare the selected IS in 10 mL of diethyl ether (*see* **Note 14**).
   (ii) Store the IS solution at 4 °C until use.
3. Extraction of SCFAs:
   (i) Weigh the optimized fermented sample in a glass test tube.
   (ii) Acidify with 200 μL of 50% (w/v) sulfuric acid and vortex for 1 min.
   (iii) Add 10 μL of IS solution to achieve a final concentration of 456.25 μM.
   (iv) Add 800 μL of ethyl ether for SCFAs extraction and centrifuge at 2800×*g* for 5 min.
   (v) Collect the upper (organic) phase and transfer it to a new tube to avoid residual sulfuric acid.
   (vi) Repeat the extraction two more times using 800 μL of ethyl ether per cycle (final volume: 2400 μL).
   (vii) Transfer the extracted samples to GC vials.
4. Chromatographic conditions:
   (i) Carrier gas: hydrogen (3.70 mL/min).
   (ii) Injection mode: splitless, 280 °C.
   (iii) Run time: 13 min.
   (iv) Oven temperature program: 40 °C for 3 min, increase at 20 °C/min to 160 °C. Then, increase ar 40 °C/min to 245 °C, hold for 1.87 min.

5. Identification and quantification of SCFAs:
   (i) Identify SCFAs by comparing the relative RT of sample peaks with those of pure standards.
   (ii) Construct a calibration curve (0.86–7000 μM) for SCFA quantification.

3.7 Quantification of Amino Acids

The following protocol for amino acid sample preparation was based on [34, 35], while quantification by HPLC is based on the methodology described by [36].

1. Sample preparation:
   (i) Precipitate 1 mL of fecal fermentation samples with 2 mL of 0.6 M perchloric acid (*see* **Note 15**).
   (ii) Homogenize for 1 h.
   (iii) Centrifuge at 4100×*g* for 15 min at 4 °C.
   (iv) Collect the supernatant and filter it through a 0.22 μm membrane for HPLC analysis.
2. Chromatographic conditions:
   (i) Internal standard: homoserine.
   (ii) λ excitation: 356 nm.
   (iii) λ emission: 445 nm.
   (iv) Sensitivity factor: 1.
   (v) Data collection rate: 2 Hz.
   (vi) Injection volume: 10 μL.
   (vii) Reagents:
   Reagent A (25 mL): 100 μL of mercaptoethanol, 0.5 g of sodium tetraphenylborate, and borate buffer; Reagent B (20 mL): 0.6 g of iodoacetic acid and borate buffer (pH = 9.5); Reagent C (25 mL): 0.175 g of OPA, 5 mL of methanol, 0.5 mL of mercaptoethanol, and borate buffer. Bubble the solution with N_2.
   (viii) Mobile phase:
   Mobile phase A: 10 mM disodium phosphate, 10 mM sodium tetraborate (pH 8.2), and 5 mM sodium azide adjusted to pH 8.4.
   Mobile phase B: acetonitrile: methanol: water (45: 45: 10, v/v).
   (ix) Flow rate: 0.9 mL/min.
   (x) Elution program: Starting with 90% of mobile phase A and 10% of mobile phase B; Linear gradient to 50% of mobile phase A during 43 min; Mobile phase B at 100% between 47 and 49 min; Return to initial conditions until 55 min.
   (xi) Column temperature: 25 °C.

3. Identification and quantification:
   - (i) Prepare standard solutions of individual amino acid standards.
   - (ii) Identify amino acids by comparing sample peaks' RT with those of pure standards.
   - (iii) Construct a calibration curve (0.05–10 mg/L) for amino acid quantification.

4 Notes

1. Although the original protocol recommends α-amylase from human saliva, α-amylase from *Bacillus licheniformis* (ref. A4551, Sigma-Aldrich) can be used as an alternative less expensive, as used by Carvalho and colleagues [27].
2. Add $CaCl_2$ directly to the sample on the assay day to avoid the precipitation associated with long-term and cold storage.
3. The use of carbonate salts in the electrolyte solutions requires the use of sealed containers with minimal headspace. In open vessels, CO_2 will be released, causing a progressive pH increase over time. If open vessels are used, such as during a 'pH-stat' approach or for sampling purposes, it is suggested to replace the sodium bicarbonate ($NaHCO_3$), the main source of carbonates, with NaCl at the same molar ratio to maintain the ionic strength of the oral, gastric and intestinal of the electrolyte solutions. Closed containers with minimal headspace are strongly recommended.
4. To calculate the reagent volumes required for the assay, use the spreadsheet available at http://www.proteomics.ch/IVD/ or calculate the volumes based on the table presented in the INFOGEST 2.0 protocol [7]. The calculations must be adjusted based on the initial mass/volume of food added (the example presented in the INFOGEST 2.0 protocol considers 5 g of food). Additionally, check and adjust the enzyme activities accordingly.
5. RGE contains both gastric lipase and pepsin, meaning that the pepsin content needs to be accounted for when calculating total pepsin activity. If using RGE, additional pepsin may not be required.
6. INFOGEST 2.0 recommended the use of membranes between 3–10 kDa [7]. You may need to adapt this based on your specific food matrix.

7. The INFOGEST intestinal endpoint contains (i) active digestive enzymes that detach these monolayers, (ii) bile salts that are toxic to cells and can shift osmolality, and (iii) food-derived compounds generated during digestion, which may also be harmful to cell cultures [21].
8. Human fecal samples must be collected fresh and maintained under anaerobic conditions for no more than 2 h before processing. Several criteria (that should be adapted to based on study objectives) must be followed for volunteer eligibility: normal omnivorous diets, had not ingested any antibiotics or other medicines known to affect the microbiota for at least 6 months before the study, and not being regular consumers of known probiotics or prebiotics.
9. The DNA extraction kit referenced in this protocol is one of several commercially available options for gut microbiota DNA extraction. While this protocol is specifically optimized for fecal fermentation fluid samples, the kit has been successfully tested on both fermentation fluids and stool samples.
10. Fecal ammonia concentration generally varies between 12 and 30 mM in humans, with safety thresholds up to 70 mM.
11. Lactic acid is produced increases with the growth of lactate-producing bacteria such as *Bifidobacterium* spp. and *Lactobacillus* spp. This increase can lead to a pH decrease. So, it is important to correlate pH values with lactic acid concentration.
12. Start by testing 250 and 500 mg of fermented samples, then adjust as necessary.
13. IS options include C5 (valeric acid) and iC6 (isohexanoic acid), depending on the fermentation medium and tested samples.
14. If selecting C5, add 120 μL of C5 in 10 mL of ethyl ether to obtain a concentration of 109.5×10^3 μM. If selecting iC6 add 4 μL of iC6 in 10 mL of ethyl ether to obtain a concentration of 19.8×10^3 μM.
15. Adjust the proportion of sample and acid, if necessary. As an alternative, SPE columns (e.g., C18 cartridges) can be used, following the manufacturer's instructions.

Acknowledgments

We would like to thank the National Funds from FCT - Fundação para a Ciência e a Tecnologia through project UIDB/50016/2020 and FCT individual PhD grant number 2020.05798.BD (H. Araújo-Rodrigues).

Supplementary Material (Table 3)

References

1. Araújo-Rodrigues H, Sousa AS, Relvas JB et al (2024) An overview on mushroom polysaccharides: health-promoting properties, prebiotic and gut microbiota modulation effects and structure-function correlation. Carbohydr Polym 333:121978
2. Niego AG, Rapior S, Thongklang N et al (2021) Macrofungi as a nutraceutical source: promising bioactive compounds and market value. J Fungi (Basel) 7:397
3. Yadav D, Negi PS (2021) Bioactive components of mushrooms: processing effects and health benefits. Food Res Int 148:110599
4. Mirończuk-Chodakowska I, Kujawowicz K, Witkowska AM (2021) Beta-glucans from fungi: biological and health-promoting potential in the COVID-19 pandemic era. Nutrients 13:3960
5. Gibson GR, Hutkins R, Sanders ME et al (2017) Expert consensus document: The International Scientific Association for Probiotics and Prebiotics (ISAPP) consensus statement on the definition and scope of prebiotics. Nat Rev Gastroenterol Hepatol 14:491–502
6. Farias DDP, de Araújo FF, Neri-Numa IA et al (2019) Prebiotics: trends in food, health and technological applications. Trends Food Sci Technol 93:23–35
7. Brodkorb A, Egger L, Alminger M et al (2019) INFOGEST static *in vitro* simulation of gastrointestinal food digestion. Nat Protoc 14: 991–1014
8. Grundy MML, Abrahamse E, Almgren A et al (2021) INFOGEST inter-laboratory recommendations for assaying gastric and pancreatic lipases activities prior to *in vitro* digestion studies. J Funct Foods 82:104497
9. Minekus M, Alminger M, Alvito P et al (2014) A standardised static *in vitro* digestion method suitable for food-an international consensus. Food Funct 5:1113–1124
10. Egger L, Ménard O, Delgado-Andrade C et al (2016) The harmonized INFOGEST *in vitro* digestion method: from knowledge to action. Food Res Int 88:217–225
11. Ma G, Xu Q, Du H et al (2022) Characterization of polysaccharide from *Pleurotus eryngii* during simulated gastrointestinal digestion and fermentation. Food Chem 370:131303
12. Zheng M, Pi X, Li H et al (2022) *Ganoderma* spp. polysaccharides are potential prebiotics: a review. Crit Rev Food Sci Nutr 64:1–19
13. Yin C, Noratto GD, Fan X et al (2020) The impact of mushroom polysaccharides on gut microbiota and its beneficial effects to host: a review. Carbohydr Polym 250:116942
14. Li X-Y, He C, Zhu Y et al (2020) Role of gut microbiota on intestinal barrier function in acute pancreatitis. World J Gastroenterol 26: 2187–2193
15. Araújo-Rodrigues H, Salsinha AS, Coscueta ER et al (2025) Understanding the impact of different post-digestion enzymatic inhibition methods and storage conditions on standardized gastrointestinal model (INFOGEST): A case study of *Pleurotus ostreatus* mushroom biomass. Food Res Int 211:116432
16. Sales PM, Souza PM, Simeoni LA et al (2012) α-Amylase inhibitors: a review of raw material and isolated compounds from plant source. J Pharm Pharm Sci 15:141–183
17. Pérez-Burillo S, Molino S, Navajas-Porras B et al (2021) An *in vitro* batch fermentation protocol for studying the contribution of food to gut microbiota composition and functionality. Nat Protoc 16:3186–3209
18. Salsinha AS, Cunha SA, Machado M et al (2023) Assessment of the bioaccessibility and bioavailability prediction of omega 3 and conjugated fatty acids by *in vitro* standardized digestion model (INFOGEST) and cell model. Food Biosci 53:102635
19. Ribeiro TB, Oliveira A, Campos D et al (2020) Simulated digestion of an olive pomace water-soluble ingredient: relationship between the bioaccessibility of compounds and their potential health benefits. Food Funct 11:2238–2254
20. Gullon B, Pintado ME, Fernández-López J et al (2015) *In vitro* gastrointestinal digestion of pomegranate peel (*Punica granatum*) flour obtained from co-products: changes in the antioxidant potential and bioactive compounds stability. J Funct Foods 19:617–628
21. Kondrashina A, Arranz E, Cilla A et al (2024) Coupling *in vitro* food digestion with *in vitro* epithelial absorption; recommendations for

biocompatibility. Crit Rev Food Sci Nutr 64(26):9618–9636

22. Antunes F, Andrade F, Araújo F et al (2013) Establishment of a triple co-culture *in vitro* cell models to study intestinal absorption of peptide drugs. Eur J Pharm Biopharm 83:427–435
23. Machado M, Costa EM, Silva S et al (2023) Understanding the anti-obesity potential of an avocado oil-rich cheese through an *in vitro* co-culture intestine cell model. Molecules 28:5923
24. Dima C, Assadpour E, Dima S et al (2020) Bioavailability and bioaccessibility of food bioactive compounds; overview and assessment by *in vitro* methods. Compr Rev Food Sci Food Saf 19:2862–2884
25. Verhoeckx K, Cotter P, López-Expósito I et al (2015) *In vitro* fermentation models: general introduction. In: Verhoeckx K, Cotter P, López-Expósito I et al (eds) The impact of food bioactives on health *in vitro* and *ex vivo* models. Springer, Cham
26. de Carvalho NM, Oliveira DL, Saleh MAD et al (2021) Preservation of human gut microbiota inoculums for *in vitro* fermentations studies. Fermentation 7:14
27. De Carvalho NM, Walton GE, Poveda CG et al (2019) Study of *in vitro* digestion of *Tenebrio molitor* flour for evaluation of its impact on the human gut microbiota. J Funct Foods 59:101–109
28. Ribeiro TB, Costa CM, Bonifácio-Lopes T et al (2021) Prebiotic effects of olive pomace powders in the gut: *in vitro* evaluation of the inhibition of adhesion of pathogens, prebiotic and antioxidant effects. Food Hydrocoll 112:106312
29. Madureira AR, Campos D, Gullon B et al (2016) Fermentation of bioactive solid lipid nanoparticles by human gut microflora. Food Funct 7:516–529
30. Gómez-García R, Sánchez-Gutiérrez M, Freitas-Costa C et al (2022) Prebiotic effect, bioactive compounds and antioxidant capacity of melon peel (*Cucumis melo* L. *inodorus*) flour subjected to *in vitro* gastrointestinal digestion and human faecal fermentation. Food Res Int 154:111045
31. Bonifácio-Lopes T, Catarino MD, Vilas-Boas AA et al (2022) Impact of circular Brewer's spent grain flour after *in vitro* gastrointestinal digestion on human gut microbiota. Foods 11:2279
32. Marques C, Meireles M, Norberto S et al (2016) High-fat diet-induced obesity rat model: a comparison between Wistar and Sprague-Dawley rat. Adipocyte 5:11–21
33. Scortichini S, Boarelli MC, Silvi S et al (2020) Development and validation of a GC-FID method for the analysis of short chain fatty acids in rat and human faeces and in fermentation fluids. J Chromatogr B Analyt Technol Biomed Life Sci 1143:121972
34. Salsinha AS, Cima A, Araújo-Rodrigues H et al (2024) The use of an *in vitro* fecal fermentation model to uncover the beneficial role of omega-3 and punicic acid in gut microbiota alterations induced by a Western diet. Food Funct. https://doi.org/10.1039/d4fo00727a
35. Hayashi T, Yamashita T, Takahashi T et al (2021) Uncovering the role of gut microbiota in amino acid metabolic disturbances in heart failure through metagenomic analysis. Front Cardiovasc Med 8:789325
36. Pripis-Nicolau L, de Revel G, Marchand S et al (2001) Automated HPLC method for the measurement of free amino acids including cysteine in musts and wines; first applications. J Sci Food Agric 81:731–738
37. Catarino MD, Marçal C, Bonifácio-Lopes T et al (2021) Impact of phlorotannin extracts from fucus vesiculosus on human gut microbiota. Mar Drugs 19:375
38. Salsinha AS, Araújo-Rodrigues H, Dias C et al (2025) Omega-3 and conjugated fatty acids impact on human microbiota modulation using an *in vitro* fecal fermentation model. Clin Nutr 49:102–117
39. Bento de Carvalho T, Barbosa JB, de Carvalho NM et al (2025) *In vitro* colonic fermentation of clean label ham formulations: GUT microbiota modulation and metabolite production. Food Res Int 209:116287

Chapter 13

Antimicrobial Compounds in Mushrooms and Methods to Evaluate Them

Sabeena Hussain Syed, Supriya Unavane, Poonam Taru, and Om Bagade

Abstract

Mushrooms are a rich source of bioactive compounds, including antimicrobial agents with significant potential against multidrug-resistant pathogens. These bioactive metabolites, such as polysaccharides, phenolic compounds, terpenoids, and sterols, exhibit diverse mechanisms of action, including disruption of microbial cell walls and interference with essential metabolic pathways. The extraction of these antimicrobial compounds requires a systematic approach involving soxhlet extraction, ultrasound-assisted extraction, microwave-assisted extraction, and supercritical fluid extraction. Standard microbiological assays, such as agar well diffusion, broth microdilution, and time-kill kinetics, are employed to assess their antimicrobial efficacy against bacterial and fungal strains. Advanced techniques including minimum inhibitory concentration (MIC) determination and TLC studies also give more insights into their respective efficacy as well as application against few infectious microbes. This chapter highlights the extraction processes and the methods used to determine antimicrobial action, along with the bioactive compounds with antimicrobial activity from mushroom.

Key words Mushroom, Bioactive compounds, Extraction, Microbiology, Antimicrobial.

1 Introduction

Microbial infections cause significant health challenges, leading to morbidity and mortality of living organisms. Additionally, these infections severely impact crops, reducing productivity and causing substantial economic losses globally [1]. Over the past decade, the widespread and inappropriate usage of antibiotics for microbial treatment has contributed to the alarming rise of multidrug-resistant infections [2]. The growing prevalence of these resistant infections has led to an increased demand for the discovery of novel antibiotics [3]. Natural products have been a key source of new therapeutic agents, and mushrooms, in particular, have proven to be valuable sources of antimicrobial agents [4].

Mushrooms, as versatile organisms within the fungal kingdom, have long been recognized for their nutritional and therapeutic

Pablo Navarro Simarro (ed.), *Fungi*, Methods and Protocols in Food Science, https://doi.org/10.1007/978-1-0716-4989-3_13,
© The Author(s), under exclusive license to Springer Science+Business Media, LLC, part of Springer Nature 2026

potential [5]. They are rich in carbohydrates (β-glucans), amino acids, minerals, vitamins (ascorbic acid, pantothenic acid, and riboflavin), fibers, and water [6]. This unique composition makes mushrooms a valuable dietary component for promoting health and well-being. Mushrooms are also prolific producers of secondary metabolites, including anthraquinones, glycosides, terpenes, steroids, phenols, flavonoids, and alkaloids [7], which exhibit a broad spectrum of therapeutic properties like anti-inflammatory, anticancer, antiviral, antioxidant, antidiabetic, antifibrotic, and both cardioprotective and hepatoprotective effects [8]. In addition to their well-documented therapeutic properties, mushrooms also exhibit significant antimicrobial activity, making them a promising source for the development of novel antimicrobial agents [9].

The bioactives from the mushroom exhibit antibacterial activity by preventing cell wall synthesis, disrupting membrane integrity, and interfering with nucleic acid synthesis [10]. Additionally, phenolic and terpenoid compounds exert antioxidant properties, including oxidative stress within bacterial cells, while certain metabolites enhance membrane permeability and interact with bacterial DNA, leading to cell death [11]. Mushrooms can also disrupt quorum sensing, a critical bacterial communication process, thereby reducing virulence and biofilm formation [12].

Antimicrobial compounds from the mushrooms are extracted by various methods like solvent extraction using water, ethanol, methanol, or their combinations. Ultrasound-assisted extraction and microwave-assisted extraction use high-frequency sound waves and microwave energy respectively, leading to breakage of cell walls, thereby enhancing the release of bioactive compounds [13]. To assess the antimicrobial activity of mushroom extract or bioactives, microdilution method, minimum inhibitory concentration test, agar well diffusion method, and synergistic test are well documented [14].

2 Materials

2.1 Extraction of Bioactive Compounds of Mushrooms

For the extraction of bioactive components from the mushroom, distilled water, methanol, ethanol, or their combinations are used as a solvent. These solvents have strong polarity and facilitate the extraction of polar constituents like phenols, saccharides, and polysaccharide-binding proteins.

For the extraction of semi-polar and nonpolar constituents, organic solvents like chloroform, acetone, ethyl acetate, hexane, and dichloromethane are used. All the solvents should be of analytical grade for the accuracy and reliability of antimicrobial activity assessment. The solvent used for the extraction not only determines the extraction of specific constituents but also determines the quality, quantity, and safety of the extracts [15].

2.2 Antimicrobial Activity

2.2.1 Microdilution Method

1. Culture medium (Tryptic Soy Broth): The concentration of broth should be 50%, ensuring adequate nutrients and reduced dilution of antimicrobial agents.
2. Mushroom extracts: Extract (10 mg/mL as initial stock solution) of mushroom prepared using appropriate solvents like water, methanol, or ethanol.
3. Microorganisms: Bacterial strains (*Escherichia coli*, *Pseudomonas aeruginosa*, *Staphylococcus aureus*, etc.) and fungal strains (*Candida albicans*, etc.) adjusted to standardized turbidity of 0.5 McFarland.
4. Microtiter plates: Sterile 96-well flat-bottomed microplates.
5. Indicator: Resazurin dye (0.01%) for colorimetric analysis.
6. Controls: Positive control (ampicillin), negative control (distilled water), blank control (broth).
7. Incubator.
8. Microplate reader.

2.2.2 Agar Diffusion Method

1. Agar plates and nutrient agar.
2. Mushroom extracts (10 mg/mL as initial stock solution).
3. Microorganisms: Bacterial strains and fungal strains adjusted to standardized turbidity of 0.5 McFarland.
4. Petri dishes: Sterile 90 mm or 100 mm diameter plates for the diffusion assay.
5. Sterile swabs: Used to apply microbial inoculum evenly on agar surface.
6. Controls: Positive control (ampicillin), negative control (distilled water), blank control (broth).
7. Incubator.

2.2.3 TLC-Bioautography

1. TLC plates: Pre-coated silica gel plates as stationary phase.
2. Solvent system: Tailored solvent system based on the bioactive compounds' polarity (e.g., ethylacetate:hexane (for nonpolar components), methanol:chloroform (for polar components)) (*see* **Note 1**).
3. Mushroom extracts dissolved in ethanol or ethanol.
4. Bacterial or fungal strains adjusted to standardized turbidity of 0.5 McFarland.
5. Growth media: Mueller–Hinton Agar (MHA) for bacterial growth and Sabouraud Dextrose Agar (SDA) for fungal growth.
6. Incubator.

2.2.4 Time-Kill Assay

1. Broth medium: Mueller–Hinton Agar (MHA) for bacterial growth and Sabouraud Dextrose Agar (SDA) for fungal growth.
2. Mushroom extracts.
3. Microbial suspension: Standardized bacterial or fungal suspension with an initial concentration of $5 \cdot 10^5$ CFU/mL.
4. Sterile tubes.
5. Incubator.

3 Methods

3.1 Solvent Extraction of Bioactive Compounds

1. Clean the mushrooms from debris and dry them in shade or in hot air oven at 50–60 °C until constant weight is achieved. Grind the dried mushrooms into coarse powder.
2. Solvent for the extraction is selected based on the polarity of the target bioactive compounds. Ethanol or methanol is commonly used for the extraction of highly polar bioactives like phenols, tannins, and polysaccharides, while nonpolar solvents like ether and benzene are used to extract nonpolar bioactives like steroids or terpenoids.
3. For extraction by simple maceration, weighed amount of mushroom (10 g) is added to container solvent (100 mL) and allowed to stand at room temperature for 24–48 h with intermittent shaking.
4. For Soxhlet extraction, dried mushroom powder is added into Soxhlet thimble, and solvent is added in the round-bottom flask. The apparatus is heated at controlled temperature for 6–8 h.
5. Filter and concentrate the extract by rotary evaporation to remove the solvent.
6. Calculate the percent yield and store the extract in amber-colored vials.

3.2 Antimicrobial Activity Determination

3.2.1 Microdilution Method

1. Prepare the microbial suspension in broth and adjust the turbidity to match the 0.5 McFarland standard (approx. $1 \cdot 10^8$ CFU/mL).
2. Dilute the suspension 1:100 in TSB to achieve a final inoculum of $1 \cdot 10^6$ CFU/mL.
3. Prepare a stock solution of mushroom extract 10 mg/mL and perform serial twofold dilution in TSB to achieve a concentration range from 10 mg/mL to 0.078 mg/mL.
4. Add 100 μL of prepared inoculum to the wells of 96-well microplate and then add 100 μL mushroom extract of the

highest concentration (10 mg/mL) to the first well. Perform serial twofold dilutions of the extract and create a range of concentrations (e.g., 10, 5, 2.5, 1.25, 0.625, 0.312, 0.156, 0.078 mg/mL) in subsequent wells.

5. Include positive control (100 μL of standard drug and 100 μL of TSB), negative control (100 μL of TSB and inoculum), and blank control (100 μL TSB).
6. Incubate at 37 °C for 18–24 h for bacteria or at 25–30 °C for fungi.
7. After incubation, visually inspect the wells for turbidity and measure the optical density of each well at 600 nm using a microplate reader. Resazurin dye is also applicable to detect microbial growth (*see* **Note 2**).
8. Identify the well with the lowest concentration of mushroom extract which inhibits the growth of microorganisms represented as Minimum Inhibitory Concentration [15].

3.2.2 Agar Diffusion Method

1. Prepare a standardized microbial suspension to achieve a final inoculum of $1 \cdot 10^6$ CFU/mL.
2. Prepare a stock solution of mushroom extract 10 mg/mL and perform serial twofold dilution in TSB to achieve a concentration range from 10 mg/mL to 0.078 mg/mL.
3. Prepare agar plates by pouring Mueller–Hinton Agar (MHA) or Nutrient Agar (NA) into sterile petri dishes and allow them to solidify.
4. Dip a sterile cotton swab into the microbial suspension and spread evenly over the surface of the agar plate in three different directions to ensure even coverage.
5. Add 20 μL of each concentration of mushroom extract into pre-made wells (using a sterile cork borer or well cutter) spaced evenly on the agar plate.
6. Include positive control (standard antibiotic) and negative control (solvent used to dissolve mushroom extract).
7. Incubate the petri plates at 37 °C for 18–24 h for bacteria or at 25–30 °C for fungi.
8. After incubation, measure the diameter of the inhibition zone (clear area around the well where growth is inhibited) in millimeters (mm) and compare against positive control [16].

3.2.3 TLC-Bioautography

Agar Diffusion Method

1. Develop the chromatogram of mushroom extract using the selected solvent system.
2. Place the developed chromatogram in direct contact with agar plate seeded with microorganisms.

3. After diffusion of few minutes to hours, remove the chromatogram and incubate the agar plate at 37 °C.
4. Zone of inhibition on the chromatogram indicates the antimicrobial activity.

Direct Bioautography

1. Develop the chromatogram of mushroom extract using the selected solvent system.
2. Dip or spray the developed chromatogram with a microbial suspension.
3. Incubate under humid conditions (25 °C for 48 h).
4. Spray tetrazolium salt on to plate, reincubate (37 °C) for 3–4 h and observe the formation of intensely colored formazan at sites of microbial growth inhibition.

Agar Overlay Assay

1. Develop the chromatogram of mushroom extract using the selected solvent system.
2. Cover the developed chromatogram with molten agar seeded with microorganisms.
3. After cooling to allow diffusion, incubate the plate at 37 °C.
4. Stain with tetrazolium dye to visualize the inhibition zone [17, 18].

3.2.4 Time-Kill Assay

1. Prepare a standardized microbial suspension to 0.5 McFarland to achieve a final inoculum of $1{\cdot}10^6$ CFU/mL.
2. Prepare mushroom extract at concentrations corresponding to MIC, 2·MIC, and 4·MIC in sterile broth medium and transfer into sterile test tubes.
3. Add an inoculum size of $1{\cdot}10^6$ CFU/mL to the test tubes.
4. Incubate the test tubes at 37 °C for 0, 6, 12, 18, 24, and 30 h (for bacteria) and at 0, 6, 12, 30, 36, 48, 54, and 72 h (for fungi).
5. Aseptically inoculate the aliquots onto sterile nutrient agar plates and incubate at 37 °C for 24 h.
6. After incubation, count the colony-forming units from each plate and plot a graph of log CFU/mL against time [19].

4 Antimicrobial Compounds in Mushrooms

Mushrooms are an exceptional source of bioactive compounds associated with their various biological activities. Mushrooms contain alkaloids, phenolics, terpenoids, proteins, peptides, lectins, fatty acids, polysaccharides, and steroids. Phytochemical profiling of different mushroom species is significant in order to ascertain

Table 1
Bioactive compounds and their bioactivity in mushrooms

S. no.	Class of bioactive compounds	Mushroom	Bioactivity	References
1	Alkaloid	*Ganoderma australe, G. cochlear, G. lucidum*	Neuroprotective and anti-inflammatory	[20]
2	Steroids	*Hypholoma lateritium*	Anti-inflammatory	[21]
3	Terpenoids	*Cyathus africanus*	Anticholinesterase	[22]
4	Phenols	*Pleurotus ostreatus*	Antimicrobial	[23]
5	Polysaccharides	*Lentinus edodes*	Anticancer	[24]
6	Fatty acids	*Morchella fluvialis*	Antidiabetic and antioxidant	[25]
7	Proteins and peptides	*Lentinus edodes*	Antidiabetic	[26]

the medicinal properties and potential uses of each species. Studying these bioactive components found in mushrooms helps us further understand their health benefits and potential clinical applications in the medical field (Table 1).

5 Notes

1. All the solvents should be of analytical grade.
2. 0.01% resazurin dye added to each well and incubated for an additional 2–4 h. The dye will turn pink in the presence of viable microorganisms or remain blue if growth is inhibited.

References

1. Salam MA, Al-Amin MY, Salam MT, Pawar JS, Akhter N, Rabaan AA, Alqumber MAA (2023) Antimicrobial resistance: a growing serious threat for global public health. Healthcare (Basel) 11:1946. https://doi.org/10.3390/healthcare11131946
2. Uddin TM, Chakraborty AJ, Khusro A, Zidan BRM, Mitra S, Emran TB, Dhama K, Ripon MKH et al (2021) Antibiotic resistance in microbes: history, mechanisms, therapeutic strategies and future prospects. J Infect Public Health 14:1750–1766. https://doi.org/10.1016/j.jiph.2021.10.020
3. Appiah T, Boakye YD, Agyare C (2017) Antimicrobial activities and time-kill kinetics of extracts of selected Ghanaian mushrooms. Evid Based Complement Alternat Med 2017:4534350. https://doi.org/10.1155/2017/4534350
4. Al Qutaibi M, Kagne SR (2024) Unearthing nature's pharmacy: exploring the antimicrobial potency of mushrooms. J Food Process Preserv 2024:8331974. https://doi.org/10.1155/2024/8331974
5. Alves MJ, Ferreira ICFR, Dias J, Teixeira V, Martins A, Pintado M (2012) A review on antimicrobial activity of mushroom (basidiomycetes) extracts and isolated compounds. Planta Med 78:1707–1718. https://doi.org/10.1055/s-0032-1315370
6. Bhambri A, Srivastava M, Mahale VG, Mahale S, Karn SK (2022) Mushrooms as

potential sources of active metabolites and medicines. Front Microbiol 13:837266. https://doi.org/10.3389/fmicb.2022.837266

7. Venturella G, Ferraro V, Cirlincione F, Gargano ML (2021) Medicinal mushrooms: bioactive compounds, use, and clinical trials. Int J Mol Sci 22:634. https://doi.org/10.3390/ijms22020634
8. Navarro-Simarro P, Gómez-Gómez L, Ahrazem O, Rubio-Moraga Á (2024) Food and human health applications of edible mushroom by-products. New Biotechnol 81:43–56. https://doi.org/10.1016/j.nbt.2024.03.003
9. Angelini P, Venanzoni R, Angeles Flores G, Tirillini B, Orlando G, Recinella L, Chiavaroli A, Brunetti L et al (2020) Evaluation of antioxidant, antimicrobial and Tyrosinase inhibitory activities of extracts from *Tricholosporum goniospermum*, an edible wild mushroom. Antibiotics 9:513. https://doi.org/10.3390/antibiotics9080513
10. Gupta R, Gupta N, Bindal S (2021) Bacterial Cell Wall biosynthesis and inhibitors. In: Gupta R, Gupta N (eds) Fundamentals of bacterial physiology and metabolism. Springer, Singapore, pp 81–98
11. Bach F, Zielinski AAF, Helm CV, Maciel GM, Pedro AC, Stafussa AP, Ávila S, Haminiuk CWI (2019) Bio compounds of edible mushrooms: *in vitro* antioxidant and antimicrobial activities. LWT 107:214–220. https://doi.org/10.1016/j.lwt.2019.03.017
12. Bains A, Chawla P, Inbaraj BS (2023) Evaluation of in vitro antimicrobial, antioxidant, and anti-quorum sensing activity of edible mushroom (*Agrocybe aegerita*). Foods 12:3562. https://doi.org/10.3390/foods12193562
13. Azmir J, Zaidul ISM, Rahman MM, Sharif KM, Mohamed A, Sahena F, Jahurul MHA, Ghafoor K et al (2013) Techniques for extraction of bioactive compounds from plant materials: a review. J Food Eng 117:426–436. https://doi.org/10.1016/j.jfoodeng.2013.01.014
14. Balouiri M, Sadiki M, Ibnsouda SK (2016) Methods for *in vitro* evaluating antimicrobial activity: a review. J Pharm Anal 6:71–79. https://doi.org/10.1016/j.jpha.2015.11.005
15. Zhang Q-W, Lin L-G, Ye W-C (2018) Techniques for extraction and isolation of natural products: a comprehensive review. Chin Med 13:20. https://doi.org/10.1186/s13020-018-0177-x
16. Magaldi S, Mata-Essayag S, Hartung de Capriles C, Perez C, Colella MT, Olaizola C, Ontiveros Y (2004) Well diffusion for antifungal susceptibility testing. Int J Infect Dis 8:39–45. https://doi.org/10.1016/j.ijid.2003.03.002
17. Marston A (2011) Thin-layer chromatography with biological detection in phytochemistry. J Chromatogr A 1218:2676–2683. https://doi.org/10.1016/j.chroma.2010.12.068
18. Choma IM, Grzelak EM (2011) Bioautography detection in thin-layer chromatography. J Chromatogr A 1218:2684–2691. https://doi.org/10.1016/j.chroma.2010.12.069
19. Sim J-H, Jamaludin NS, Khoo C-H, Cheah Y-K, Halim SNBA, Seng H-L, Tiekink ERT (2014) In vitro antibacterial and time-kill evaluation of phosphanegold(I) dithiocarbamates, R3PAu[S2CN(iPr)CH2CH2OH] for R = Ph, cy and et, against a broad range of gram-positive and gram-negative bacteria. Gold Bull 47:225–236. https://doi.org/10.1007/s13404-014-0144-y
20. Lu S-Y, Peng X-R, Dong J-R, Yan H, Kong Q-H, Shi Q-Q, Li D-S, Zhou L et al (2019) Aromatic constituents from *Ganoderma lucidum* and their neuroprotective and anti-inflammatory activities. Fitoterapia 134:58–64. https://doi.org/10.1016/j.fitote.2019.01.013
21. Ványolós A, Muszyńska B, Chuluunbaatar B, Gdula-Argasińska J, Kała K, Hohmann J (2020) Extracts and steroids from the edible mushroom *Hypholoma lateritium* exhibit anti-inflammatory properties by inhibition of COX-2 and activation of Nrf2. Chem Biodivers 17:e2000391. https://doi.org/10.1002/cbdv.202000391
22. Yu M, Kang X, Li Q, Liang Y, Zhang M, Gong Y, Chen C, Zhu H, Zhang Y (2022) Thirteen cyathane diterpenoids with acetylcholinesterase inhibitory effects from the fungus *Cyathus africanus*. Phytochemistry 193:112982. https://doi.org/10.1016/j.phytochem.2021.112982
23. Yakobi S, Zuma L, Gasa N, Pooe O (2024) Antimicrobial potential of organic phenolic compounds from wild mushroom extracts: impact on proliferation and kinetic growth of multidrug-resistant Neisseria gonorrhoeae strains. J Food Biochem 2024:2336255. https://doi.org/10.1155/2024/2336255
24. Ponnusamy C, Uddandrao VVS, Sasikumar V, Prabhu P, Saravanan G (2024) Partially purified polysaccharides from *Lentinus edodes* (mushroom) scavenge free radicals and induce apoptosis in MCF-7 cancer cells by regulating apoptotic genes. Indian J Pharm Edu Res 58:535–545. https://doi.org/10.5530/ijper.58.2.60

25. Turk A, Lee S, Lee HH, Yeon SW, Ryu SH, Seo GH, Chang HY, Hwang BY, Lee MK (2023) α-Glucosidase inhibitory fatty acids from *Morchella fluvialis* mushroom. Folia Hortic 35:369–379. https://doi.org/10.2478/fhort-2023-0026

26. Zhang Y, Chen Y, Liu X, Wang W, Wang J, Li X, Sun S (2023) Preparation and identification of peptides with α-glucosidase inhibitory activity from shiitake mushroom (*Lentinus edodes*) protein. Foods 12:2534. https://doi.org/10.3390/foods12132534

Part III

Development of Novel Foods and Ingredients

Chapter 14

Functional Flours from *Agaricus bisporus* Stem Waste

Miguel Ángel Gallardo, Manuel Álvarez-Ortí, José Emilio Pardo, and M. Esther Martínez-Navarro

Abstract

Mushroom stems are the main waste generated during mushroom production, which are still treated as residue, although their nutritional and bioactive composition is very similar to that of the fruiting body. It is therefore a waste with great potential to be transformed into a high-value-added by-product as a functional ingredient or incorporated directly into formulations to develop new functional foods. This chapter provides a guide for the production of functional flours from mushroom stem residues, through a drying process (dehydration or freeze-drying), to be incorporated into basic bakery doughs, thus enhancing their nutritional profile and health-promoting properties.

Key words *Agaricus bisporus*, Stem, Mushroom waste, By-products, Functional food

1 Introduction

The global market for edible mushroom production has shown steady growth since the late twentieth century. Annual production is estimated at 12.74 million tonnes, generating over 46.6 billion USD in 2022 [1, 2]. Moreover, global output is projected to reach 20.84 million tonnes by 2026 [3]. The most widely cultivated variety is *Agaricus bisporus*, also commonly known as the button mushroom or white button mushroom, which accounts for approximately 54% of the total edible mushroom production [4]. The popularity of mushrooms, particularly *A. bisporus*, is attributed to their organoleptic characteristics, culinary versatility, and well-established nutritional and bioactive properties. Numerous studies have shown that *A. bisporus* contain a wide range of bioactive compounds (including protein–polysaccharide complexes, β-glucans, and polyphenols), which confer various medicinal properties, including antioxidant, antimicrobial, antiviral, immunomodulatory, antitumor, and cholesterol-lowering effects, among others [5]. These attributes position *A. bisporus* as both a promising source for pharmaceuticals and nutraceuticals and as a

Pablo Navarro Simarro (ed.), *Fungi*, Methods and Protocols in Food Science, https://doi.org/10.1007/978-1-0716-4989-3_14,
© The Author(s), under exclusive license to Springer Science+Business Media, LLC, part of Springer Nature 2026

functional food, i.e., a food that not only provides basic nutrition but also offers additional health benefits beyond its essential nutrients [6].

During mushroom harvesting, a large amount of waste is generated, consisting of mushroom stems and mushrooms that do not meet the shape or size requirements for supermarket distribution. Mushroom stems can make up more than 20% of the total yield of mushroom production [7]. Currently, these wastes are primarily used for composting or disposed of by incineration—a practice that can produce unpleasant odors and attract insects, which conflicts with the goals of the European Green Deal's Farm to Fork Strategy and the United Nations' 2030 Agenda for Sustainable Development [8, 9]. Nevertheless, *A bisporus* stem wastes have a bioactive composition very similar to that of the remaining fruiting body [10], highlighting their potential to be transformed from waste into value-added ingredients suitable for functional food production.

In recent years, numerous authors have developed various techniques, both eco-friendly and conventional, to extract bioactive compounds from different mushrooms [5, 9, 11, 12]. However, these extraction processes often involve high costs and the use of solvents that generate additional waste. As an alternative, this book chapter proposes transforming *A. bisporus* stem wastes into by-products by incorporating them directly into flours, resulting in functional flours for the preparation of basic bakery doughs, composed predominantly of flour and water, which provide enhanced nutritional value and health benefits.

2 Materials

1. Water.
2. Wheat flour.
3. Olive oil.
4. Dried or lyophilized white button mushroom stem waste.

3 Methods

3.1 Stem Preparation

1. Cut the mycelium contained in the stem of mushroom.
2. Clean the stems by soaking it in cold water and then dry them (*see* **Note 1**).
3. Remove the moisture content from the stems using a dehydrator or freeze-dryer (prior freezing) (*see* **Notes 2–4**).
4. Store in a dry and dark place until use.

3.2 Stem Flour Production

1. Grind the dehydrated or freeze-dried stem waste in a blade mill (*see* **Note 5**).
2. Sieve the ground powder to a particle size <1 mm (*see* **Note 6**).
3. Keep the stem powder (stem flour) in a dry place, protected from light (*see* **Note 7**).

3.3 Dough Preparation

1. Mix 25 g of wheat flour with 25 g of stem flour dehydrated or freeze-dried (*see* **Note 8**).
2. Add 4 g olive oil and then 27.5 mL of water (*see* **Note 9**).
3. Knead the mixture until a homogeneous dough is obtained.
4. The dough can be baked immediately (200 °C, 20 min) or stored frozen (≤ -18 °C).

4 Notes

1. Cold water is recommended to remove the remains of the casing layer (Fig. 1), although brushing and air blowing are also effective methods for cleaning.
2. Moisture less than 10%.
3. Previous sensory studies ($n \geq 100$) showed that consumers preferred products made from freeze-dried *Agaricus bisporus* stem waste over those made from dehydrated stem waste. Although the advantages of freeze-dried products are well known, this may be a major economic disadvantage compared

Fig. 1 *Agaricus bisporus* stems unwashed (left) and washed in cold water (right)

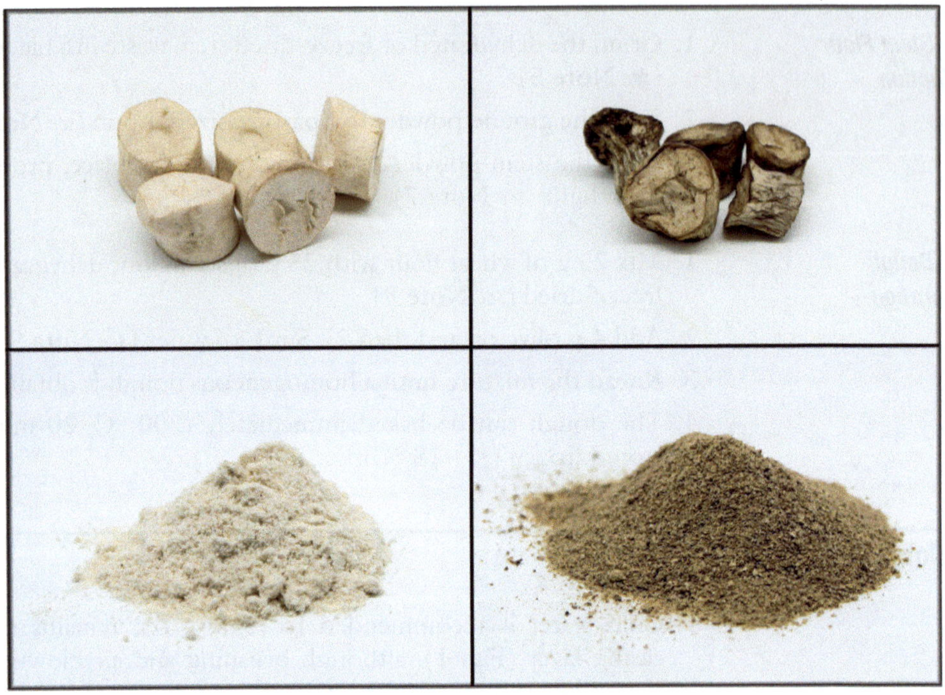

Fig. 2 Difference between the color of freeze-dried (top-left) and dehydrated (top-right) stems and their corresponding freeze-dried (bottom-left) and dehydrated (bottom-right) flours

to dehydrated alternatives, which also achieved high acceptability values at a lower cost.

4. The stem flour obtained through dehydration is darker than the freeze-dried stem flour. This consideration is crucial when developing the product, as the color difference may directly influence consumer perception and acceptance (Fig. 2).
5. Mill at 10,000 rpm for 10 s.
6. Freeze-dried stem powder does not need to be sieved due to its small particle size.
7. Stem flours should be stored in airtight containers due to their high hygroscopicity, which makes them prone to absorbing moisture from the environment.
8. Proportions to make a dough with 50% stem mushroom flour. Higher concentrations make the dough very brittle and difficult to handle.
9. Higher amounts of mushroom flour can be added to prepare the dough. In this case, the amounts of oil and water remain constant for the different concentrations of stem mushroom

dough, except for the 100% dehydrated or freeze-dried stem dough, where the water to be added is 32.5 mL, due to the higher water absorption capacity.

References

1. Bermúdez-Gómez P, Fernández-López J, Pérez-Clavijo M, Viuda-Martos M (2024) Optimization of mushroom (*Agaricus bisporus* and *Pleurotus ostreatus*) by-products processing for prospective functional flour development. Foods 13(24):4046. https://www.mdpi.com/2304-8158/13/24/4046
2. Wang X, Han Y, Li S, Li H, Li M, Gao Z (2024) Edible fungus-derived bioactive components as innovative and sustainable options in health promotion. Food Biosci 59:104215. https://doi.org/10.1016/j.fbio.2024.104215
3. Atallah E, Zeaiter J, Ahmad MN, Leahy JJ, Kwapinski W (2021) Hydrothermal carbonization of spent mushroom compost waste compared against torrefaction and pyrolysis. Fuel Process Technol 216:106795. https://doi.org/10.1016/j.fuproc.2021.106795
4. Wang R, Yun J, Wu S, Bi Y, Zhao F (2022) Optimisation and characterisation of novel angiotensin-converting enzyme inhibitory peptides prepared by double enzymatic hydrolysis from *Agaricus bisporus* scraps. Foods 11(3):394. https://www.mdpi.com/2304-8158/11/3/394
5. Ramos M, Burgos N, Barnard A, Evans G, Preece J, Graz M, Ruthes AC, Jiménez-Quero A, Martínez-Abad A, Vilaplana F, Ngoc LP, Brouwer A, van der Burg B, del Garrigós MC, Jiménez A (2019) *Agaricus bisporus* and its by-products as a source of valuable extracts and bioactive compounds. Food Chem 292:176–187. https://doi.org/10.1016/j.foodchem.2019.04.035
6. Ferreira ICFR, Fernandes Â, Heleno SA (2017) Chemical, nutritional, and bioactive potential of mushrooms. In: Edible and medicinal mushrooms. Wiley, pp 455–501. https://doi.org/10.1002/9781119149446.ch21
7. Papoutsis K, Grasso S, Menon A, Brunton NP, Lyng JG, Jacquier J-C, Bhuyan DJ (2020) Recovery of ergosterol and vitamin D2 from mushroom waste—potential valorization by food and pharmaceutical industries. Trends Food Sci Technol 99:351–366. https://doi.org/10.1016/j.tifs.2020.03.005
8. De Cianni R, Varese GC, Mancuso T (2023) A further step toward sustainable development: the case of the edible mushroom supply chain. Foods 12(18):3433. https://www.mdpi.com/2304-8158/12/18/3433
9. Guo J, Zhang M, Fang Z (2022) Valorization of mushroom by-products: a review. J Sci Food Agric 102(13):5593–5605. https://doi.org/10.1002/jsfa.11946
10. Gallardo MA, Vieira Júnior WG, Martínez-Navarro ME, Álvarez-Ortí M, Zied DC, Pardo JE (2024) Impact of button mushroom stem residue as a functional ingredient for improving nutritional characteristics of pizza dough. Molecules 29(21):5140. https://www.mdpi.com/1420-3049/29/21/5140
11. Joradon P, Rungsardthong V, Ruktanonchai U, Suttisintong K, Thumthanaruk B, Vatanyoopaisarn S, Uttapap D, Mendes AC (2024) Extraction of bioactive compounds from lion's mane mushroom by-product using supercritical CO2 extraction. J Supercrit Fluids 206:106162. https://doi.org/10.1016/j.supflu.2023.106162
12. Sun M, Zhuang Y, Gu Y, Zhang G, Fan X, Ding Y (2024) A comprehensive review of the application of ultrasonication in the production and processing of edible mushrooms: drying, extraction of bioactive compounds, and post-harvest preservation. Ultrason Sonochem 102:106763. https://doi.org/10.1016/j.ultsonch.2024.106763

Chapter 15

Mushroom and Their Coproducts as Flavor Enhancers in Salt-Reduced Food

Patricia Bermúdez-Gómez, Manuel Viuda-Martos, Juana Fernández-López, and Margarita Pérez-Clavijo

Abstract

Edible mushrooms have been an important part of the human diet for centuries, with their stems accounting for up to 20% of total production. Excessive salt (NaCl) intake is a major global health concern due to its association with cardiovascular diseases, prompting the WHO to recommend food reformulation as a reduction strategy. One approach involves using flavor enhancers, such as umami compounds naturally present in mushrooms. Therefore, mushroom powder and umami-rich extracts have been incorporated into reduced-sodium foods. Additionally, stem flours of *Agaricus bisporus* (ABSF) and *Pleurotus ostreatus* (POSF) contain MSG-like amino acids. This chapter outlines a step-by-step protocol for preparing low-sodium bread and sodium-reduced mortadella using POSF and ABSF stem flours.

Key words Mushroom stem, Functional food, Sodium-reduce, By-products, Bread, Mortadella

1 Introduction

Edible mushrooms have been an important part of the human diet for centuries due to their nutritional and health benefits, complemented by their appealing flavor and aroma [1, 2]. It is broadly acknowledged that mushrooms are rich in proteins, fiber, vitamins, and minerals and contain bioactive compounds with antioxidant, anti-inflammatory, and anticancer properties [1–3]. Consequently, the global edible mushroom market has witnessed considerable growth in recent decades, with production estimated to exceed 20 million tons annually by 2026 [4, 5]. Among the most widely cultivated and consumed species worldwide are *Agaricus bisporus* (common button mushroom), *Lentinula edodes* (shiitake), and various species of the *Pleurotus* genus [1, 2, 6]. Although their cultivation process could reduce agricultural waste by using them as substrate, the rapid expansion of the market has led to the generation of approximately 60 million tons of other waste annually, posing a significant management challenge for producers [3, 5,

Pablo Navarro Simarro (ed.), *Fungi*, Methods and Protocols in Food Science, https://doi.org/10.1007/978-1-0716-4989-3_15,
© The Author(s), under exclusive license to Springer Science+Business Media, LLC, part of Springer Nature 2026

7–9]. Mushroom by-products mainly consist of spent mushroom substrate, mushroom stems, and mushrooms that do not meet the required size or shape for commercial sale [5]. The mushroom stem is functionally the pathway to absorb nutrients from the growth substrate and represents up to 20% of the total production yield [10]. This waste is typically managed through composting; however, it could also be applied in the food, energy, pharmaceutical, and cosmetic industries [5, 11]. As well as the fruiting bodies, mushroom stems have a valuable healthy nutritional profile and contain numerous bioactive compounds [7, 10, 12–15]. Overall, mushrooms and their by-products have been explored as a source for extracting valuable biomolecules such as ergosterol, peptides, phenolic compounds, glucans, and chitin [16–18]. Additionally, several mushroom species, their stems, and some isolated bioactive agents from both fruiting bodies and stems have been proven in meat and meat analogs, dairy products, and bakery products [11, 19, 20].

Excessive consumption of salt (NaCl) is a global public health concern and represents the primary dietary factor associated with mortality from noncommunicable diseases, with an estimated three million deaths annually attributed to this risk factor [21–23]. Various scientific studies have demonstrated an association between excessive sodium intake and high blood pressure (BP), thereby increasing the risk of cardiovascular disease (CVD) [21, 24, 25]. Epidemiological evidence suggests that reducing sodium ingestion leads to a significant decrease in blood pressure and, consequently, a lower incidence of cardiovascular events [25]. For these reasons, the World Health Organization (WHO) has established a recommended sodium intake limit of ≤2 g/day (equivalent to 5 g/day of NaCl) [21, 24]. However, sodium consumption in various regions significantly exceeds this recommendation, with estimated values ranging between 3.9 and 5.6 g/day in populations from East Asia, Central Asia, Eastern Europe, Central Europe, and the Middle East/North Africa [21–23]. In Western countries, approximately 77% of dietary sodium comes from processed foods, with the main contributors being bakery products, processed meats and fish, dairy products, snacks, and sauces [23–25]. In this context, the World Health Organization (WHO) has advocated several strategies to reduce global salt intake, with the primary measures including mandatory food reformulation, food labeling, taxes or subsidies, and communication campaigns [21, 24]. The reformulation of food products has involved the development of reduced, low, very low, and sodium-free formulations [26]. However, the production of certain food products without salt is not a straightforward process. The primary challenge associated with salt reduction in processed foods lies in its role in preservation and, consequently, food safety, as it helps lower water activity [27]. Other important factors include changes in flavor and

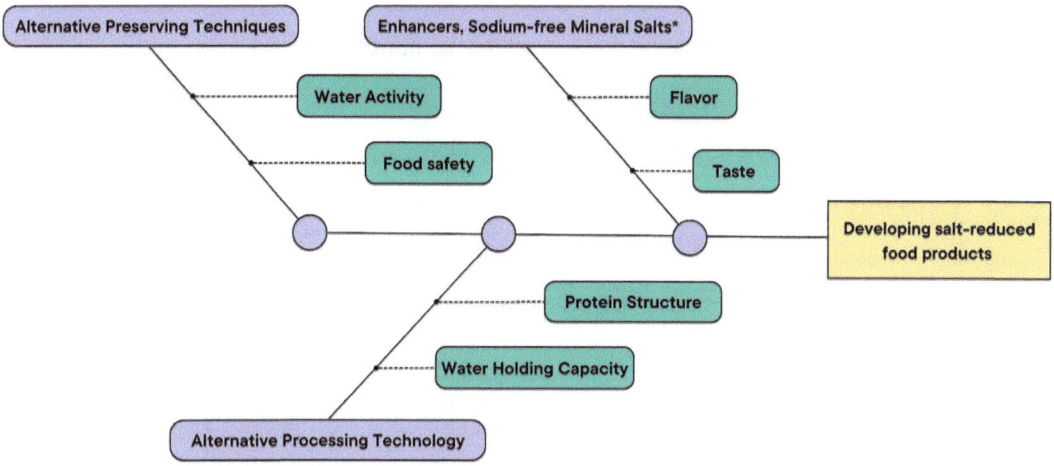

Fig. 1 Challenges to reduce sodium in food and their technological strategy solutions

the technological functions of salt, which enhance water-holding capacity and facilitate the structural modifications of proteins required to achieve the desired texture. These functions encompass gluten network formation in bread and emulsion development in meat products [26–28] (Fig. 1).

The primary strategies developed to mitigate these challenges related to food safety, texture, and flavor involve alternative preservation techniques and processing technologies, such as ultrasound, high hydrostatic pressure, and irradiation [27–29]. Finally, to minimize changes in the flavor, aroma, and texture of foods, the use of sodium-free mineral salts such as potassium chloride (KCl), magnesium chloride ($MgCl_2$), calcium chloride ($CaCl_2$), and magnesium sulfate ($MgSO_4$) has been explored. Among these, KCl has emerged as a promising alternative; however, at high concentrations, it has been reported to result in a metallic taste [30]. Another key area of research in this field is the use of flavor enhancers/boosters that amplify the perception of salt, such as umami compounds, including monosodium glutamate (MSG), 5′-nucleotides, and organic acids (OAs) [31] (Fig. 1). It is widely accepted that these compounds are naturally present in certain foods, such as edible mushrooms, tomatoes, and spices [30–33]. In recent years, there has been a growing interest in obtaining umami substances from natural sources, leading to experimentation with the extraction of MSG-like amino acids and 5′-mononucleotides from mushrooms. Additionally, this extraction, as well as fresh mushroom and mushroom powder, has been incorporated into reduced-sodium foods due to its potential as a flavor enhancer (Table 1) [34–45]. Monosodium glutamate (MSG)-like amino acids, particularly L-Glu, have also been found in stem flours of *Agaricus bisporus* and *Pleurotus ostreatus* [15].

Table 1
Reduced-sodium foods by incorporating fresh mushrooms, powder, and extracts

Foods	Mushroom type	Format inclusion	Formulation	Main results	References
Corn extruded snacks	*Lentinula edodes* (shiitake mushroom)	Extract from shiitake by-products (umami ingredient (UI))	**Formulations:** Control: 2.8% salt, 0.6% MSG, 0% UI 0.8% salt, 0.6% MSG, 0% UI 0.8% salt, 0% MSG, 1.0% and 1.5% UI	UI enhanced umami and salty taste similar to MSG Lower sensory acceptance than the control but were not rejected No significant differences in color and texture	[35]
Chicken sausage	*Flammulina velutipes* (winter mushroom)	Mushroom powder (freeze-dried and ground) (WMP)	**Formulations:** Control: 0.3% sodium pyrophosphate 0% WMP 0% sodium pyrophosphate 0.5% WMP 0% sodium pyrophosphate 1% WMP 0% sodium pyrophosphate	Improved emulsion stability and lower oxidation No significant effect on color Reduced hardness with the increasing amount of WMP No negative impact on sensory acceptance	[36]
Beef burgers	*Lentinula edodes* (shiitake mushroom) by-products (stipes)	Extract from shiitake by-products (umami ingredient (UI))	**Formulations:** Control: 1.9% NaCl, 0% UI 1.24% NaCl +1% UI or 2% UI 0.91% NaCl +1% UI or 2% UI 0.57% NaCl +1% UI or 2% UI	Burgers with 70% salt reduction less hard Increase proline and phenylalanine The volatile profile exhibited a slight increase in compounds (e.g., 3-methylbutanal, (S)-(+)-1,2-propanediol, and dimethyl sulfide)	[37]
Chicken broth	*Agaricus bisporus* (button mushroom)	Enzymatically hydrolyzed mushroom protein (eHMP)	**Formulations:** eHMP alone or with cysteine Cysteine alone	Odorants generated from the eHMP + cysteine treatment significantly enhanced the perceived saltiness of the broth	[38]

(continued)

Table 1 (continued)

Foods	Mushroom type	Format inclusion	Formulation	Main results	References
Beef burgers	*Lentinula edodes* (shiitake mushroom)	Water extract from dried shiitake mushroom (E)	10 **formulations** with 0–75% NaCl reduction and E1 = 5%, E2 = 12.5%, E3 = 20% shiitake extract concentrations	E3 enhanced salty taste perception in burgers with 50% NaCl reduction but not with 75% reduction E3 also improved sensory scores for color, aroma, texture, and flavor	[39]
Beef taco filling	*Agaricus bisporus* (white button mushroom)	Fresh	**Formulations**: 0% mushroom +0.97% salt 15% mushroom +0.82% salt 30% mushroom +0.68% salt 45% mushroom +0.53% salt	A 45% mushroom substitution level maintained physical properties Consumers liked the control equally as much as the 45% mushroom	[40]
Beef burgers	*Pleurotus ostreatus* (oyster), *Agaricus bisporus* (button), *Agaricus brunnescens* (portobello)	Hot-air dried, ground flour	**Formulations:** Control: 1.8% NaCl 1.35% mushroom flour from each species +0.45% NaCl	Consumers did not detect 75% sodium Chloride reduction Burgers with mushroom flour had higher cooking loss and shrinkage than control Minor changes in pH and water activity; no significant impact on texture	[41]
Liver pâté	*Pleurotus ostreatus* (oyster mushroom) (PO), *Agaricus bisporus* (white button mushroom) (AB)	Hot-air dried, ground flour	**Formulations:** Control: 2% NaCl 50% fat and salt using AB and PO (7.5% hydrated 50/50 w/w AB and PO) 50% fat and salt	Increased moisture, fiber, and protein content, improving nutritional value AB darkened color, PO increased yellowness 10% resulted in a	[42]

(continued)

Table 1
(continued)

Foods	Mushroom type	Format inclusion	Formulation	Main results	References
			using AB and PO (10% hydrated 70/30 w/w)	harder texture and increased lipid oxidation Pâtés with 7.5% mushroom flour were sensorial acceptable	
Frankfurter sausages	*Pleurotus ostreatus* (oyster mushroom) (PO), *Agaricus bisporus* (white button mushroom) (AB)	Hot-air dried, ground flour	**Formulations:** Control: 1.5% NaCl MF-Ab: 50% salt reduction, 2.5% AB MF-Po: 50% salt reduction, 2.5% PO LF-Ab: 50% salt reduction, 5% AB LF-Po: 50% salt reduction, 5% PO MF-PoAb: 50% salt reduction, 2.5% AB +2.5% PO	Increased moisture, dietary fiber, and lipid oxidation 5% AB results in higher protein content AB darkened the color, while PO resulted in a softer texture Sensory analysis showed that mushroom-containing sausages were acceptable despite lower scores in color and flavor	[43]
Beef burgers	*Agaricus bisporus* (white button mushroom)	Fresh, finely chopped and heated	**Formulations**: Combinations of 0.5%, 1.25%, 2.0% of salt and 0%, 15%, 30% of mushroom	30% mushroom significantly softened texture Sensory analysis showed that mushrooms improved umami perception and preserved overall acceptance despite salt reduction	[44]
Beef patties	*Pleurotus ostreatus* (oyster mushroom) (PO), *Agaricus bisporus* (white button	Hot-air dried, ground flour	**Formulations:** Control: 1.2% NaCl 2.5% *A. bisporus* and *P. ostreatus* flours, 50% salt reduction 5.0% *A. bisporus*	Increased protein and dietary fiber content; modified color (AB darkened color, PO increased yellowness)	[45]

(continued)

Table 1 (continued)

Foods	Mushroom type	Format inclusion	Formulation	Main results	References
	mushroom) (AB)		and *P. ostreatus* flours, reducing salt by 50%	Slight antimicrobial effect against *pseudomonas* Patties with 2.5% mushroom flour maintained acceptable sensory properties	

Given the urgent need to reduce sodium intake while maintaining the sensory and technological qualities of food products, exploring natural flavor-enhancing ingredients has become a key area of research. Due to the promising role of natural flavor enhancers from mushrooms and their coproducts, this chapter aims to describe, step by step, the protocol for preparing low-sodium bread and sodium-reduced mortadella through the incorporation of flours derived from the stems of *Pleurotus ostreatus* (POSF) and *Agaricus bisporus* (ABSF).

2 Materials

2.1 Ingredients

The ingredients used in the production of bread and mortadella can vary depending on the specific objectives of the product, whether related to nutritional profile, sensory characteristics, or physicochemical properties. However, certain essential ingredients are commonly included in their formulations.

Bread Ingredients:

1. Wheat flour (*see* **Note 1**).
2. Water (*see* **Note 2**).
3. Salt.
4. Dry yeast as a powder (*see* **Note 3**).
5. Stems of *Agaricus bisporus* and *Pleurotus ostreatus* (*see* **Note 4**).

Mortadella Ingredients:

6. Lean meat.
7. Pork dewlap.
8. Fat: pork backfat (*see* **Note 5**).
9. Ice or water.

10. Salt.
11. Preservative: sodium nitrite.
12. Stabilizer: sodium tripolyphosphate.
13. Antioxidant: sodium ascorbate.
14. Potato starch (*see* **Note 6**).
15. Seasoning: pepper, nutmeg, garlic powder.
16. Artificial casings.
17. Stems of *Agaricus bisporus* and *Pleurotus ostreatus*.

2.2 Equipment

First, the processing of *Agaricus bisporus* and *Pleurotus ostreatus* stems requires (*see* **Note** 7):

1. Laminator.
2. Dehydrator.
3. Mill.
4. Sieve 0.18 mm (*see* **Note 8**).

   The following is a list of essential equipment required to produce bread:
5. Weighing balance.
6. Thermometer.
7. Glass beaker.
8. Automatic bread maker (*see* **Note 9**).
9. Freezer.
10. Oven (*see* **Note 10**).

    On the other hand, the production of mortadella requires:
11. Cutter.
12. Stuffer.
13. Clipping machine.
14. Cooking kettle.

3 Methods

3.1 Bread Production

3.1.1 Low-Sodium Bread Formulation

The first step before starting the production process is calculating the concentration of each ingredient in the formulation. The experimental design to produce low-sodium bread is indicated in Table 2. The percentage of stem flour added was quantified based on the substitution level of salt. The increase in ABSF and POSF concentrations from 1% to 3% and 6% in the 100% salt replacement is based on findings from the literature and is adjusted by reducing the amount of wheat flour (*see* **Note 11**).

Table 2
Formulations of bread prototypes with low-sodium content by the addition of *Agaricus bisporus* stem flour (ABSF) and *Pleurotus ostreatus* stem flour (POSF)

Ingredient g/100 g bread	Control	$BAB_{0.5}/BPO_{0.5}$	BAB_1/BPO_1	BAB_3/BPO_3	BAB_6/BPO_6
Wheat flour	60	60	60	58	55
Water	37	37	37	37	37
Salt	1	0.5	0	0	0
ABSF/POSF	0	0.5	1	3	6
Yeast	2	2	2	2	2

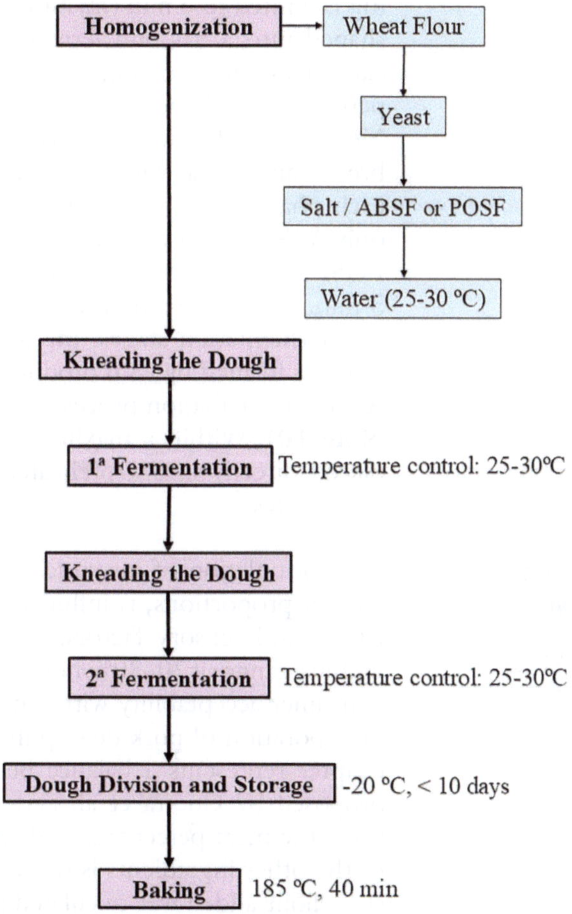

Fig. 2 Low-sodium bread production flowchart

3.1.2 Low-Sodium Bread Processed

The production of bread involves several critical steps that contribute to the quality of the product. Figure 2 shows the low-sodium bread flowchart. After determining the proportion of each ingredient in the formulation, it is essential to add them to the bowl, bread maker, or kneading machine in the correct order. First, wheat flour is added, a hole is made in the center, the yeast is placed inside, and then covered (*see* **Note 12**), Next, salt and/or mushroom stem flours are incorporated, followed by water at a temperature between 25 °C and 30 °C (*see* **Note 2**). Once all ingredients are included, they are mixed until a homogeneous dough is obtained; this process takes no more than 2 minu. Once homogenized, the kneading procedure begins, with its duration varying depending on the level of salt substitution and the nature of the ingredients used for this purpose (*see* **Note 13**). Kneading is carried out for 10 min for both the control formulation and those with 50% salt substitution, as well as for prototypes with 100% salt replacement using only ABSF. However, when POSF is incorporated in bread without salt, kneading is limited to 6 min (*see* **Note 14**). After kneading, the dough is shaped into a ball and fermented at 25–30 °C. Fermentation is carried out in two stages, with duration depending on the salt percentage and the specific characteristics of the flours used (*see* **Note 15**). The first and second fermentations for the control breads and those with 50% salt substitution require 15 min for each phase. In contrast, bread with 100% salt substitution using only ABSF requires just 10 min for each stage. More notably, when POSF is used in bread without salt, the first fermentation lasts only 5 min, while the second takes 10 min (*see* **Note 16**). Between the two fermentation stages, the dough is stretched to break the air bubbles formed by CO_2 production and then reshaped into a ball. After the incubation process, the dough is divided and frozen (*see* **Note 10**). Within a maximum of 10 days, the frozen dough is placed directly into a preheated oven at 185 °C and baked for 40 minutes.

3.2 Mortadella Production

3.2.1 Reduced-Sodium Mortadella Formulation

The formulation of mortadella, particularly concerning the meat and fat proportions, is influenced by various technological, regulatory, and sensory factors. Generally, fat content is maintained within a range of 20–30%; however, studies have shown even higher consumer acceptability with a meat/fat ratio of 90/10 [46]. The incorporation of pork dewlap in the formulation described in this chapter represents a balance between the 90/10 meat/fat ratio proposed by Guerra et al. and the typical fat range of 20–30%. Only the meat percentage adds up to 100%, while the percentage of the other ingredients is related to the meat. The fraction of the stem flour added was calculated based on the substitution of both salt and potato starch, and in both cases, it was within the ranges previously reported in the literature (Table 3).

3.2.2 Reduced-Sodium Mortadella Processed

The flowchart presented in Fig. 3 indicates step-by-step instructions to produce mortadella reduced in sodium. The first step is to chop and grind the meat in the cutter to obtain a size of 5 mm. Once the meat has been gridded, the fat is added to the cutter and

Table 3
Formulations of mortadella prototypes with reduced-sodium content by the addition of *Agaricus bisporus* stem flour (ABSF) and *Pleurotus ostreatus* stem flour (POSF)

Ingredient g/100 g bread	Control	$MAB_{2.75}$	$MPO_{2.75}$	$MAB_{4.25}$	$MPO_{4.25}$
Lean meat	65	65	65	65	65
Pork dewlap	25	25	25	25	25
Pork backfat	10	10	10	10	10
Salt	2.5	1.25	1.25	1.25	1.25
Ice	12.5	12.5	12.5	12.5	12.5
Potato starch	3	1.5	1.5	0	0
Pepper	0.1	0.1	0.1	0.1	0.1
Nutmeg	0.05	0.05	0.05	0.05	0.05
Garlic powder	0.2	0.2	0.2	0.2	0.2
Ascorbate	0.05	0.05	0.05	0.05	0.05
Tripolyphosphate	0.03	0.015	0.015	0.015	0.015
Nitrite	0.015	0.015	0.015	0.015	0.015
ABSF	0	2.75	0	4.2.5	0
POSF	0	0	2.75	0	4.25

homogenized. After that, the salt and half of the total amount of ice are incorporated and mixed until a homogeneous dough is obtained. This homogenization stage in the cutter takes a few minutes to be accomplished. Then, the remaining ice, seasoning, and additives are included in the paste. Again, the dough is mixed in the cutter for a few minutes to obtain a homogeneous aspect. Finally, starch and/or its substitutes (ABSF and POSF) are integrated (*see* **Note 17**), and the mixture is stirred again until a uniform mass is achieved, ensuring that the temperature remains lower than 12 °C to prevent emulsion breakdown (*see* **Note 18**). After that, the mortadella dough is stuffed into artificial casings and clipped to separate it into pieces up to 10 cm in length (*see* **Note 19**). For thermal processing, the mortadella is placed in the cooking kettle and kept submerged in water at 80 °C until it reaches an internal temperature of 72 °C. As soon as the thermal treatment is completed, the mortadella is cooled in ice until it reaches 15 °C and then stored at a refrigeration temperature of 0–4 °C.

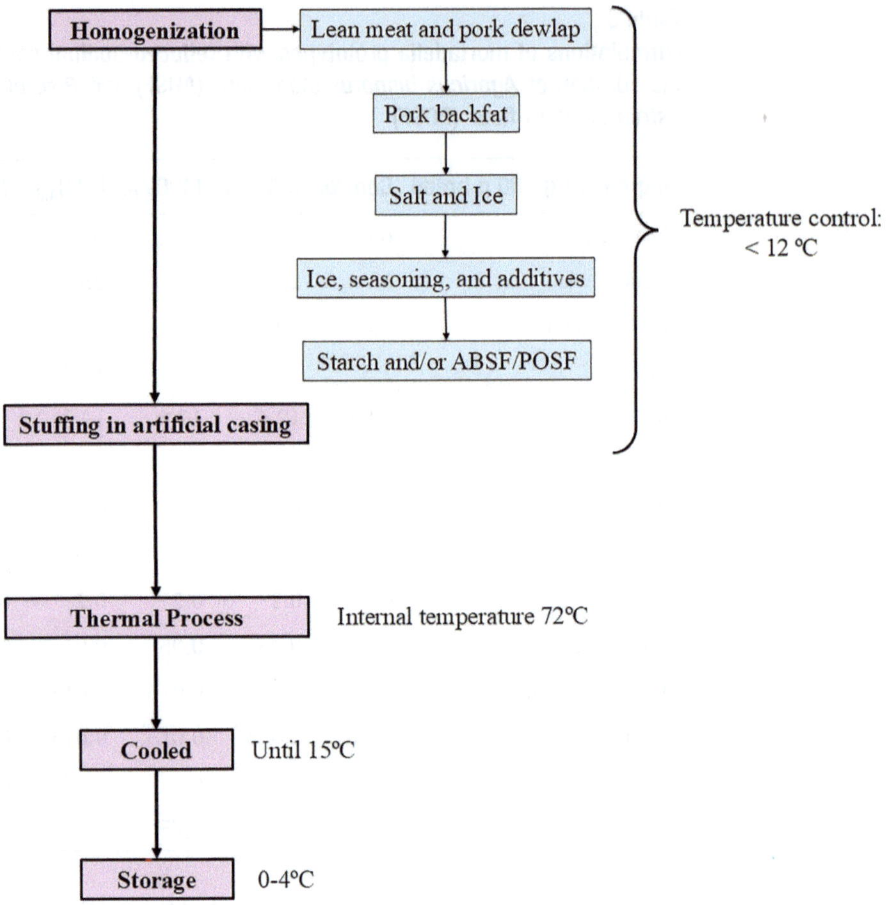

Fig. 3 Sodium-reduced mortadella production flowchart

4 Notes

1. Wheat flour is the most commonly used in bread-making due to its high gluten content, which provides the necessary elasticity and structure [47]. However, alternative flours such as rice, corn, and oats can also be used [48, 49]. Oat flour is increasingly used in bread-making due to its nutritional benefits, such as high fiber and beta-glucan content [49]. Although rice and corn can alter dough viscoelasticity, resulting in distinct textures, these flours enable the production of gluten-free bread suitable for individuals with celiac disease [48].
2. Water is essential for flour hydration and gluten activation. It is recommended to add water at the optimal temperature for yeast activation (between 25 °C and 30 °C) to prevent delays in its activity or potential inactivation if the temperature exceeds 35 °C [50].

3. The microorganism *Saccharomyces cerevisiae* is used in bread-making. Yeast can be added either fresh or as a dry powder and is responsible for fermentation and the development of the bread's structure.
4. It is recommended to store the stems at a temperature between 0 °C and 4 °C for no more than 5 days before processing to prevent degradation, the growth of undesirable microorganisms, or mycelium development. This is particularly important for *Agaricus bisporus* stems, which can form a compact, difficult-to-handle mass with the mycelium.
5. Added as bacon or animal fat to enhance juiciness, texture, and overall palatability.
6. Starch is added to meat products as it enhances emulsion quality by increasing viscosity due to its swelling capacity when heated. It has numerous functional applications, including adhesion, texture improvement, viscosity enhancement, water retention, and emulsion stabilization, among others [51]. The incorporated stem flours, used as salt substitutes, exhibit techno-functional properties similar to those of starch, mainly due to their high fiber content. Therefore, they could also be included in formulations to reduce potato starch addition [15].
7. The flour production process has been previously described and involves an initial washing step for *Agaricus bisporus* stems due to the presence of adhered peat, followed by drying in dehydrators at 50 °C for a variable duration depending on the species [52].
8. It is essential to consider their particle size when incorporating ingredients to enrich, fortify, or replace another component in formulations. Particle size is critical due to its impact on the techno-functional properties of the ingredient. According to a previous study, a particle size of <0.18 mm for both POSF and ABSF appears to be suitable to produce foods requiring hydration properties and, as a result, for the development of functional foods with low sodium content [15]. Therefore, this particle size will be the recommended standard for the flours derived from the by-products of both species in the formulation of reformulated foods incorporating this ingredient.
9. A dough mixer can be used instead, followed by a stainless-steel bowl for fermentation. The specific equipment used in the preparation described below was an 850 W automatic bread maker, which allows for adjustments in kneading and resting times.
10. It can be baked directly in the automatic bread maker; however, for logistical reasons, once the dough was prepared, it was

frozen and baked within 10 days to perform the necessary analytical determinations.

11. The incorporation of mushroom powder into bakery product formulations as a replacement for wheat flour has been extensively studied [20]. The findings of these studies revealed an improvement in the nutritional profile of various bakery products, with higher levels of vitamins, minerals, polyphenols, fiber, and proteins. In this context, increasing the concentration of ABSF and POSF at the expense of wheat flour was considered a promising alternative for upcycling mushroom stems.
12. To prevent direct contact with salt and thereby avoid potential inactivation or slowdown of its activity [26].
13. The kneading time is determined by the time required for the dough to achieve its optimal viscoelastic properties, ensuring effective gas retention essential for bread-making [53]. This occurs when a stable gluten network is formed. In the presence of salt, gluten hydration is delayed due to the interaction between the charged amino acids of gluten and salt, thereby increasing the kneading time [26]. In the absence of salt and with the incorporation of flours rich in insoluble fiber, which competes with gluten for water absorption, the kneading time is reduced [15, 54].
14. The difference in kneading time between ABSF and POSF is attributed to their compositional differences, as POSF has a higher insoluble fiber content [15]. Several tests were conducted, and longer kneading times resulted in dough stickiness with lower viscoelastic properties, leading to operational challenges.
15. The irregular gluten network caused by the high insoluble fiber content in the flours, particularly in POSF, hinders air incorporation during mixing and further retention and uniform distribution of air bubbles [54]. This, combined with the absence of salt—an essential ingredient for controlling yeast activity—explains the different fermentation times selected for each mushroom species and level of salt. Salt inhibits yeast activity, thereby reducing the formation of gas bubbles (CO_2), which are responsible for weakening the gluten network in bread with 100% salt replacement [26].
16. Fermentation times exceeding the 15 minutes specified in this chapter result in weak, sticky, and difficult-to-handle doughs.
17. Although mushroom stem flours, as well as fruiting bodies, have been linked to flavor-enhancing compounds, such as amino acids related to umami taste, they are also added in these prototypes to reduce starch. ABSF and POSF have

suitable water absorption, emulsifying, and gelling capacities to substitute starch in the formulations, especially ABSF [15].

18. During the cutting process, the temperature should not exceed 12 °C to prevent the separation of fat and water, which would result in an unstable emulsion and alterations in the texture of the mortadella. To avoid this, it is essential to add the meat at 4 °C, incorporate ice to dissipate the heat generated, and carefully control both the cutting speed and processing time [55].
19. If analyses such as emulsion stability, pH, or emulsion color are required, it is recommended to extract a sample of the dough before the mortadella is stuffed.

The production of bread and mortadella prototypes described in the Methods section was funded by Grants for Industrial Doctorates from the Autonomous Community of La Rioja, Spain.

References

1. Raman J, Jang KY, Oh YL, Oh M, Im JH, Lakshmanan H et al (2021) Cultivation and nutritional value of prominent *Pleurotus* Spp.: an overview. Mycobiology 49:1–14
2. Guillamón E, García-Lafuente A, Lozano M, Ďarrigo M, Rostagno MA, Villares A et al (2010) Edible mushrooms: role in the prevention of cardiovascular diseases. Fitoterapia 81: 715–723
3. Kakon AJ, Bazlul M, Choudhury K, Saha S (2012) Mushroom is an ideal food supplement. J Dhaka Natl Med Coll Hos:18
4. Kousar A, Khan HA, Farid S, Zhao Q, Zeb I (2024) Recent advances on environmentally sustainable valorization of spent mushroom substrate: a review. Biofuels Bioprod Biorefin 18:639–651
5. Guo J, Zhang M, Fang Z (2022) Valorization of mushroom by-products: a review. J Sci Food Agric 102(13):5593–5605
6. Cağlarirmak N (2011) Edible mushrooms: an alternative food item. In: Proceedings of the 7th international conference on mushroom biology and mushroom products (ICMBMP7), Turkey, pp 548–554
7. Ramos M, Burgos N, Barnard A, Evans G, Preece J, Graz M et al (2019) *Agaricus bisporus* and its by-products as a source of valuable extracts and bioactive compounds. Food Chem 292:176–187
8. Grimm D, Wösten HAB (2018) Mushroom cultivation in the circular economy. Appl Microbiol Biotechnol 102:7795–7803
9. Silva M, Ramos AC, Lidon FJ, Reboredo FH, Gonçalves EM (2024) Pre- and postharvest strategies for *Pleurotus ostreatus* mushroom in a circular economy approach. Foods 13
10. Papoutsis K, Grasso S, Menon A, Brunton NP, Lyng JG, Jacquier JC et al (2020) Recovery of ergosterol and vitamin D2 from mushroom waste—potential valorization by food and pharmaceutical industries. Trends Food Sci Technol 99:351–366
11. Navarro-Simarro P, Gómez-Gómez L, Ahrazem O, Rubio-Moraga Á (2024) Food and human health applications of edible mushroom by-products. New Biotechnol 81:43–56. https://doi.org/10.1016/j.nbt.2024.03.003
12. Valchev N (2020) Nutritional and amino acid content of stem and cap of *Agaricus bisporus*, Bulgaria. Bulg J Agric Sci 26:192–201
13. Serafín Muñoz AH, Corona FG, Wrobel K, Soto GM, Wrobel K (2005) Subcellular distribution of aluminum, bismuth, cadmium, chromium, copper, iron, manganese, nickel, and lead in cultivated mushrooms (*Agaricus bisporus* and *Pleurotus ostreatus*). Biol Trace Elem Res 106(3):265–277
14. Zawadzka A, Janczewska A, Kobus-Cisowska J, Dziedziński M, Siwulski M, Czarniecka-Skubina E et al (2022) The effect of light conditions on the content of selected active ingredients in anatomical parts of the oyster mushroom (*Pleurotus ostreatus* L.). PLoS One 17:e0262279
15. Bermúdez-Gómez P, Fernández-López J, Pérez-Clavijo M, Viuda-Martos M (2024)

Evaluation of sample size influence on chemical characterization and in vitro antioxidant properties of flours obtained from mushroom stems coproducts. Antioxidants 13(3):349. https://doi.org/10.3390/antiox13030349

16. Wang R, Yun J, Wu S, Bi Y, Zhao F (2022) Optimisation and characterisation of novel angiotensin-converting enzyme inhibitory peptides prepared by double enzymatic hydrolysis from *Agaricus bisporus* scraps. Foods 11(3):394–417
17. Wu QC, Zhang YY, Li YB, Alitongbieke G, Xue Y, Li XM et al (2023) A novel cell-wall polysaccharide derived from the stipe of *Agaricus bisporus* inhibits mouse melanoma proliferation and metastasis. Arch Biochem Biophys 744:109678–109693
18. Sirohi R, Negi T, Rawat N, Sagar NA, Sindhu R, Tarafdar A (2023) Emerging technologies for the extraction of bioactives from mushroom waste. J Food Sci Technol
19. Wang L, Brennan MA, Guan W, Liu J, Zhao H, Brennan CS (2021) Edible mushrooms dietary fibre and antioxidants: effects on glycaemic load manipulation and their correlations pre- and post-simulated *in vitro* digestion. Food Chem 351:129320–129330
20. Salehi F (2019) Characterization of different mushrooms powder and its application in bakery products: a review. Int J Food Prop 22(1): 1375–1385
21. Vargas-Meza J, Gonzalez-Rocha A, Campos-Nonato I, Nilson EAF, Basto-Abreu A, Barquera S et al (2023) Effective and scalable interventions to reduce sodium intake: a systematic review and meta-analysis. Curr Nutr Rep 12:486–494
22. Toft U, Riis NL, Lassen AD, Trolle E, Andreasen AH, Frederiksen AKS et al (2020) The effects of two intervention strategies to reduce the intake of salt and the sodium-to-potassium ratio on cardiovascular risk factors. A 4-month randomised controlled study among healthy families. Nutrients 12(5)
23. Riis NL, Lassen AD, Bjoernsbo K, Toft U, Trolle E (2022) Dietary effects of introducing salt-reduced bread with and without dietary counselling—a cluster randomized controlled trial. Nutrients 14(18)
24. Riis NL, Bjoernsbo KS, Toft U, Trolle E, Hyldig G, Hartley IE et al (2021) Impact of salt reduction interventions on salt taste sensitivity and liking: a cluster randomized controlled trial. Food Qual Prefer 87
25. Dunteman A, Yang Y, McKenzie E, Lee Y, Lee SY (2021) Sodium reduction technologies applied to bread products and their impact on sensory properties: a review. Int J Food Sci Technol 56:4396–4407
26. Silow C, Axel C, Zannini E, Arendt EK (2016) Current status of salt reduction in bread and bakery products—a review. J Cereal Sci 72: 135–145
27. Toldrá F, Barat JM (2015) Low-salt foods: types and manufacture. In: Encyclopedia of food and health. Elsevier Inc., pp 576–578
28. Wang J, Huang XH, Zhang YY, Li S, Dong X, Qin L (2023) Effect of sodium salt on meat products and reduction sodium strategies—a review. Meat Sci 205:109296
29. Monteiro MLG, Mársico ET, Conte-Junior CA (2024) Reduced-sodium meat products, pp 137–149
30. Pashaei M, Zare L, Khalili Sadrabad E, Hosseini Sharif Abad A, Mollakhalili-Meybodi N, Abedi AS (2022) The impacts of salt reduction strategies on technological characteristics of wheat bread: a review. J Food Sci Technol 59: 4141–4151
31. Ma F, Li Y, Zhang Y, Zhang Q, Li X, Cao Q et al (2024) Effects of umami substances as taste enhancers on salt reduction in meat products: a review. Food Res Int 185:114248
32. Sun L, Zhang Z, Xin G, Sun B, Bao X, Wei Y et al (2020) Advances in umami taste and aroma of edible mushrooms. Trends Food Sci Technol 96:176–187
33. Dos Santos FF, Dantas NM, Simoni NK, Pontes LS, Pinto-E-silva MEM (2020) Are foods naturally rich in glutamic acid an alternative to sodium reduction? Food Sci Technol (Brazil) 40:190–196
34. Poojary MM, Orlien V, Passamonti P, Olsen K (2017) Improved extraction methods for simultaneous recovery of umami compounds from six different mushrooms. J Food Compos Anal 63:171–183
35. dos Harada-Padermo SS, Dias-Faceto LS, Selani MM, Conti-Silva AC, S d VTMF (2021) Umami ingredient from shiitake byproducts in low-sodium products: a case study in corn extruded snacks. LWT 138:110806
36. Jo K, Lee J, Jung S (2018) Quality characteristics of low-salt chicken sausage supplemented with winter mushroom powder. Korean J Food Sci Anim Resour 38(4):768–779
37. França F, dos Harada-Padermo SS, Frasceto RA, Saldaña E, Lorenzo JM, S d VTMF et al (2022) Umami ingredient from shiitake (*Lentinula edodes*) by-products as a flavor enhancer in low-salt beef burgers: effects on

physicochemical and technological properties. LWT 154:112724

38. Lopez J, Kerley T, Jenkinson L, Luckett CR, Munafo JP (2019) Odorants from the thermal treatment of hydrolyzed mushroom protein and cysteine enhance saltiness perception. J Agric Food Chem
39. Mattar TV, Gonçalves CS, Pereira RC, Faria MA, de Souza VR, de Carneiro JDS (2018) A shiitake mushroom extract as a viable alternative to NaCl for a reduction in sodium in beef burgers: a sensory perspective. Br Food J 120(6):1366–1380
40. Wong KM, Decker EA, Autio WR, Toong K, DiStefano G, Kinchla AJ (2017) Utilizing mushrooms to reduce overall sodium in taco filling using physical and sensory evaluation. J Food Sci 82(10):2379–2386
41. Botella-Martínez C, Muñoz-Tebar N, Lucas-González R, Pérez-Álvarez JA, Fernández-López J, Viuda-Martos M (2023) Assessment of chemical, physico-chemical and sensory properties of low-sodium beef burgers formulated with flours from different mushroom types. Foods 12(19)
42. Cerón-Guevara MI, Santos EM, Lorenzo JM, Pateiro M, Bermúdez-Piedra R, Rodríguez JA et al (2021) Partial replacement of fat and salt in liver pâté by addition of *Agaricus bisporus* and *Pleurotus ostreatus* flour. Int J Food Sci Technol 56(12):6171–6181
43. Cerón-Guevara MI, Rangel-Vargas E, Lorenzo JM, Bermúdez R, Pateiro M, Rodríguez JA et al (2020) Reduction of salt and fat in frankfurter sausages by addition of *Agaricus bisporus* and *Pleurotus ostreatus* flour. Foods 9(6)
44. Patinho I, Saldaña E, Selani MM, de Camargo AC, Merlo TC, Menegali BS et al (2019) Use of Agaricus bisporus mushroom in beef burgers: antioxidant, flavor enhancer and fat replacing potential. Food Prod Process Nutr 1(1)
45. Cerón-Guevara MI, Rangel-Vargas E, Lorenzo JM, Bermúdez R, Pateiro M, Rodriguez JA et al (2020) Effect of the addition of edible mushroom flours (*Agaricus bisporus* and *Pleurotus ostreatus*) on physicochemical and sensory properties of cold-stored beef patties. J Food Process Preserv 44(3)
46. Guerra ICD, Félex SSS, Meireles BRLM, Dalmás PS, Moreira RT, Honório VG et al (2011) Evaluation of goat mortadella prepared with different levels of fat and goat meat from discarded animals. Small Rumin Res 98(1–3): 59–63
47. Goesaert H, Brijs K, Veraverbeke WS, Courtin CM, Gebruers K, Delcour JA (2005) Wheat flour constituents: how they impact bread quality, and how to impact their functionality. Trends Food Sci Technol 16:12–30
48. Sabanis D, Tzia C (2009) Effect of rice, corn and soy flour addition on characteristics of bread produced from different wheat cultivars. Food Bioprocess Technol 2(1):68–79
49. Majzoobi M, Raiss Jalali A, Farahnaky A (2016) Impact of whole oat flour on dough properties and quality of fresh and stored part-baked bread. J Food Qual 39(6):620–626
50. Struyf N, Van der Maelen E, Hemdane S, Verspreet J, Verstrepen KJ, Courtin CM (2017) Bread dough and baker's yeast: an uplifting synergy. Compr Rev Food Sci Food Saf 16(5):850–867
51. Gençcelep H (2012) The effect of using dried mushroom (*Agaricus bisporus*) on lipid oxidation and color properties of sucuk. J Food Biochem 36(5):587–594
52. Bermúdez-Gómez P, Fernández-López J, Pérez-Clavijo M, Viuda-Martos M (2024) Optimization of mushroom (*Agaricus bisporus* and *Pleurotus ostreatus*) by-products processing for prospective functional flour development. Foods 13(24)
53. Lu X, Brennan MA, Serventi L, Brennan CS (2018) Incorporation of mushroom powder into bread dough—effects on dough rheology and bread properties. Cereal Chem 95(3): 418–427
54. Djordjević M, Djordjević M, Šoronja-Simović D, Nikolić I, Šereš Z (2022) Delving into the role of dietary fiber in gluten-free bread formulations: integrating fundamental rheological, technological, sensory, and nutritional aspects. Polysaccharides 3:59–82
55. Cenci DF, Kilian J, Janeczko MU, Manzoli A, Rigo E, Soares MBA (2018) Effect of meat and water temperature and emulsion speed on the industrial process for chicken mortadella. J Food Process Eng 41(8)

Chapter 16

Edible Mushroom-Based Films and Coatings: Properties, Development, and Applications

Aijaz Hussain Soomro, Asadullah Marri, and Nida Shaikh

Abstract

Conventional food packaging materials, such as polyethylene, polystyrene, metal foils, and glass, have long been used to preserve food quality, ensure safety, extend shelf life, and reduce food loss. Despite their effectiveness, growing consumer concerns over the negative environmental and health impacts such as microplastic accumulation in food, metal erosion, and the impermeability of glass have led to an increasing demand for environmentally friendly packaging solutions. Biodegradable packaging, particularly edible food films and coatings (EFFC), is emerging as a sustainable alternative, driven by the need for safer, cost-effective, and eco-friendly options. The scientific literature has explored various agro-materials for developing EFFC, but research on edible and medicinal mushrooms as a source of EFFC remains scarce. These mushrooms, rich in bioactive compounds such as polysaccharides (β-glucan, hemicellulose, chitin, and mannans), proteins, antimicrobial agents, antioxidants, and nutraceutical components, offer a promising alternative for developing innovative and functional packaging materials. Additionally, residues and by-products from mushroom cultivation can be repurposed to produce EFFC. This chapter will explore the potential of edible mushrooms and their by-products in developing EFFC, with a focus on their application in preserving perishable food commodities. This will also highlight the advantages of mushroom-based EFFC in extending shelf life, reducing food waste, and addressing environmental concerns, making them an ideal candidate for sustainable food packaging solutions.

Key words Edible mushrooms, Food films, Coatings

1 Food Packaging-Edible Food Films and Coatings

The emergence of food packaging brought a revolution in food technology. In the beginning, humans relied on natural materials for food packages (i.e., leaves, animal skins, shells) for food storage and transport [1]. In the twentieth century, the plastic revolution embarked on the use of polyethylene and polystyrene as food packages and transformed the entire scenario of food packaging technology. Plastic-based synthetic food packages were nonrenewable materials and had multiple pros (low-weight, low-cost, durable, waterproof, and transparent) over natural food packages. However, they also created a variety of deleterious impacts on

Pablo Navarro Simarro (ed.), *Fungi*, Methods and Protocols in Food Science, https://doi.org/10.1007/978-1-0716-4989-3_16,
© The Author(s), under exclusive license to Springer Science+Business Media, LLC, part of Springer Nature 2026

Table 1
Differentiation among various classes of food packaging materials

	Primary packaging	Secondary packaging	Tertiary packaging
Definition	Packaging that is directly in contact/adhered at food	It covers or packages food that is already packed with primary packaging materials	The packaging is used for bulk handling, storage, distribution, and transportation of food commodities
Examples	Tins, cans, bottles, jar, wrappers, films, coatings, etc.	Boxes, plastic shrink wraps, trays, etc.	Wooden or plastic/fiber crates, cardboard boxes, etc.
Purposes	Protects food from contamination, delays fruit ripening, and enhances shelf life	Facilitates food handling, loading, storage, and food display in retail market	Enables protection of food from damage during shipment or transport and facilitates convenience during loading and unloading of food

environment and public health. Alongside, the poor biodegradability of plastic packages lifted the challenge of proper waste management and contributed to global climate change. These crises shaped the need to develop some suitable and sustainable alternatives against plastic-based food packages.

Food packaging continues to become a crucial segment within the food supply chain [2]. A sustainable food package must be biodegradable, recyclable, compostable, cost-effective, and durable. In food technology, packaging performs critical functions, such as serving as a container for food; preventing food contamination and spoilage; protecting food from physical damage during handling, storage, transportation, or distribution; and delivering vital information regarding packaged food through food labels [3]. Based on the specific role of food packaging materials in the field of food technology, they are classified as primary, secondary, and tertiary packaging materials (Table 1).

Food packaging materials are mainly obtained from various natural or synthetic sources. Both natural and synthetic food packaging materials differ in several aspects to each other such as origin, environmental effect, degree of resilience against food perishability, durability, and disposing system. Some common food packaging materials and their sources are discussed below:

1. **Plant leaves:** The leaves from plants (mainly from edible plants) are used as food packages with or without processing.
2. **Plastic:** It is derived from petrochemicals such as natural gas or crude oil.
3. **Glass:** It is composed of natural raw materials, namely, silica, soda ash, and limestone.

4. **Cardboard/paper:** Wood pulp and some plant fibers are used to develop cardboard and paper.
5. **Metal:** Some natural mineral ores are used to synthesize metal packages.
6. **Polystyrene foam:** It is a petroleum-based plastic material.
7. **Biodegradable packages:** They are compostable, renewable plant-based substances immensely being utilized in modern era packages.
8. **Edible packages:** These packages are food-grade materials composed of edible food components.

Basically, all types of food packages can be categorized as biodegradable (e.g., plant leaves, cardboard/paper, plant-based plastic, edible packages) or nonbiodegradable (plastic, polystyrene foam, metal, etc.) materials. Biodegradable packages are eco-friendly, and renewable materials naturally decomposed by microbial activity without leaving any residual harm in the environment [2]. In contrast, nonbiodegradable packages are not decomposable and environmentally persistent for longer periods depending upon their type, composition, and processing.

1.1 Edible Films and Coatings

The food films and coatings are types of primary packaging materials. They are flexible, thin-layered packages responsible for providing strong barrier functions against food quality deteriorating agents (i.e., moisture, oxygen, contaminants, and light). Edible food films and coatings (EFFCs) are also associated with performing similar functions; however, they have components that have been defined by the Food and Drug Administration as GRAS (generally recognized as safe for human consumption) [4]. Over time, food films and coatings significantly evolved their forms with progressions in materials science and packaging requirements. Nowadays, food scientists are extensively experimenting with natural biomaterials for the development of food films and coatings due to their sustainability, environment-friendly properties, and frequent degradability [5]. EFFCs are eco-friendly food packaging alternate to conventional synthetic packaging materials. They are significantly involved in protecting food from microbial contamination, preserving food from quality destruction, and significantly enhancing food shelf life.

In most instances, terms "film" and "coating" are used indistinguishably and meant for covering food surfaces from thin layers formulated with certain natural materials. It is to be considered that films and coatings are variable from each other (Table 2) in several prospects such as in composition, method of development, mode of application, and performance. Any EFFCs are comprised of three main components: (i) film-forming material, (ii) solvent, and (iii) additive (Table 3). EFFCs can be mainly formulated from the

Table 2
Difference between edible food films and coatings

	Films	Coatings
Definition	Thin solid layer preformed from film forming material that can be wrapped around food surfaces	Thin liquid layer of solution prepared from different solid, semi-solid, or liquid constituents that can be directly coated on the food surfaces
Form	Solid, self-supporting membranes or sheets that are not strongly adhered to the surface of food	Liquid solution that is directly adhered to the surface of food
Mode of application	Wrapped or covered around food surfaces	Brushed, sprayed over the surfaces while foods may also be immersed in the coating solution, removed after certain period, and dried
Composition	Composed of synthetic or natural or edible materials	Usually composed of natural or edible materials
Removability	Removed/unwrapped prior to consumption	Removed by plain water wash or consumed along with food

Table 3
Major components of edible food film and coatings and their roles, examples, and sources

Components	Purpose	Examples	Sources
Film-forming material	Forms a thin, continuous, and uniform layer Moisture and gas barrier Carrier of functional additives	Starch, chitin, beta-glucan, chitosan, cellulose, pectin, gelatin, whey protein, soy protein, waxes, and fatty acids, etc.	Corn, rice, potato, sweet potato, tapioca, edible mushrooms, crustacean shells, fruit peels, animal connective tissues, seaweed, soybean, animal tissues etc.
Solvent	Serves as a medium to thoroughly dissolve film-forming material and functional additives Evaporates during film or coating drying or dehydration Allows solidification of film-forming material in thin continuous layer	Water, alcohol, organic solvents acidic or basic solutions, etc.	Deionized or distilled water, natural or synthetic solvents, etc.
Functional additive	Adds antimicrobial and antioxidant activity in film and coatings Decreases film brittleness and enhances film flexibility	Essential oils, vitamins, emulsifiers, glycerol, minerals, proteins, nanoparticles, etc.	Plant materials, animal tissues, fresh edible mushrooms and their industrial residues, etc.

components extracted or taken from edible biological sources; therefore, they are safer and may be consumed along filmed or coated foods [6]. According to Thakur et al. [7], underutilized animal by-products (bones, skin, feathers, hairs, horns, fish scales, fins, whey proteins, etc.) from meat- and milk-processing industries are significant sources of biopolymers (chitosan, keratin, gelatin, etc.) that can serve as valuable food film and coating materials. Plant-based compounds (polysaccharides, proteins, waxes, fats, and oils) are used for formulating EFFCs to acquire desirable food-preserving properties [8]. Edible mushroom-sourced compounds are also gaining popularity in the field of food film and coating development [9].

1.2 Current Trends in Edible Food Films and Coatings

1.2.1 Agro-Based Edible Food Films and Coatings

Agro-based EFFCs are developed from the components derived from agricultural produce (i.e., polysaccharides, proteins, fats and oils, waxes, gums, resins). By-products or wastes generated from food industries or agro-industries can be used for developing EFFCs. Manufacturing EFFCs from food industry by-products or waste not only limits greenhouse gas emissions but also reduces the cost of manufacturing synthetic food packages and tends to develop innovative food packaging materials [10].

1.2.2 Nanoparticle-Based Edible Food Films and Coatings

Nanoparticle-based EFFCs are developed from ultra-small-sized naturally occurring particles (organic and inorganic nanoparticles). The nanoparticles containing EFFCs add more functional attributes to conventional EFFCs by improving food quality, safety, and shelf life [11].

1.2.3 Mushroom-Based Edible Food Films and Coatings

Mushroom-based EFFCs are rapidly gaining traction in the food packaging sector, thanks to their outstanding functional properties. These innovative, eco-friendly, and bioactive packaging materials are derived from edible mushrooms and their industrial residues or by-products [9]. Mushroom-derived valuable compounds are becoming popular because of their abundant availability in nature, biodegradability, and immense functional attributes.

2 Materials

The promising role of edible mushrooms and their industrial residues in forming EFFCs is mainly due to their nutritional composition, presence of valuable bioactive compounds, and food film strengthening structural constituents. Several species of edible mushrooms and their industrial residues have been studied to evaluate their food-preserving capacities as EFFCs [12–17].

2.1 Film-Forming Materials

2.1.1 Agaricus bisporus

Agaricus bisporus or button mushroom remains the most extensively studied edible mushroom for the development of EFFCs, owing to its higher availability, simple processing, and effective performance. This mushroom is rich in active compounds that exert structural, functional, and food-preserving roles, making it a strong candidate for developing EFFCs. Raw A. *bisporus* as well as its residual remains after processing can be utilized for developing EFFCs. Chitin and β-glucan are chief polysaccharides in *A. bisporus* [12]. Chitin is a structural polysaccharide that serves stability and mechanical strength in food films. Water-holding capacity is also one main characteristic of β-glucan [18] that adds value in EFFCs. Chitosan is basically deacetylated chitin. Chitosan derived from *A. bisporus* possesses great ability to enhance the shelf life of perishable fruits and vegetables [17]. *A. bisporus* also contains a smaller proportion of hemicellulose and pectin-like compounds that are responsible for improving film binding property. Proteins that occur in *A. bisporus* are mainly water-soluble; however, tyrosinases and lectin proteins [19] are responsible for exerting tensile strength and enhancing mechanical strength in food films. A small concentration of unsaturated fatty acid is present in *A. bisporus* that serve food film hydrophobicity and limits the permeability of water vapors. Furthermore, dietary fiber in this mushroom reinforces mechanical durability in EFFCs [20].

2.1.2 Lentinula edodes

Lentinula edodes is commonly known as shiitake mushroom. Its exceptional flavor profile, nutritional composition, and numerous medicinal benefits make it one of the most widely consumed and studied edible mushrooms [21]. *L. edodes* is a rich source of dietary fiber, particularly β-glucan. A reasonable proportion of protein and unsaturated fatty acids is also found in *L. edodes.* This mushroom supplies many important polysaccharides that can be adequately used for developing antimicrobial and biodegradable food packages. Lentinan is the main bioactive polysaccharide in *L. edodes*; this is a glucan with potent anticancer properties [22]. This glucan is water-soluble and has high molecular weight. Lentinan has recently gained attention for its immune-boosting actions and other important bio-functionalities. It may be used as an exceptional film-forming agent and may form thinner, transparent, nontoxic, and biodegradable food films for direct food surface contact due to its strong cohesive properties [23]. Moreover, Zhang et al. [24] suggest that lentinan exhibits excellent barrier properties against water vapors and gases for maintaining food freshness and limiting oxidative spoilage of fruits, vegetables, meat, seafood, and bakery products. Some naturally occurring phenolic compounds (e.g., gallic and ferulic acid) in *L. edodes* are associated with antioxidant role in coated foods [25]. *L. edodes* can be a subject of interest in the field of EFFCs due to compounds such as ergosterol and its derivatives. Ergosterol is a vitamin D precursor that may provide protection against UV radiation and thus could be used to develop EFFCs that prevent light-dependent food spoilage [26]. Proteins in *L. edodes* tend to play an efficient role in developing EFFCs. *L. edodes* primarily contain proteins such as lectins, enzymes, glycoproteins, and fibrous proteins. Lectins contribute to antimicrobial activity, antioxidative properties, and mechanical strength. It also provides elasticity in the film by forming non-covalent bonds with polysaccharides. The enzymes in the mushroom also contribute to excellent forming properties in EFFCs. The laccases are copper-containing oxidases in *L. edodes* that facilitate the oxidative cross-linking of phenolic compounds to increase film moisture resistance and mechanical strength. Additionally, the bioactive nature of other proteins of *L. edodes* offers antimicrobial and antioxidant properties, which can extend the shelf life of packaged foods [27].

2.1.3 *Pleurotus ostreatus*

P. ostreatus itself and its industrial residues may be efficiently utilized for formulating environmentally friendly EFFCs against synthetic films and coatings [28, 29]. β-Glucans, chitin, chitosan, xylans, and glucomannans are among the chief polysaccharides in *P. ostreatus*. These polysaccharides serve as natural biopolymers that facilitate film-forming properties through strong matrix cross-linking and cohesiveness. Among the different polysaccharides, beta-glucans are present in major proportion in *P. ostreatus*. β-Glucan plays a crucial role in film cohesiveness, flexibility, and elasticity. Chitin and chitosan impart antimicrobial actions and barrier functions in EFFCs, protecting coated foods against microbial spoilage and oxidative damage [30]. In addition, these compounds further facilitate the formation of a rigid and stable matrix system in EFFCs, however, glucomannans help retain moisture in coated foods. An abundance of essential fatty acids is found in *P. ostreatus* [31], such as linoleic acid (50–60% of total fatty acids), linolenic acid (1–2% of total fatty acids), and oleic acid (15–25% of total fatty acids). These fatty acids aid hydrophobicity in EFFCs by lowering the permeability of water vapors [29]. Phospholipids and other fat soluble compounds in *P. ostreatus* may further contribute to plasticizing effects in EFFCs.

2.1.4 *Ganoderma lucidum*

Ganoderma lucidum (or Reishi mushroom) is an edible, wood-decaying mushroom rich in valuable edible food film-forming biopolymers and numerous health-promoting bioactive compounds that exhibit antitumor and immunomodulatory effects [32, 33]. β-Glucans and other heteropolysaccharides can facilitate flexibility, cohesiveness, and transparency in EFFCs. Moreover, these polysaccharides also contribute to antioxidative actions, antimicrobial activities, biodegradability, water resistance etc. in EFFCs. However, the bioactivity of these polysaccharides is often compromised due to limited optimized extraction methods. Therefore, it is necessary to explore sustainable, cost-effective, and simple methods for extracting polysaccharides from these mushrooms while preserving the bioactivity of the extracted biopolymers. The lipids in *G. lucidum* are also valuable for developing edible packages, moreover, Tangaraju et al. [34] stated that more than 20% lipids may be extracted from the mycelial biomass of G. *lucidum*. Essential fatty acids, sterols, phospholipids, and some other fat-soluble compounds in this mushroom perform an efficient role in water impermeability, mechanical strength, and emulsification. One limitation is the bitter taste of *G. lucidum*, primarily due to its triterpenoid content [35], which may hinder its use in the development of EFFCs.

G. *lucidum* industrial residues are comprised of mycelial biomass, remains from its fruiting bodies, and spores. These residual substances have a plentiful concentration of biopolymers, lipids, and several phenolic compounds. Chen et al. [36] suggest that formulating EFFCs from the residues of this mushroom is a suitable waste management option, which may provide nontoxic and biodegradable packaging solutions.

2.1.5 Other Mushroom Species

Alongside the abovementioned types of edible mushrooms, some other lesser-known species of mushrooms (e.g., *Auricularia auricula-judae*, *Flammulina velutipes*, *Volvariella volvacea*, *Grifola frondosa*, *Hypsizygus tessellatus*, *Tremella fuciformis*, *Hericium erinaceus*, and *Coprinus comatus*) may also serve as efficient food film- and coating-forming agents.

2.2 Other Food Film- or Coating-Forming Materials

2.2.1 Solvents

Water, alcohol, organic solvents, and acidic or basic solutions can serve as media to thoroughly dissolve film-forming materials and functional additives.

2.2.2 Functional Additives

Essential oils, vitamins, emulsifiers, glycerol, minerals, proteins, and nanoparticles add functionality (e.g., antimicrobial or antifungal activity, etc.) in EFFCs.

3 Methods

Edible mushrooms can serve as excellent choices for developing EFFCs due to their unique bioactive compounds [37]. Typically, the development of EFFCs from mushrooms involves isolation of film-formulating bioactive compounds, preparation of food film or coating solutions from isolated bioactive compounds, food film casting, drying of food film, and characterization of formulated films or coatings for their quality attributes, such as physical, chemical, and rheological attributes. Finally, EFFCs are applied on the surface of food products to assess their suitability and efficiency in enhancing the shelf life of filmed or coated food products.

3.1 Extraction of Mushroom's Bioactive Compounds

3.1.1 Hot Water Extraction

This extraction method is more suitable for extracting β-glucan from mushrooms [38]. Hot water facilitates the destruction of the tough mushroom cell walls that are composed of valuable bioactive compounds. The elevated temperature diffuses and hydrolyzes the water-soluble bioactive compounds. However, the kinetic movement (agitation) of hydrolyzed bioactive compounds contributes to the separation of extracted compounds. Hot water extraction is ineffective at extracting nonpolar compounds.

3.1.2 Solvent Extraction

Solvents, mainly ethanol and methanol, are used in this extraction method to especially isolate nonpolar and semipolar bioactive compounds, such as phenolics.

3.1.3 Ultrasonic Extraction

Ultrasonication is a widely studied and employed method for the extraction of bioactive compounds from edible mushrooms due to its efficient rate of extracting polysaccharides and phenolic compounds [39, 40]. This technology is based on the combination of both cavitation and thermal effects. During ultrasonication, ultrasonic waves are released, and their cavitation effect produces shock waves that enhance the extraction of bioactive compounds by the rupture of mushroom cell walls. The efficiency of this extraction may be further enhanced when assisted with some other extraction methods.

3.1.4 Supercritical Carbon Dioxide Extraction

Supercritical carbon dioxide extraction is an attractive, modern, green technology-based method for extracting bioactive compounds from edible mushrooms. This technology is economical, nontoxic, highly efficient, and eco-friendly, and it damages mushroom cells by physical means [41]. Supercritical carbon dioxide extraction does not require solvents for extracting compounds. However, pressure and temperature conditions adjust the density and other physicochemical properties of CO_2 [42, 43], which influence the yield and composition of the extracted compounds.

3.1.5 Microwave-Assisted Extraction

Microwaves are electromagnetic waves that can influence the extraction of bioactive compounds from biological samples [44]. It is a cost-effective, energy-efficient, and environmentally friendly extraction method that is applicable for isolating polysaccharides, antioxidants, and other phenolic compounds [45].

3.2 Preparation of Mushroom-Based Food Films or Coating Solutions

Film-forming materials are mixed with additives in a solvent and stirred vigorously to acquire uniform solubilization of all film-forming components. The resulting solution may be served as food coatings while for developing films the solution may be cast (wet method) or extruded (dry method). Pérez-Bassart et al. [13] developed biodegradable food-preserving films from mushroom waste biomass by casting film solution on Teflon plates. The casted films tend to dry under different drying temperatures. After drying, the thin-layered films or sheets may be used as primary food packages. Thin and continuous food films can be developed from extrusion technology in which the film-forming material is fed on the screw extruder to obtain food films.

3.3 Characterization of Mushroom-Based Edible Films and Coating Solutions

Prior to food films and coating applications on food products, it is important to characterize the newly developed mushroom-based food film or coating solutions for their physical properties (thickness, transparency, color, solubility, wettability, surface roughness), chemical properties (water activity, moisture content, chemical composition), mechanical properties (tensile strength, elongation, Young's modulus), thermal properties (thermal stability, glass transition temperature), rheological properties (viscosity, flow behavior index, thixotropy), and microscopy to estimate their overall performance efficiency.

3.4 Application of Mushroom-Based Edible Food Films and Coatings

3.4.1 Modes of Application of Edible Coating on Food Surfaces

Adhesion is accompanied by the connection of food coating solution on the surface area of the food products [46]. The mushroom-based food coatings can be applied on food surfaces by spraying, dipping, panning, brushing, and fluidized-bed methods [45]. The spraying method of coating is appropriate for large-scale preservation of food commodities and employs spraying devices to apply coating solutions. Dipping is a cost-efficient method of coating that involves submerging the food in the coating solution, followed by removal, dripping of excess solution, and drying. It is suitable for small-scale preservation of fruits, vegetables, and nuts. The panning method is done through tilted/rotating pans or cylinders (filled with coating solutions), in which the foods to be coated (mainly candies, nuts, other dried fruits) are loaded in rotating pans or cylinders. Food products may be coated through brushing, in which brushes are manually used to apply coating solutions on the food surfaces. The fluidized-bed method is another coating method that requires specialized equipment. In the fluidized-bed method, the foods to be coated are hung in an upward airflow and coated by spraying. It facilitates uniform and even coatings on irregular food surfaces [47].

3.4.2 Modes of Application of Edible Films on Food Surfaces

Food films developed from mushroom-based compounds can be applied to the surface of food products using different methods, such as wrapping, layering, heat sealing, vacuum filming, and preformed mold fitting. Wrapping facilitates thorough surface coverage to individual food products. It can be done manually or mechanically to film food products such as fruits, vegetables, bread, cakes, or processed meat products. Cling and shrink wrapping are two major types of film wrapping in the field of food packaging. The layering on the other side is one important food film application method, in which films or sheets are directly positioned on the top of food surfaces to attain food preservation. Heat sealing is a heat-dependent method in which film edges are adhered to the surfaces of foods (e.g., frozen or ready-to-eat foods) by heating. In vacuum filming, the food films are enveloped on the food surface, while air is vacuumed prior to sealing to ensure tight adhesion of the film to the food surface. Preformed mold fitting involves placing food into molded films, followed by sealing for a compact fit.

3.5 Determining the Role of Mushroom-Based Edible Films or Coatings in Food Preservation

1. Shelf life: The specimens of filmed or coated food products are stored under controlled atmospheric conditions for certain periods, alongside control (non-filmed, non-coated) food specimens. During intervals, the specimens are examined for quality characteristics (e.g., nutritional, sensorial, microbial analyses) to compare the performance of control and treated food products in enhancing shelf life.
2. Physical appearance: The physical appearance of the filmed or coated food products is compared with that of the control using some visual parameters (e.g., size, shape, surface shrink, surface gloss or transparency, etc.).
3. Sensory assessment: Trained experts evaluate the sensorial attributes, such as taste, texture, flavor, and aroma, of the filmed or coated food products compared with their control counterparts at different intervals during the storage period. Moreover, in advanced methods, the sensory assessment of the food specimen is performed based on scientific instruments (sensor-based electronic nose and tongue). Smart EFFCs should be odorless and tasteless.
4. Biodegradability is the most attractive attribute of food-grade films or coatings. The biodegradable package naturally breaks down into the environment without causing pollution. The rate of EFFC biodegradability can be determined by examining their degradation under natural environmental conditions [47].
5. Stability of nutritional and functional properties: An effective film or coating developed from mushroom-based residues also reduces the oxidation or rancidity of the lipid-containing foods. However, lipid oxidation in filmed or coated foods can

be examined through peroxide value or thiobarbituric acid reactive substance test.

6. Determining factors such as water activity, moisture retention, and weight loss of filmed or coated food products reveal a clear understanding about mushroom-based EFFCs. The water activity within the safe limit reduces microbial activity, retention of moisture for longer periods maintains food freshness, and a slow rate of food weight loss indicates that the filming or coating material contributes to maintaining the shelf life of filmed or coated food products.

References

1. Verma MK, Shakya S, Kumar P, Madhavi J, Murugaiyan J, Rao MVR (2021) Trends in packaging material for food products: historical background, current scenario, and future prospects. J Food Sci Technol 58:4069
2. Kumar L, Ramakanth D, Akhila K, Gaikwad KK (2022) Edible films and coatings for food packaging applications: a review. Environ Chem Lett 20:1–26
3. Sarkar S, Aparna K (2020). Food packaging and storage. Research trends in home science and extension AkiNik Pub, pp. 27–51
4. Ribeiro IS, Maciel GM, Bortolini DG, Fernandes IDAA, Maroldi WV, Pedro AC et al (2024) Sustainable innovations in edible films and coatings: an overview. Trends Food Sci Technol 143:104272
5. An Z, Yuan M, Xu X, Huang Z, Zhu L, Cai Z, Shen Y (2024) Active pullulan-based coatings incorporated with *Auricularia auricular* extracts for preserving potato fresh-cuts. Food Sci Biotechnol 33(5):1147–1161
6. de Oliveira Júnior SD, de Araújo JS, de Asevedo EA, de Medeiros FG, dos Santos VS, de Sousa Júnior FC et al (2021) Exploiting films based on pectin extracted from yellow mombin (*Spondias mombin* L.) peel for active food packaging. *Biomass Conv Bioref*:1–15
7. Thakur R, Santhosh R, Kumar Y, Suryavanshi VR, Singhi H, Madhubabu D et al (2023) Characteristics and application of animal byproduct-based films and coatings in the packaging of food products. Trends Food Sci Technol:104143
8. Gupta D, Lall A, Kumar S, Patil TD, Gaikwad KK (2024) Plant-based edible films and coatings for food-packaging applications: recent advances, applications, and trends. Sustain Food Technol 2(5):1428–1445
9. Alimi BA, Pathania S, Wilson J, Duffy B, Frias JMC (2023) Extraction, quantification, characterization, and application in food packaging of chitin and chitosan from mushrooms: a review. Int J Biol Macromol 237:124195
10. Choudhary U, Bhinchhar BK, Paswan VK, Kharkwal S, Yadav SP, Singh P (2021) Utilization of agro-industrial wastes as edible coating and films for food packaging materials. J Food Sci Technol 58(9):3344–3356
11. Ruiz-Velducea HA, Moreno-Vásquez MDJ, Guzmán H, Esquer J, Rodríguez-Félix F, Graciano-Verdugo AZ, Tapia-Hernández JA (2024) Valorization of agave Angustifolia bagasse biomass from the Bacanora industry in Sonora, Mexico as a biochar material: preparation, characterization, and potential application in ibuprofen removal. Sustain Chem 5(3): 196–214
12. Hernández JA (2024) The application of organic and inorganic nanoparticles incorporated in edible coatings and their effect on the physicochemical and microbiological properties of seafood. PRO 12(9):1889
13. Pérez-Bassart Z, Reyes A, Martínez-Abad A, López-Rubio A, Fabra MJ (2023) Feasibility of *Agaricus bisporus* waste biomass to develop biodegradable food packaging materials. Food Hydrocoll 142:108861
14. Liu T, Tang Q, Lei H, Zhen X, Zheng N, Qiu P, Zhao J (2024) Preparation, physicochemical and biological evaluation of chitosan *Pleurotus ostreatus* polysaccharides active films for food packaging. Int J Biol Macromol 254: 127470
15. Du H, Hu Q, Yang W, Pei F, Kimatu BM, Ma N et al (2016) Development, physiochemical characterization and forming mechanism of *Flammulina velutipes* polysaccharide-based edible films. Carbohydr Polym 152:214–221

16. Shao G, Xu D, Xu Z, Jin Y, Wu F, Yang N, Xu X (2024) Green and sustainable biomaterials: edible bioplastic films from mushroom mycelium. Food Hydrocoll 146:109289
17. Zhao A, Berglund L, Rosenstock Völtz L, Swamy R, Antonopoulou I, Xiong S et al (2024) Fungal innovation: harnessing mushrooms for production of sustainable functional materials. Adv Funct Mater 24:12753
18. Kumar S, Mishra A, Pandey S (2021) Development and characterization of chitosan-based edible coatings from *Agaricus bisporus* for food preservation. Food Packag Shelf Life 29: 100645
19. Berggren S (2018) Water holding capacity and viscosity of ingredients from oats: the effect of β-glucan and starch content, particle size, pH and temperature. Food Hydrocoll 82:161–167
20. Usman M, Murtaza G, Ditta A (2021) Nutritional, medicinal, and cosmetic value of bioactive compounds in button mushroom (*Agaricus bisporus*): a review. Appl Sci 11(13): 5943
21. Tirta Ismaya W, Tjandrawinata RR, Rachmawati H (2020) Lectins from the edible mushroom *Agaricus bisporus* and their therapeutic potentials. Molecules 25(10):2368
22. Abdelshafy AM, Luo Z, Belwal T, Ban Z, Li L (2023) A comprehensive review on preservation of shiitake mushroom (*Lentinus edodes*): techniques, research advances and influence on quality traits. Food Rev Int 39(5):2742–2775
23. Chakraborty S, Beura M, Sharma SK, Singh A, Dahuja A, Krishnan V (2023) Lentinan, β-glucan from shiitake (*Lentinula edodes*): a review on structure, conformational transition, and gastro-intestinal interaction contributing towards its anti-diabetic potential. Trends Food Sci Technol 142:104224
24. Wang H, Wang X, Zhao J (2021) Mechanical and barrier properties of lentinan-enhanced edible films. Carbohydr Polym 255:117498
25. Zhang R, Sun J, Huang Y (2019) Water vapor barrier properties of lentinan-based coatings for fresh produce. Postharvest Biol Technol 157:110950
26. Xu W, Zhang X, Zhang H (2020) Antimicrobial properties of lentinan incorporated into food packaging materials. Int J Biol Macromol 160:1096–1103
27. Liu Y, Wang Y, Feng X (2020) Lentinan-based edible coatings: antioxidant and antimicrobial effects on food preservation. J Food Sci 85(8): 2156–2164
28. Wu J, Ge S, Liu H, Wang Y (2020) Functional properties of edible films made from mushroom protein combined with polysaccharides. J Food Sci Technol 57(5):1578–1588
29. Corrêa RCG, Brugnari T, Bracht A, Peralta RM, Ferreira IC (2016) Biotechnological, nutritional and therapeutic uses of *Pleurotus* spp. (oyster mushroom) related with its chemical composition: a review on the past decade findings. Trends Food Sci Technol 50:103–117
30. Petrović J, Džinić N, Filipović J (2021) Potential use of mushroom-derived polysaccharides in edible films. Polysaccharide Innov 14(3): 223–230
31. Yang S, Liu Y, Chen W (2020) Antimicrobial and physical properties of fungal-residue-based edible films. J Biodegrad Mater 37(4):154–162
32. Karunarathna SC, Ediriweera A, Prasannath K, Mingfei Y, Hapuarachchi KK (2024) Exploring the health benefits of *Ganoderma*: bioactive compounds and mechanisms of action; immunomodulatory and anti-tumour activities. N Z J Bot:1–85
33. Chen H, Liu X, Zhang Z (2022) Development of edible films from *Ganoderma lucidum* polysaccharides and their food applications. Carbohydr Polym 286:119270
34. Tangaraju M, Ilham Z, Adlim M, Wan WAQI, Jamaludin AA (2024) Fueling the future with fungi: efficient lipid extraction from *Ganoderma* mycelium biomass. Prog Energy Environ 2:16–25
35. Chuensun T, Chewonarin T, Laopajon W, Samakradhamrongthai RS, Chaisan W, Utama-Ang N (2024) Evaluation of the phytochemical, bioactive compounds and descriptive sensory of encapsulated lingzhi (*Ganoderma lucidum*) extracts with combined wall materials for masking effect on the perception of off-flavour and bitterness. Heliyon 10(21): e26132
36. Chen Y, Zhang J, Liu X (2021) Application of *Ganoderma lucidum* waste residues in biodegradable film production. J Appl Polym Sci 138(22):50562
37. Li X, Luo L, Wang X, Zhu M (2025) Further insights into the molecular mechanisms underlying tobacco straw cultivation of *Pleurotus ostreatus* by comparative transcriptome analyses. Genomics:110992
38. Losoya-Sifuentes C, Cruz M, del Refugio R-PM, Loredo-Treviño A, Belmares R (2025) Edible mushrooms: a nutrient-rich ingredient for healthier food products–a review. Curr Nutr Rep 14(1):9
39. Saetang N, Rattanapot T, Manmai N, Amornlerdpison D, Ramaraj R, Unpaprom Y (2024) Effect of hot water extraction process

on schizophyllan from split gill mushroom. Biomass Convers Biorefinery 14(1): 1017–1026

40. Aliaño-González MJ, Barea-Sepúlveda M, Espada-Bellido E, Ferreiro-González M, López-Castillo JG, Palma M et al (2022) Ultrasound-assisted extraction of total phenolic compounds and antioxidant activity in mushrooms. Agronomy 12(8):1812
41. Krivošija S, Jerković I, Nastić N, Zloh M, Jokić S, Banožić M, Aladić K, Senka V (2023) Green pathway for utilisation of orange peel dust and *in silico* evaluation of pharmacological potential. Microchem J 193:109132
42. Attard TM, Bainier C, Reinaud M, Lanot A, McQueen-Mason SJ, Hunt AJ (2018) Utilisation of supercritical fluids for the effective extraction of waxes and cannabidiol (CBD) from hemp wastes. Ind Crop Prod 112:38–46
43. Krivošija S, Nastić N, Karadžić Banjac M, Kovačević S, Podunavac-Kuzmanović S, Vidović S (2025) Supercritical extraction and compound profiling of diverse edible mushroom species. Foods 14(1):107
44. Sirohi R, Negi T, Rawat N, Sagar NA, Sindhu R, Tarafdar A (2024) Emerging technologies for the extraction of bioactives from mushroom waste. J Food Sci Technol 61(6): 1069–1082
45. Xiaokang W, Lyng JG, Brunton NP, Cody L, Jacquier JC, Harrison SM, Papoutsis K (2020) Monitoring the effect of different microwave extraction parameters on the recovery of polyphenols from shiitake mushrooms: comparison with hot-water and organic-solvent extractions. Biotechnol Rep (Amst) 27:e00504
46. Senturk Parreidt T, Schmid M, Müller K (2018) Effect of dipping and vacuum impregnation coating techniques with alginate-based coating on physical quality parameters of cantaloupe melon. J Food Sci 83(4):929–936
47. Nawawi WFW, Lee KY, Kontturi E, Murphy RJ, Bismarck A (2019) ACS Sustain Chem Eng 7(7):6492–6496

Chapter 17

Mushroom-Based Edible Films and Coatings with Nanointegration

Sumaiya Hassan, Saima Fariha, Suraiya Hassan Richi, and A. T. M. Mijanur Rahman

Abstract

The growing concern over the environmental impact of conventional food packaging has driven the shift towards biodegradable alternatives, with edible films and coatings emerging as a promising solution. This chapter provides a comprehensive overview of the materials and methodologies of mushroom-derived edible films and coatings, integrating advanced extraction and nanotechnology techniques. Subcritical water extraction (SWE) is employed to isolate bioactive polysaccharides from mushrooms under controlled temperature and pressure, ensuring high extraction efficiency without the use of organic solvents. The extracted compounds are then utilized to formulate biodegradable edible films and coatings, enhanced with silver nanoparticles (AgNPs) to improve mechanical strength, antimicrobial activity, and barrier properties. The incorporation of AgNPs further enhances food protection by inhibiting microbial growth and reducing oxygen permeability.

Key words Mushroom, Polysaccharides, Subcritical water extraction, Silver nanoparticles, Edible films, Edible coatings

1 Introduction

Food preservation is a methodical procedure involving several meticulously designed steps to slow down spoilage and improve quality [1]. Packaging is among the integral parts of preservation, having characteristics tailored to specific food products and designated to preserve, protect, and facilitate commercialization and distribution. To date, different conventional materials have been employed as packaging materials which, however, have been linked to the upsurge production of urban solid waste (USW). This issue has fostered a global transition towards biodegradable alternatives, aiming to prioritize sustainability [2]. Research suggests that transitioning from complex multilayer or multilevel packaging systems

Authors Sumaiya Hassan and Saima Fariha have equally contributed to this chapter.

Pablo Navarro Simarro (ed.), *Fungi*, Methods and Protocols in Food Science, https://doi.org/10.1007/978-1-0716-4989-3_17,
© The Author(s), under exclusive license to Springer Science+Business Media, LLC, part of Springer Nature 2026

to a single-component simplified packaging approach, which can be done by edible packaging, has the potential to enhance recyclability with uncompromised protective qualities [3].

Edible packaging refers to a category of biopolymer-based packaging materials—namely, polysaccharides, proteins, lipids, or their composites—engineered to function as a protective layer. When applied in liquid form via coating, spraying, or immersion, they undergo a drying process and form an ultrathin layer (coating) with a thickness below 0.025 mm, which may either remain on the food surface for consumption or be removed. Alternatively, they can be fabricated into structurally independent, solid laminated sheets (films/sheets) exceeding 0.050 mm in thickness, serving as encapsulating or enveloping mediums, functioning analogously to conventional food packaging or wraps, which can also be removed or consumed along with the food [4–7]. Meanwhile, edible packaging derived from naturally sourced materials represents a specialized category within the broader spectrum of bio-based and biodegradable packaging [8]. Numerous biologically produced polymers have been the subject of extensive investigation to create protective coatings or continuous membrane constructions. Among these, polysaccharide hydrocolloids, sourced from plants, crustaceans, and microorganisms, are the most widely utilized class of biopolymers to produce edible coatings and films due to their favorable physicochemical properties and film-forming capabilities [9].

As reported by Scopus database, more than half of the research findings on biodegradable films are centered on plant-derived polysaccharides, with cellulose emerging as the predominant polymer. In contrast, biodegradable films derived from fungal and bacterial sources account for less than 10% of the published research. This is because, for a high production yield, bacteria and fungi require strictly regulated conditions, including precise control over temperature, substrate availability, and exposure duration, to ensure optimal yield [10]. Nevertheless, due to their natural origin, mushroom-derived polysaccharides inherently possess biodegradable properties [11]. Numerous bioactive substances, including various polysaccharides, proteins, and phenols, are usually found in mushrooms [12]. However, the most notable component is chitin, a naturally occurring biopolymer found in the cell walls of fungi and yeasts, was initially identified in both edible and nonedible mushroom species. Its partial deacetylation leads to the formation of chitosan, a highly versatile polysaccharide, which is extensively utilized in the development of edible films and coatings due to its remarkable properties, including strong antimicrobial activity, excellent biocompatibility, biodegradability, and superior film-forming capacity. Additionally, it exhibits a nontoxic profile, high mechanical strength comparable to many commercial polymers, hydrophilicity, and suitability for a diverse range of food applications, making it an invaluable material in food packaging and preservation [13, 14].

The escalation of the quality of coatings or films can be done by the incorporation of nanosystems, which enhances the surface area, thereby improving the loading capacity of bioactive compounds, stability, and mechanical properties of the films or coatings [15]. This highly promising technology has indisputable advantage over edible packaging by developing protection towards light-sensitive foods, accelerating food stability, and inhibiting the proliferation of undesirable microorganisms. Among the various inorganic nanoparticles utilized in edible packaging, titanium dioxide, silver, and zinc oxide are the most prevalent [16]. However, among them silver nanoparticles (AgNPs) are particularly well documented, with extensive research supporting their antimicrobial efficacy and safety for human consumption. The antimicrobial activity of AgNPs is primarily attributed to the release of Ag^+ ions, which interact with electron donor groups in molecules containing sulfur, oxygen, or nitrogen, thereby disrupting microbial cellular processes [17]. Beyond their antimicrobial benefits, silver nanoparticles contribute to the enhancement of key physical properties in food packaging, particularly by improving oxygen and water vapor barrier properties, which in turn contribute to extending food shelf life [18]. Regulatory authorities, such as the European Union, have established a safety threshold for silver in food at 0.05 mg/kg. Additionally, research suggests that a silver concentration of 0.06 mg/L is deemed safe for application in the coating of fruits and vegetables [17]. However, the permitted concentration/safe limit of AgNPs in edible coatings is suggested by Siqueira [19] which is below 1 mgL^{-1}, as concentrations at this level can cause liver cell degeneration in rats.

Subcritical water extraction or SWE is an emerging technique for segregating bioactive compounds from plant materials, employing water as a solvent under subcritical parameters [20]. This process is carried out at temperatures ranging between 100 °C and 374 °C and pressures ranging from 1 to 22.1 MPa, ensuring that water remains in its liquid phase [21]. SWE enhances the extraction of plant polysaccharides by increasing the ionization constant, leading to a greater availability of H^+ and OH^- ions, thereby improving the efficiency of extracting both polar and nonpolar compounds. Compared to conventional extraction techniques, SWE offers significant advantages, including eliminating the need for organic solvents [22], making it an environmentally sustainable method aligned with the principles of green chemistry [21]. Additionally, the increased diffusivity efficiency of water under subcritical conditions enhances the recovery yield of plant polysaccharides while reducing extraction time [22]. Due to its high efficiency and ability to preserve biological activity and stability, SWE is widely applied in the food and pharmaceutical industries for extracting polysaccharides, proteins, antioxidants, and polyphenols [21].

Hence, the current chapter will discuss the SWE method for extracting bioactive compounds from mushrooms and the formulation of edible films and coatings using nanotechnology.

2 Materials

1. Mushroom fruiting bodies.
2. Hot-air oven; hot-air performance incubator.
3. Mechanical grinder; sieve (≤0.45 mm mesh).
4. Distilled or deionized water.
5. High-pressure extraction reactor; nitrogen gas (N_2); magnetic stirrer; electric heater.
6. Water bath; ice bath.
7. Whatman No.1 filter paper (11 μm pore size).
8. Centrifuge; vacuum rotary evaporator.
9. 95% ethanol; absolute ethanol; acetone; diethyl ether.
10. Silver nitrate ($AgNO_3$).
11. UV-Vis spectrophotometer.
12. 2 g each of starch and agar; glycerol.
13. Flat Teflon film-coated glass plate; desiccator.
14. Dark-colored sprayer (for protecting AgNPs from light exposure).
15. Freeze dryer; refrigerator.

3 Methods

3.1 Drying and Grinding

1. Place mushroom fruiting bodies in a hot-air oven at 60 °C for 24 h to remove moisture [23]. If dealing with highly fibrous mushrooms, *see* **Note 1**.
2. Ground the dried mushrooms using a mechanical grinder until the powder size is less than 0.45 mm, to improve extraction efficiency [24]. To ensure uniform particle size, pass the ground material through a sieve (≤0.45 mm mesh) (*see* **Note 2**).

3.2 Extraction Process

1. Weigh the powdered mushroom sample and mix it with purified water in a solid-to-liquid ratio of 1:25 g/mL in a high-pressure extraction vessel.

2. Seal the vessel tightly and purge with nitrogen gas (N) for 5 min to prevent the oxidation of bioactive compounds (*see* **Note 3**).
3. Maintain the extraction pressure at 1.5 MPa for optimal extraction efficiency and gradually heat the reactor to 140 °C using an electric heater (*see* **Notes 4** and **5**).
4. Continuously agitate the mixture at 300 rpm with a magnetic stirrer to ensure uniform heat and mass transfer.
5. Monitor both temperature and pressure inside the reactor in real time using a pressure gauge and temperature controller.
6. Record the extraction time (15 min) once the interior temperature reached the target range.

3.3 Cooling, Filtration, and Storage

1. After the extraction process, use controlled cooling—gradually in a water bath at room temperature, then in an ice bath—to halt further extraction (*see* **Note 6**).
2. Pass the extracted mixture through vacuum filtration using Whatman No. 1 filter paper (11 μm pore size) to remove solid residues. If necessary, use centrifugation (4000 rpm, 1800×*g*, 10 min) for additional purification (*see* **Note 7**).
3. Collect the filtered extract and store at 4 °C for further processing.

3.4 Concentration and Separation

1. The mixtures were combined and concentrated under reduced pressure using a rotary vacuum evaporator at 45 °C, 50 mbar. Avoid exceeding 50 °C to prevent heat damage to polysaccharides.
2. The concentrated solution was then centrifuged at 4000 rpm for 10 min, and the supernatant was carefully collected.

3.5 Precipitation and Purification

1. Add four volumes of 95% ethanol to the supernatant and incubate at 4 °C overnight to precipitate polysaccharides.
2. Centrifuge the resulting precipitate at 4000 rpm for 10 min.
3. Wash the precipitate sequentially with absolute ethanol, acetone, diethyl ether, and dissolve in water.

3.6 Lyophilization and Storage

1. Freeze the purified polysaccharides at −20 °C overnight, then freeze-dry under vacuum (≤0.1 mbar) for 24–48 hours to obtain a fine, dry powder (*see* **Note 8**).
2. Store the powder in airtight glass vials at 4 °C in a desiccator (50% RH) to prevent moisture absorption.

3.7 Biosynthesis and Characterization of AgNPs

1. Biosynthesize the AgNPs by mixing 10% mushroom extract with silver nitrate ($AgNO_3$) solution while keeping the final concentration of the mixture always at 1 mM.
2. Heat the mixture in a hot-air performance incubator until the color changes. The color change of the reaction mixture from the initial colorless to yellowish color and finally to red brown color primarily indicates the completion of the reduction of silver ion (Ag^+) to silver particle (Ag^0) and the biosynthesis of AgNPs (*see* **Note 9**).
3. Use a UV-Vis spectrophotometer to further confirm the formation of AgNPs by scanning 3 mL of the reaction mixtures in the range between 300 and 700 nm.
4. Record the absorption spectrum for each reaction mixture as a function of reaction time with a resolution of 1 nm.
5. Freeze-dry the solution to a powder form and store at 4 °C for 1 week.

3.8 Preparation and Application of AgNPs-Loaded Film and Coating

1. Mix 2 g each of starch and agar with 200 mL of distilled water, including 1.2 g of glycerol (30 wt% of biopolymer) as a plasticizer, and dissolve thoroughly by strong agitation at 90 °C for 1 h (*see* **Note 10**).
2. Blend AgNPs into the mixture and deposit the resulting blend onto a glass plate coated with a thin layer of Teflon (*see* **Note 11**).
3. Once dry, peel off the film and condition it in a desiccator (50% RH, 25 °C) to stabilize moisture content.
4. For coating, transfer the AgNP-blended mixture into any dark-colored sprayer and apply the coating to the desired food products.

4 Notes

1. For highly fibrous mushrooms, drying at 60 °C for 24 h may not fully remove the moisture, leading to inconsistent grinding. Extend drying time by 6–12 h if moisture remains high or use a vacuum drying method to accelerate the process while preventing heat degradation.
2. Uneven grinding may result in inconsistent extraction efficiency, as larger particles release fewer bioactive compounds. If a mechanical grinder is unavailable, ball milling can be used to achieve finer particle sizes.
3. Incomplete nitrogen purging may lead to oxidation of bioactive compounds, reducing the antioxidant properties of the film. Verify purging by checking for oxygen-free conditions

using an oxygen sensor if available. You may remove this step unless the mushroom species is highly prone to oxidation.

4. Maintaining pressure can be challenging, especially with improperly sealed extraction vessels. Regularly check gasket integrity and apply gradual pressure increase to prevent sudden fluctuations.
5. Temperatures above 160 °C may degrade heat-sensitive polysaccharides, leading to molecular breakdown. Conduct preliminary tests at incremental temperatures (120 °C, 140 °C, 160 °C) to determine optimal extraction without degradation.
6. Rapid cooling may cause precipitation of unwanted compounds, leading to the loss of valuable polysaccharides. Use gradual cooling (ice bath with controlled stirring) instead of direct immersion to prevent precipitation of impurities.
7. Incomplete solvent removal may lead to poor film formation. Ensure solvent residue is <5%.
8. Rapid freezing can lead to structural issues.
9. Optimize the temperature accordingly. Start with 60 °C and observe the reaction rate. If necessary, increase the temperature until uniform, stable AgNPs are achieved. Avoid exceeding 90 °C, as high heat may degrade key bioactives.
10. Excess glycerol (above 30% of polysaccharide weight) may lead to sticky or weak films.
11. It is mandatory to evaluate the safety of AgNP-incorporated edible films or coating mixtures after formulation. An in vivo toxicity assessment should be conducted using a rat model. Healthy adult Wistar rats should be divided into control (receiving distilled water) and experimental groups (receiving AgNP-incorporated edible film extracts or coating mixture at low and high doses, based on estimated dietary exposure). The test substance should be administered orally daily for 14–28 days while monitoring body weight, behavioral changes, food and water intake, and any signs of toxicity. At the end of the study, blood samples should be collected for biochemical (ALT, AST, creatinine, urea) and hematological (RBC, WBC, hemoglobin) analyses, while oxidative stress markers and inflammatory cytokines should be assessed to detect systemic effects. Tissues of organs such as the liver, kidneys, and intestines should undergo hematoxylin–eosin (H&E) staining to assess morphological alterations and potential nanoparticle deposition. If significant toxicity is observed, adjustments in AgNP concentration or synthesis parameters may be necessary to ensure the film's safety for food applications. Ethical approval must be obtained before the study, and results should be compared to control values to determine the

suitability of AgNP-incorporated edible films for safe consumption.

References

1. Amit SK, Uddin MM, Rahman R, Islam SR, Khan MS (2017) A review on mechanisms and commercial aspects of food preservation and processing. Agric Food Secur 6:1–22
2. Díaz-Montes E, Castro-Muñoz R (2021) Edible films and coatings as food-quality preservers: an overview. Foods 10(2):249
3. Bharti SK, Pathak V, Alam T, Arya A, Basak G, Awasthi MG (2020) Materiality of edible film packaging in muscle foods: a worthwhile conception. J Packag Technol Res 4(1):117–132
4. Hasan M, Kumar A, Maheshwari C, Mangraj S (2020) Biodegradable and edible film: a counter to plastic pollution. Int J Chem Stud 8:2242–2245
5. Janowicz M, Galus S, Ciurzyńska A, Nowacka M (2023) The potential of edible films, sheets, and coatings based on fruits and vegetables in the context of sustainable food packaging development. Polymers 15(21):4231. https://doi.org/10.3390/polym15214231
6. Nair SS, Trafiałek J, Kolanowski W (2023) Edible packaging: a technological update for the sustainable future of the food industry. Appl Sci 13(14):8234
7. Shah YA, Bhatia S, Al-Harrasi A, Khan TS (2024) Advancements in the biopolymer films for food packaging applications: a short review. Biotechnol Sustain Mater 1(1):2
8. Teixeira-Costa BE, Andrade CT (2021) Natural polymers used in edible food packaging—history, function and application trends as a sustainable alternative to synthetic plastic. Polysaccharides 3(1):32–58
9. Kocira A, Kozłowicz K, Panasiewicz K, Staniak M, Szpunar-Krok E, Hortyńska P (2021) Polysaccharides as edible films and coatings: characteristics and influence on fruit and vegetable quality—a review. Agronomy 11(5):813
10. Díaz-Montes E (2022) Polysaccharide-based biodegradable films: an alternative in food packaging. Polysaccharides 3(4):761–775
11. Anis A, Pal K, Al-Zahrani SM (2021) Essential oil-containing polysaccharide-based edible films and coatings for food security applications. Polymers 13(4):575
12. Yu Y, Liu Z, Song K, Li L, Chen M (2023) Medicinal value of edible mushroom polysaccharides: a review. J Future Foods 3(1):16–23
13. Alimi BA, Pathania S, Wilson J, Duffy B, Frias JMC (2023) Extraction, quantification, characterization, and application in food packaging of chitin and chitosan from mushrooms: a review. Int J Biol Macromol 237:124195
14. Muñoz-Tebar N, Pérez-Álvarez JA, Fernández-López J, Viuda-Martos M (2023) Chitosan edible films and coatings with added bioactive compounds: antibacterial and antioxidant properties and their application to food products: a review. Polymers 15(2):396
15. Koirala P, Nirmal NP, Woraprayote W, Visessanguan W, Bhandari Y, Karim NU et al (2023) Nano-engineered edible films and coatings for seafood products. Food Packag Shelf Life 38:101135
16. Tauferová A, Javůrková Z, Pospiech M, Koudelková Mikulášková H, Těšíková K, Dordevic D et al (2022) Nanoparticles and plant by-products for edible coatings production: a case study with zinc, titanium, and silver. Polymers 14(14):2837
17. Odetayo T, Tesfay S, Ngobese NZ (2022) Nanotechnology-enhanced edible coating application on climacteric fruits. Food Sci Nutr 10(7):2149–2167
18. Alfei S, Marengo B, Zuccari G (2020) Nanotechnology application in food packaging: a plethora of opportunities versus pending risks assessment and public concerns. Food Res Int 137:109664
19. Siqueira MC, Aouada MRM, Castro VLS, Brandão HM, Rech RR, Marconcini JM, Mattoso LHC (2013) Caracterização e avaliação da toxicidade de nanopartíulas de prata para incorporação em matriz polimérica para uso em embalagens de alimentos. VII workshop de nanotecnologia aplicada ao agronegócio, São Carlos, SP, Brazil, 10–13 June 2013. Retrieved from https://bit.ly/2IDMtPd
20. Atanasova A, Petrova A, Teneva D, Ognyanov M, Georgiev Y, Nenov N, Denev P (2023) Subcritical water extraction of rosmarinic acid from lemon balm (*Melissa officinalis* L.) and its effect on plant cell wall constituents. Antioxidants 12(4):888
21. Zhao H, Yang J, Zeng J, Zhou B, Yang M, Yang X, Sun R (2025) Development of subcritical water extraction for areca alkaloids and its influence on the structure of areca nut husk. Molecules 30(4):886

22. Rizkyana AD, Ho TC, Roy VC, Park JS, Kiddane AT, Kim GD, Chun BS (2022) Sulfation and characterization of polysaccharides from oyster mushroom (*Pleurotus ostreatus*) extracted using subcritical water. J Supercrit Fluids 179:105412
23. Yang L, Qu H, Mao G, Zhao T, Li F, Zhu B et al (2013) Optimization of subcritical water extraction of polysaccharides from *Grifola frondosa* using response surface methodology. Pharmacogn Mag 9(34):120
24. Liu J, Li Y, Liu W, Qi Q, Hu X, Li S et al (2019) Extraction of polysaccharide from *Dendrobium nobile* Lindl. by subcritical water extraction. ACS Omega 4(24):20586–20594

Chapter 18

Fungal-Based Foods Beyond Mushrooms: Traditional Practices and Modern Innovations

Antoni Gandia, Natalie Nussbaum, Carole Zermatten, and Patrick A. Rühs

Abstract

Fungal-based foods and fungal fermentation processes have played a crucial role in global food traditions for millennia, contributing to preservation, nourishment, and flavor enhancement. Beyond foraged and cultivated mushrooms, traditional fungal fermentations, such as tempeh, sufu, koji-based products, and mold-ripened cheeses and sausages, demonstrate how fungi can enhance both nutritional and sensory properties of foods. Biotechnological advances in cell culture and bioprocess engineering have further expanded the range of fungal-based food applications, now including mycoprotein-based meat alternatives, realistic whole cuts derived from self-structured fibrous mycelia, and a plethora of fungal-derived functional ingredients. Filamentous fungi such as *Aspergillus*, *Rhizopus*, *Mucor*, *Neurospora*, *Fusarium*, *Penicillium*, and many others, including basidiomycetes, have become staples in precision fermentation, biowaste valorization, and animal feed supplementation, demonstrating their versatility in modern food production. As the industrial applications of fungi continue to expand, AI-assisted fermentation systems, synthetic biology, materials science, and metabolic engineering are now at the forefront of this innovative industry, aiming to optimize scale-up efforts and bring new, sustainable products to our tables.

Key words Fungal-based foods, Mycelium, Fermentation, Mycoprotein, Tempeh, Koji, Food biotechnology, Precision fermentation, Food structuring, Novel foods, Functional foods

1 Introduction

Fungi—including yeasts, molds, and mushroom-forming species—have played a fundamental role in human diets for thousands of years, serving both as direct sources of food and as essential agents in fermentation processes. Archaeological evidence suggests that fermented foods using fungi date back to ancient civilizations, where fermentation was adopted to improve flavor, texture, and preservation [1–3]. Beyond their historical applications, fungi are now at the forefront of food biotechnology and innovation, with recent progress leading to the development of realistic meat alternatives and precision-fermented fungal ingredients.

Advances in fermentation technology, metabolic engineering, and bioprocess automation have significantly improved the

Pablo Navarro Simarro (ed.), *Fungi*, Methods and Protocols in Food Science, https://doi.org/10.1007/978-1-0716-4989-3_18,
© The Author(s), under exclusive license to Springer Science+Business Media, LLC, part of Springer Nature 2026

scalability and nutritional quality of fungal-based foods, allowing them to emerge as viable and sustainable alternatives to traditional animal-derived proteins [4, 5]. The structural properties of filamentous fungal mycelia, particularly their natural fiber-like architecture and ability to retain water and fat, are especially valuable in the development of meat analogs that mimic the fibrous texture and mouthfeel of animal tissues, such as skin and muscle [6–9].

In the West, companies such as Marlow Foods Ltd. (Quorn™, UK), MycoTechnology Inc. (USA), The Better Meat Co. (USA), 3F BIO Ltd. (Enough, UK), MyForest Foods Co. (USA), Planetary (Switzerland), and Nature's Fynd (USA), to cite a few, are pioneering commercial innovations in fungal food production, leveraging solid- and liquid-state fermentation techniques to improve protein yield, texture, and environmental sustainability. Meanwhile, in East and Southeast Asia, new and mature enterprises such as Emmay (Vietnam), 70/30 Food Tech (Singapore), Yukiguni Maitake Co. Ltd. (Japan), OmniFoods (China), and Mycovation (Singapore) are emerging as key players, using regional expertise in fungal biology to develop novel, culturally tailored products that align with growing consumer demand for sustainable protein alternatives.

The study of fungal-based foods extends beyond their role as protein alternatives. The fermentation process itself is responsible for the production of numerous bioactive compounds, including peptides, polyphenols, and polysaccharides, many of which have been associated with improved gut health, immune function, and metabolic regulation [10–12]. As food systems evolve toward sustainability and holistic nurturing, the potential of fungi in food production continues to expand, offering solutions that align with environmental concerns and the growing demand for healthy plant-based nutrition [13, 14].

This chapter explores the historical and modern applications of the filamentous fungi in food production, detailing some prominent traditional fermentation methods, industrial innovations, nutritional benefits, and emerging trends in fungal-based foods. Special emphasis is placed on the transformation of fungi into structured foods, the production of mycoproteins, and the nutritional benefits of a fungal-based diet. In addition, future directions in fungal biotechnology, AI-assisted optimization, consumer trends, and market insights are briefly discussed.

2 Traditional Fungal-Based Foods

The use of fungi in food production is deeply embedded in culinary traditions around the world (Fig. 1). Fungal fermentation, in particular, plays a key role in improving the flavor, digestibility, and nutritional quality of staple foods [8]. In East Asia, the deliberate

Fig. 1 Traditional fungal fermentations: (**a**) black bean tempeh, (**b**) Gorgonzola blue-veined cheese, (**c**) *pehtzes* of sufu (also known as hairy tofu), and (**d**) a block of koji rice. (Images taken by Antoni Gandia, © 2025)

cultivation of the fungus *Aspergillus* spp. on grains and legumes led to the development of famous koji-based products such as miso, soy sauce, and sake [1, 15]. Meanwhile, in Southeast Asia, the practice of solid-state fermentation with *Rhizopus* species gave rise to tempeh, a protein-rich, easily digestible soybean product [2, 16]. In Europe, spontaneous colonization of cheeses by *Penicillium* species resulted in the distinctive characteristics of blue-veined and Camembert-like cheeses (Fig. 1b), which are now recognized as integral components of the European gastronomical pride [17]. These traditional fungal-fermented foods, developed through centuries of culinary experimentation, showcase the biochemical arsenal of fungi used to break down complex food matrices, improve palatability, and extend shelf life [14, 18].

2.1 Tempeh, Sufu, and Oncom: Fungal-Fermented Soy and By-product Foods

Tempeh, sufu, and oncom are three prominent examples of fungal-fermented foods, each utilizing different filamentous fungi to transform plant-based substrates into nutrient-dense, microbiologically stable products. These traditional foods, originating in Southeast Asia and China, illustrate how controlled fungal fermentation improves digestibility and enriches the flavor profiles of otherwise bland foodstuffs.

Tempeh, which originates from Indonesia and is currently widely produced around the globe, is made by solid-state fermentation of cooked soybeans with *Rhizopus oligosporus* or *R. oryzae* for a short period of 30–40 h. The fungal mycelium binds the soybeans into a compact cake-like structure while secreting enzymes that degrade complex carbohydrates and proteins, increasing bioavailability and improving digestibility [2, 19]. Fermentation also reduces antinutritional compounds such as phytic acid and enhances the production of bioactive peptides with potential cholesterol-lowering and antihypertensive effects [14, 18]. The mild, nutty flavor and firm texture of tempeh make it a widely accepted plant-based protein alternative on the global market. Recent iterations in tempeh production have introduced other grains and legumes, such as black beans (Fig. 1a), chickpeas, and quinoa, which are combined with soybeans or used as main substrates, expanding the range of flavors and textures traditionally attributed to tempeh.

Sufu, often referred to as "Chinese cheese" (Fig. 1c), is a fermented soy product traditionally prepared by inoculating tofu blocks with molds such as *Actinomucor elegans*, *Mucor* spp., or *Rhizopus* spp. [20]. During fermentation, these fungi hydrolyze soy proteins and lipids, creating a creamy texture and developing strong umami flavors. Sufu fermentation also produces bioactive compounds with potential antioxidant and antimicrobial properties [21]. The final maturation stage, involving brine or rice wine, contributes to the intense aroma and taste, making sufu a distinctive fermented delicacy in Chinese cuisine [22, 23].

Oncom, a lesser known but important Indonesian fermented food, is produced by fermenting agricultural by-products such as peanut press cake, coconut residue, or cassava pulp with *Neurospora intermedia* or *Rhizopus* species [24]. The fungal fermentation process degrades fibrous material, reduces antinutritional factors, and improves the protein content of the substrate, making oncom a valuable food source in regions where food security and nutrient deficiencies are concerns. Unlike tempeh, which is grown primarily on split soybeans, oncom is grown on agricultural by-products, which are then converted into an edible, protein-rich food, representing an early example of biowaste valorization in traditional food systems [25, 26].

These three Asian fermented products demonstrate the versatility of fungal fermentation in the transformation of plant-based substrates into highly nutritious and flavorful foods. As global interest in fermented foods and flavors grows, these traditional techniques are being further explored for their general simplicity, functional properties, and potential applications in the production of novel and sustainable vegan food [27–31].

2.2 Koji and Koji-Fermented Foods and Beverages

Koji fermentation, a process that relies on the filamentous fungus *Aspergillus oryzae*, is a cornerstone of Japanese, Chinese, and Korean cuisine [32–34]. Koji generally refers to rice, barley, or soybeans that have been inoculated with *A. oryzae* spores and cultivated until the fungus' mycelium covers the grains (Fig. 1d), producing a characteristic sweet and nutty aroma. Historical records indicate that koji originated from Chinese Qu fermentation, which has been documented for more than two millennia [1]. Qu fermentation, first described during the Zhou Dynasty (1046–256 BCE), used mixed microbial communities—including *Aspergillus*, *Rhizopus*, yeasts, and lactic acid bacteria—to produce fermented beverages and condiments [35–37]. The technique was later spread to Japan, where it was adapted and refined into koji, with *A. oryzae* becoming the dominant mold and rice replacing the mixed grain substrates traditionally used in Qu [15, 38]. The first recorded mention of koji in Japan appears in the Engishiki (927 CE), an ancient government document detailing brewing techniques [39]. Koji fermentation is central to the production of miso, soy sauce, mirin, and sake [15, 40, 41]. The enzymatic activity of *A. oryzae* and related fungi, such as *A. sojae* or *A. luchuensis*, is essential in the fermentation process, as it secretes amylases, proteases, and lipases that hydrolyze complex carbohydrates, proteins, and fats into more accessible nutrients. This biochemical transformation raises the umami profile of koji-fermented foods, contributes to preservation, and facilitates alcoholic fermentation in sake and other traditional beverages [42–44].

Modern advancements in koji fermentation have expanded its applications beyond traditional products. The ability of *A. oryzae* to hydrolyze plant-based proteins has led to its incorporation into alternative protein processing and mycoprotein production [45]. Industrial fermentation facilities now optimize koji growth using temperature-controlled trays and automated aeration systems to ensure consistent enzyme production. Additionally, researchers are exploring the potential of koji-fermented substrates as prebiotic ingredients, for example, in amazake, a traditional Japanese koji-fermented drink capable of modulating the composition of the gut microbiota and promoting digestive health [46].

2.3 European Fungal-Ripened Sausages and Cheeses

In European gastronomy, fungi have played a critical role in the ripening and preservation of animal flesh and dairy, giving birth to celebrated cured sausages and cheeses. The controlled growth of filamentous fungi on these foods contributes to their characteristic flavors, textures, and microbial stability, making them highly valued traditional products [47].

The production of mold-ripened sausages, such as French saucisson sec, Italian salami, and Spanish fuet, involves inoculating the surface of the sausages with *Penicillium nalgiovense* or *Penicillium chrysogenum*. These fungi form a protective white mycelial layer that prevents the growth of spoilage microorganisms and pathogenic bacteria while also contributing to the development of distinct aroma compounds [48–50]. The enzymatic activity of these fungi leads to the breakdown of lipids and proteins, enhancing the umami taste and contributing to the characteristic mouthfeel associated with fermented sausages.

Mold-ripened cheeses such as Camembert, Brie, Roquefort, Gorgonzola, and Stilton rely on specific fungal strains to develop their unique textures and flavors. For example, Camembert and Brie are inoculated with *Penicillium camemberti*, which produces a bright rind on the surface that facilitates proteolytic breakdown of casein, resulting in a soft and creamy interior [51]. In contrast, blue-veined cheeses such as Roquefort and Gorgonzola (Fig. 1b) utilize *Penicillium roqueforti*, which colonizes both the surface and the interior of the cheese, forming characteristic blue-green veins. The metabolic activity of *P. roqueforti* generates methyl ketones and free fatty acids, contributing to the complex and pungent flavor profiles of these cheeses [17].

The historical development of mold-ripened cheeses and sausages showcases the traditional use of fungi in food preservation before the advent of refrigeration. Modern production methods now optimize microbial communities and controlled conditions to ensure product consistency and safety while preserving the authentic qualities that have made these foods culinary staples for centuries. The continued popularity of fungal-ripened cheeses and sausages highlights the enduring value of these fermentation techniques in the global food culture. One notable development in this regard is the emergence of fungal-fermented vegan cheeses that use the same fungi and bacterial strains as their animal-derived counterparts, now offered by companies such as New Roots (Switzerland), Mondarella (Germany), and Veganz (Germany) across European supermarkets, which reflects a growing interest in plant-based adaptations of traditional fermented products.

3 Novel Fungal-Based Foods

The application of fungi in food biotechnology has expanded beyond traditional fermentation, leading to the development of novel fungal-based foods (*see* Figs. 2 and 3). These include meat alternatives based on mycoproteins, self-structured mycelium-based whole cuts, and precision-fermented fungal ingredients, all of which serve as sustainable solutions to conventional animal-derived products. Furthermore, innovations in material science and cell culture techniques are paving the way for hyperrealistic microbial tissue biofabrication, blurring the limits of meat-free product applications [52–55].

At the core of these innovations are two key fermentation approaches: liquid-state fermentation (LSF) and solid-state fermentation (SSF) (*see* Fig. 2). In LSF, fungi are grown in

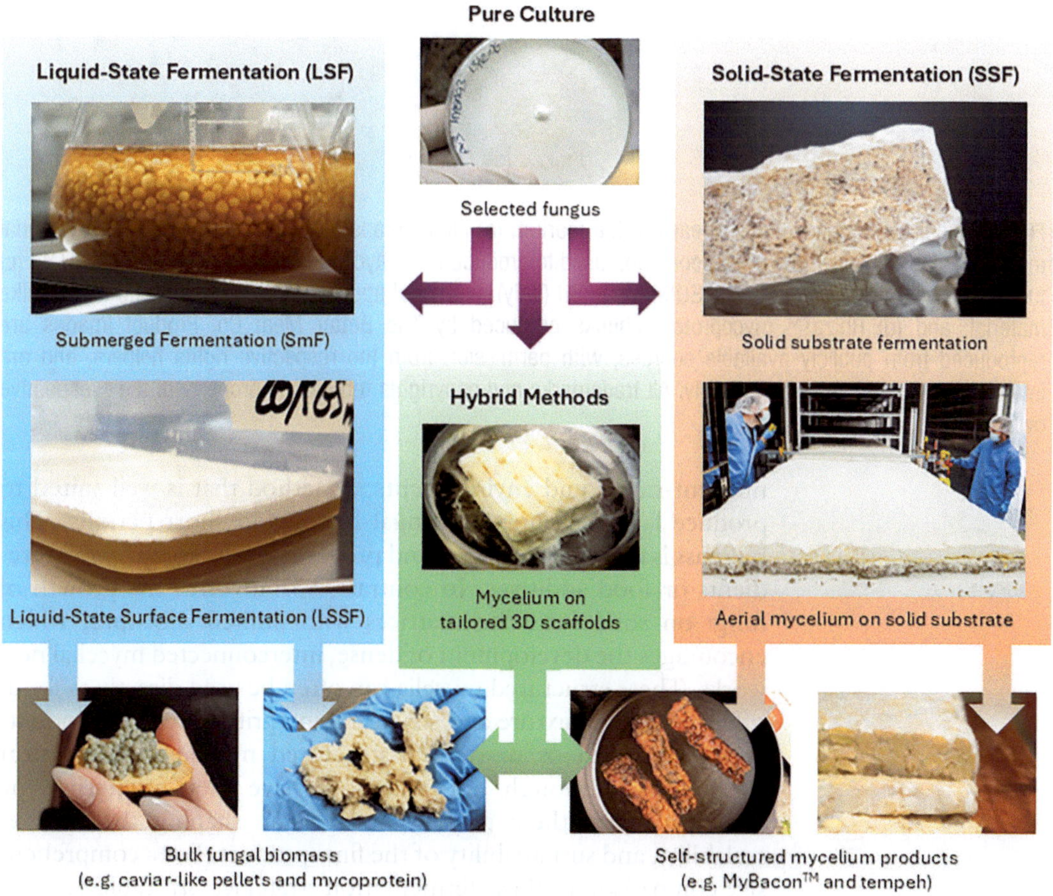

Fig. 2 Production systems for novel fungal-based foods, including liquid-state fermentation (LSF), solid-state fermentation (SSF), and emerging hybrid methods (e.g., hydrogel scaffolds). The image depicting aerial mycelium was provided by Ecovative LLC (USA), and the image of the self-structured mycelium bacon (MyBacon™) is courtesy of MyForest Foods Co. (All the other images were taken by Antoni Gandia, © 2025)

Fig. 3 Fungal-based innovations already in the market: (**a**) Lion's Mane steak by OmniFoods; (**b**) raw aerial mycelium slab grown by MyForest Foods Co., used to produce their mycelium-based bacon (MyBacon™); (**c**) fungal mat grown through LSSF by SQIM/Mogu Srl (Italy), with dual application as foodstuff and leather-like material; and (**d**) Rhiza™ mycoprotein chunks produced by The Better Meat Co. Product images are reproduced from publicly available sources, with permission from the respective rights holders, and are intended for illustrative purposes only. All trademarks and copyrights remain the property of their respective owners

nutrient-rich liquid environments, a method that is well suited to produce large amounts of fungal biomass in short periods. This biomass is typically harvested and processed into protein-rich ingredients or food additives. In contrast, SSF involves the growth of fungi on solid substrate matrices with limited moisture, which encourages the development of dense, interconnected mycelial networks. These structured mycelia can often be used directly as food, offering natural texture and structural integrity as seen in products such as tempeh or newer self-structured mycelium-based meat alternatives. Although both methods have their strengths, the choice between them plays a vital role in shaping the texture, scalability, and sustainability of the final product. For a comprehensive comparison of these fermentation strategies, *see* Table 1.

Within LSF, two main approaches can be distinguished. The first is submerged fermentation (SmF), where fungal biomass grows dispersed or pelleted in suspension within a nutritious liquid medium. This method is widely used in industrial applications for

Table 1
Comparison of liquid-state fermentation (LSF) and solid-state fermentation (SSF) in fungal-based food production

Feature	Liquid-state fermentation (LSF)	Solid-state fermentation (SSF)
Definition	Fungal cultivation in aqueous media; includes SmF and LSSF	Fungal growth on moist solid substrates with limited free water
Growth morphology	Pelleted or dispersed in suspension (SmF); surface mycelial mat formation (LSSF)	Dense, intact filamentous mycelial networks on or within substrate
Typical substrates	Liquid nutrient broths (e.g., glucose, molasses, yeast extract), potato starch, and vegetable waste slurries	Cooked grains, legumes, agricultural by-products (e.g., soy, rice, cassava); lignocellulosic waste for aerial mycelium
Example fungi	*Fusarium venenatum*, *Aspergillu soryzae*, *Rhizopus* spp.	*Rhizopus* spp., *Mucor* spp., *Neurospora intermedia*, *Pleurotus* spp.
Example products	Quorn™ fillets, Rhiza™ chunks, fungal protein powders, enzymes, organic acids	Tempeh, oncom, sufu, mushrooms, MyForest Foods' mycelium bacon (MyBacon™)
Applications	Mycoprotein, enzyme production, flavor/aroma compounds, organic acids (SmF); structured mats for layered or whole-cut foods (LSSF)	Self-structured mycelium-based whole cuts, traditional fermented foods (e.g., tempeh, oncom, mold-ripened cheeses), bioactive compounds
Advantages	High process control and scalability; suitable for continuous operation (SmF); enhanced surface growth and texture potential in LSSF; lower capital investment in LSSF	Inherently fibrous texture; low-tech, low-cost setup; minimal downstream processing for whole-cut formats; existing operational infrastructure (mushroom farms, tray systems)
Disadvantages	SmF biomass requires post-processing for texture; capital-intensive (large volume bioreactors, CIP); high water use and foaming issues (SmF)	Difficult to scale uniformly; lower process control (O_2, pH, temp.); limited heat and mass transfers; lower growth rates; solid waste

its scalability and process control, particularly in the production of fungal biomass for ingredients such as mycoproteins, enzymes, and flavor compounds. The second approach, introduced more recently, is liquid-state surface fermentation (LSSF), where fungi grow at the air–liquid interface of shallow liquid cultures [7]. This surface-oriented configuration enables greater oxygen availability and promotes the formation of cohesive and structured mycelial mats. LSSF retains many of the scalability advantages of SmF while enabling the production of food-grade mycelium with structural integrity, potentially suitable for self-structured whole-cut applications or layered textures without intensive downstream processing.

More recently, hybrid fermentation strategies have emerged that take advantage of the strengths of both LSF and SSF. One promising direction involves cultivating fungal mycelium within engineered hydrogel scaffolds, which can be customized in composition and geometry [9, 53, 56, 57]. These scaffolds can not only mimic the porous structure of natural substrates but also support nutrient flow and controlled growth conditions. When combined with 3D printing technologies, they allow the creation of finely tuned architectures, potentially allowing producers to "design" the texture, shape, and mouthfeel of mycelial foods from the ground up. Such approaches offer a new level of control over fungal growth and product development and may drive the next generation of functional, high-fidelity meat alternatives.

3.1 Mycoprotein Production and Related Advancements

Mycoprotein, a filamentous fungal biomass used as an alternative protein source, was first commercialized in the 1980s by Marlow Foods Ltd. (UK) under the commercial brand Quorn™, following extensive research on the fungus *Fusarium venenatum* [14, 58, 59]. Mycoprotein is notable for its high protein and fiber content, as well as essential micronutrients. These attributes make it a healthy and sustainable alternative to animal meat [60]. The production of mycoprotein involves submerged fermentation (SmF) and is performed in highly controlled airlift or stirred bioreactors, where fungal cultures are grown submerged in a liquid nutritious medium under carefully regulated conditions of pH, oxygenation, axenicity, and nutrient availability. Following fermentation, the biomass is harvested, heat-treated to reduce RNA content, and processed into high-protein food products with a fibrous texture resembling muscle tissue [13, 59].

Recent progress has further optimized mycoprotein production by integrating metabolic engineering to enhance amino acid balance and overall protein content [61, 62]. Additionally, researchers are exploring alternative carbon sources, such as agricultural by-products, to replace glucose-based feedstocks and increase sustainability [61, 63–73]. Machine learning and AI-assisted fermentation control are also promising tools capable of optimizing fungal growth conditions, allowing fully automatic control of fermentation parameters [74, 75]. These innovations contribute to improving the efficiency of large-scale fungal protein production while maintaining high nutritional quality.

The flexibility, bioreactor control improvements, and growing expertise in the fungal biomass production process allowed younger companies such as Planetary (Switzerland), Nosh.bio (Germany), and The Better Meat Co. (USA) to develop their own mycoprotein production technology, some of them using other well-known filamentous fungi species such as *Aspergillus oryzae* [76] and *Rhizopus oryzae* [77], which are both generally recognized as safe (GRAS) fungi, and accepted under the EU Novel Food Regulation (NFR) [78–80]. In addition, new post-processing techniques emerged, allowing muscle-like fiber alignment [81, 82] and the obtention of fat analogs and dairy-like alternatives [83–86], bringing mycoprotein-based products to the next level. Furthermore, on the bioprocess technology side, companies such as Nature's Fynd have pushed the boundaries by developing LSSF tray cultivation systems that are apparently less capital-intensive than classical SmF bioreactors. Nature's Fynd LSSF tray system is reportedly designed to produce fast-growing fungal mats with novel *Fusarium* strains that are later post-processed as bulk mycoprotein and converted to familiar products such as yogurt, patties, and bites [87, 88]. Interestingly, these tray LSSF systems are also being used for the production of fungal-based skins and leather analogs (Fig. 3c).

3.2 Self-structured Mycelium-Based Meat Analogs

Unlike traditional mycoprotein-based products such as Quorn™, which undergo mechanical processing and mixing with binding agents to achieve a cohesive and structured texture, whole self-structured mycelium-based foods leverage the natural filamentous growth and architecture of the fungal mycelium to create whole-cut meat analogs that realistically mimic animal muscle fiber. For instance, the dense interwoven hyphal networks of fungi such as *Rhizopus* spp., *Ganoderma* spp., and *Pleurotus* spp. can render meaty textures without the need for extrusion or extensive processing [9].

Innovations in the cultivation of self-structured mycelium for whole-cut alternatives are being led by companies such as MyForest Foods Co. (previously known as Atlast), which has developed a proprietary SSF method to grow the mycelium of basidiomycete species under specific conditions that promote aerial vegetative growth in the form of slabs, with fibrous structure and good retention of fat and moisture [89, 90]. This approach has led to the creation of realistic structured meat alternatives, including mycelium-based bacon (MyBacon™, *see* Figs. 2 and 3b). Another interesting approach to obtain self-structured whole cuts, specifically targeting seafood analogs, was developed by Aqua Cultured Foods (USA). Their initial approach consisted of a hybrid technique that cultivates filamentous fungi within a bacterial cellulose scaffold, also known as kombucha SCOBYs, aiming to improve their texture, overall appeal, and nutritional characteristics [91, 92]. However, in 2024 the company switched exclusively to cellulose-based platforms, abandoning fungal integration after 2 years of development challenges [93].

Another interesting and emerging trend that combines simplicity and innovation involves producing whole-cut meat analogs by directly cultivating and post-processing mushrooms to replicate traditional meat cuts and seafood. Notable examples include the use of the lion's mane mushroom (*Hericium erinaceus*) to create steak-like products (Fig. 3a), leveraging its fibrous and stringy basidiocarps to mimic the texture of meat. Similarly, the versatility of the king oyster mushroom (*Pleurotus eryngii*) allows the creation of bacon-like slices and scallop substitutes due to its dense stem, large size, and delicate flavor. Additionally, oyster mushrooms (*Pleurotus ostreatus*) are widely used to reproduce chicken crisps and pulled meat, capitalizing on their tender fibrous texture and mild taste. Apparently, there is also growing interest in the cultivation of species such as chicken of the woods (*Laetiporus sulphureus*) and beefsteak mushroom (*Fistulina hepatica*) to develop products that emulate their namesake meats.

These developments and meat-like innovations underscore the potential of mushrooms and self-structured fungal mycelia as sustainable, scalable, and nutritious alternatives to traditional animal-derived products, offering diverse textures and flavors that cater to a wide range of culinary applications.

3.3 Fungal-Derived Functional Ingredients and Precision Fermentation

Precision fermentation is an emerging biotechnological platform that utilizes genetically optimized fungal strains to produce targeted functional compounds for food applications. This approach enables the synthesis of proteins, lipids, and bioactive metabolites with high specificity and efficiency, offering viable alternatives to animal-derived ingredients [94–96].

Fungi can serve as particularly effective microbial hosts for molecular manufacturing due to their eukaryotic cellular machinery, which facilitates proper protein folding and posttranslational modifications, making them suitable for producing complex enzymes and other high-functionality ingredients [97].

One of the most industrially relevant fungal species in precision fermentation is *Aspergillus niger*, which is used to produce citric acid and hydrolytic enzymes such as glucoamylase and pectinases, which play an essential role in food processing applications [98, 99]. Similarly, *Aspergillus oryzae* has been the subject of recent synthetic biology efforts; Maini Rekdal et al. (2024) engineered this species to overproduce ergothioneine and heme pigments, rendering red-colored, flavor-enriched fungal biomass ready to use in meat analogs [96].

In addition to enzymes and pigments, precision fermentation is being leveraged to produce fungal-derived fats and volatile flavor compounds that mimic the sensory characteristics of animal-based foods [100, 101]. Fungi are also being investigated as biofactories for natural food additives, such as exopolysaccharides—including betaglucans and galactomannans—that act as thickeners, emulsifiers, and stabilizers. These compounds often confer additional health benefits, including cholesterol-lowering effects and immune modulation [102].

Moreover, fungal enzymes are increasingly being applied to enhance the organoleptic properties of plant-based products through the selective hydrolysis of plant proteins. This enzymatic modification can improve the solubility, mouthfeel, and flavor of alternative protein formulations, thus expanding the sensory appeal of next-generation vegan foods.

A notable example of commercial application is provided by MycoTechnology Inc., which uses proprietary mycelial fermentation platforms to develop flavor-modulating and sweetening ingredients. Their flagship product, ClearIQ™, is produced via SmF and serves as a natural flavor clarifier. It reduces bitterness and astringency while increasing positive flavor notes, including umami and citrus, in a wide range of food matrices such as plant-based proteins, coffee, and cocoa [103]. Their latest and *second-generation* product is derived via precision fermentation from a sweet protein found in honey truffles (*Heimioporus* spp.). This new ingredient received GRAS approval from FEMA (USA) and functions as a flavor enhancer and bitterness masker in sugar-reduced applications, with a commercial launch planned for 2026 [104]. MycoTechnology Inc. previously also offered FermentIQ™, a fermented plant protein product (e.g., pea and rice) using the basidiomycete *Lentinula edodes* (shiitake), which reportedly improved protein digestibility and reduced off-flavors. However, this line has been discontinued as the company consolidates its focus on flavors and sweeteners [105]. MycoTechnology's innovations manifest the practical application of fungal fermentation to improve sensory profiles and nutritional functionality in modern food systems.

Looking ahead, the integration of AI-assisted bioprocess optimization and synthetic biology is expected to accelerate the development of fungal-based functional products. These tools will enable more precise control over metabolic pathways, improved production yields, and greater customization of food properties. Future innovations are likely to focus on improved nutritional profiles, solving scalability bottlenecks, and expanding the range of mycelium-derived specialty ingredients tailored for plant-based and hybrid food systems.

4 Industrial-Scale Fermentation: Challenges, Market Insights, and Consumer Trends

Fungal-based foods have transitioned from traditional small-scale fermentation to industrial-scale commercial production in the span of a century, fueled by fast progress in bioprocess engineering, synthetic biology, and fermentation control. Although mycoprotein-based foods have been commercially available since the 1980s, the recent innovations in mycelium-structured meat analogs and precision fermentation briefly commented on in the previous sections are rapidly reshaping the alternative protein industry, leading to a rapid expansion of the fungal-based product catalog.

However, turning these innovations into reliable manufacturing processes introduces a different set of technical and operational challenges. While fungal biomass production appears to be a straightforward process, scaling up fungal-based food production requires precise environmental control, strict regulatory compliance, and high initial capital investment—points that are repeatedly emphasized in this chapter. Industrial-scale mycelium cultivation in SmF bioreactors, in particular, requires large-volume and well-optimized fermentation systems that maintain consistent quality and yield while minimizing contamination risks. Fungal morphology, whether suspended mycelium or pelletized forms, critically impacts growth, metabolism, and downstream processing, with factors such as agitation intensity, dissolved oxygen, pH, and temperature, requiring careful and skilled regulation for optimal production [106, 107].

Beyond the complexities of managing the fungal cultivation process itself, downstream processing introduces more critical considerations and costs. For instance, while the established mycoprotein heat treatment (described earlier in this chapter) effectively reduces RNA content to levels compliant with safety guidelines, the residual RNA still imposes a practical limitation on maximum daily consumption [108]. This highlights an ongoing challenge and the need for research into more advanced RNA reduction strategies, aiming not only to improve dietary flexibility but also to potentially improve overall process efficiency. This requirement for specialized postharvest treatment connects with another often overlooked challenge in LSF production systems: the need to process the fungal biomass immediately at the production site due to its susceptibility to microbial degradation, requiring substantial ancillary infrastructure.

The high capital and operating costs related to SmF are in fact considered one of the biggest limitations of this technology, restricting its applicability to developed nations and well-funded enterprises. For context, a Quorn™ facility built in 2018 with an output of 20,000 tons per year was reported to cost approximately USD 200 million [8, 109]. However, several newer companies appear to have developed their initial pilot- or small- to medium-scale commercial facilities with considerably less capital. For example, the Swedish company Mycorena raised €7.5 M in Pre-Series A funding (2021) [110], later securing €24 M in a record European alt-protein Series A (2022) [111], totaling €35 M+ before its 2024 bankruptcy and acquisition by Naplasol/VEOS Group [112]. In Switzerland, Planetary received USD 8 M (€7.3 M) in their 2022 seed funding for R&D [113], followed by CHF 3 M (2024) from the international agronomical cooperative Royal Cosun to scale modular fermentation [114]. Similarly, The Protein Brewery (NL) collected €22 M in Series A funding in 2020, with a portion allocated to its pilot plant [115].

Although these investment figures represent total funding, not only the direct costs of a pilot plant construction, they are significantly lower than the investment reportedly required for a large-scale Quorn™ facility. This suggests that the capital outlay to establish initial production capacity among emerging players was likely modest, potentially facilitating market entry and innovation in the mycoprotein sector by prioritizing incremental scaling and strategic partnerships while minimizing capital expenditure. However, as exemplified by Mycorena's bankruptcy, even reduced CAPEX cannot guarantee scalability in the absence of market readiness.

Reflecting the growing interest and positive investment in the sector, the global fungal protein market is projected to expand from an estimated USD 4.2 billion in 2024 to USD 6.7 billion by 2034, growing at a compound annual growth rate (CAGR) of 6.1% [116], indicating a robust market for fungal-based products. In addition, the last decade (2015–2025) has seen the formation of dozens of companies focused on alternative proteins enabled by fermentation, several of which have product pipelines based on filamentous fungi [8]. This robust growth is primarily driven by the growing interest in alternative proteins; concerns about the environmental impact of traditional livestock farming, with life cycle evaluations consistently showing better performance for mycoprotein [13, 14, 60, 72, 80, 94]; and the recognized nutritional benefits of fungal-based foods, which are reviewed in the next section of this chapter.

This surge in interest is not limited to mycoprotein alone; the global market for cultivated mushrooms shows significant commercial value, reaching USD 46.1 billion back in 2020 [8, 117]. Current projections indicate a robust expansion, with the market expected to grow from USD 66 billion reported in 2024 to USD 105.54 billion by 2029, reflecting a compound annual growth rate (CAGR) of 10.2% [118].

Consumer trends also reveal a growing preference for minimally processed whole food alternatives to conventional meat substitutes [119]. Mycelium-structured whole-cut meats, traditional fungal-based products such as tempeh, and mushrooms align perfectly with this preference, as they generally require less processing than extruded products. As discussed previously, beyond classic meat alternatives, new ventures are exploring diverse innovative applications for fungi, including fungal-fermented beverages (e.g., low-sugar and high-protein milk analogs), dairy-free cheeses and yogurts, fat replacements, and novel functional ingredients enriched with bioactive compounds, which may provide gut health and immune-supporting benefits [14, 94].

However, the positive market outlook is tempered by ongoing challenges in terms of regulatory approval and achieving large-scale production cost-efficiency. In Europe, novel fungal strains used in food applications require extensive safety evaluations prior to commercialization [78, 79]. Moreover, public perception of fungi-based proteins is still evolving, underscoring the critical role of consumer education and targeted marketing in broader market adoption. As technological advancements continue to tackle all these challenges, fungal-based proteins are poised to gain greater acceptance and integration into mainstream diets [60, 120].

5 Nutritional and Health Implications

Fungal-based foods offer a nutritionally rich and sustainable alternative to traditional animal-derived proteins. The protein quality of fungal-based products is particularly notable, as many species provide all nine essential amino acids required for human nutrition [5, 88]. In addition, fungal fermentation improves food digestibility, reduces antinutritional factors, and generates bioactive compounds that may contribute to metabolic health and disease prevention [8, 18].

In terms of nutritional aspects, fungal-based proteins, including mycoproteins and fermented soy products, exhibit high protein quality and digestibility. Mycoprotein, derived from *Fusarium venenatum*, has a protein digestibility-corrected amino acid score (PDCAAS) comparable to high-quality animal proteins, with an optimal balance of essential amino acids [59, 60]. Its protein and fiber content also offers functional benefits, such as promoting satiety, which may support weight management [13, 121]. Furthermore, the fermentation process involved in tempeh and koji-based foods significantly improves protein bioavailability by breaking down complex protein structures and eliminating antinutritional compounds such as phytic acid and trypsin inhibitors. As a result, fermented fungal-based foods are more easily absorbed by the body, making them particularly valuable in plant-based diets where protein quality can be a limiting factor [14, 18, 121].

Beyond protein, the consumption of fungal-fermented foods has been associated with positive effects on the composition of the gut microbiota. The presence of fungal-derived beta-glucans and prebiotic polysaccharides supports the proliferation of beneficial gut bacteria, including *Bifidobacterium* and *Lactobacillus* species, which are known to promote intestinal health and immune function [8, 46, 122, 123]. Studies also suggest that regular consumption of these foods improves short-chain fatty acid (SCFA) production, which, in turn, strengthens the intestinal barrier and reduces inflammation [124]. Besides, the enzymatic transformation of soy isoflavones during fungal fermentation may also contribute to improved hormone regulation, particularly in postmenopausal women [125, 126].

The benefits of incorporating fungal-based foods into the diet extend to cardiovascular and metabolic health. Mycoprotein consumption has been associated with reduced LDL ("bad") cholesterol and improved blood lipid profiles, attributed in part to its fiber content and lipid-lowering effects [5, 127, 128]. Moreover, fungal-derived proteins show a hypoglycemic effect, with studies indicating improved insulin sensitivity and glycemic control in individuals with type 2 diabetes [13, 129]. Furthermore, fermentation of soy-based foods with filamentous fungi has been shown to enhance the bioavailability of polyphenols and peptides with angiotensin-converting enzyme (ACE) inhibitory properties, suggesting a potential role in the regulation of blood pressure [130, 131]. These same fungal polyphenols and peptides have also been shown to modulate inflammatory pathways and reduce pro-inflammatory cytokine levels. For instance, koji-fermented foods contain ferulic acid derivatives that exhibit endothelial protective effects, while tempeh-derived isoflavones have been linked to reduced systemic inflammation [46, 126, 132–136].

In addition to cardiovascular effects, fungal polysaccharides and ergothioneine—a naturally occurring amino acid derivative found in fungi—may offer antioxidant protection [137–139]. Another important area of research emphasizes the immunomodulatory properties of some of these compounds; in particular, beta-glucans present in fungal cell walls have been widely studied in this regard. These polysaccharides interact with immune receptors such as Dectin-1 and Toll-like receptors (TLRs), increasing macrophage activation and promoting adaptive immune responses [8, 140, 141].

Although the health benefits are promising, it is important to consider potential safety considerations and allergenicity. Commercialized fungal-based foods are generally recognized as safe (GRAS), but some individuals may exhibit sensitivity to fungal proteins or chitin, a component of fungal cell walls. Mild gastrointestinal discomfort has been reported in some consumers of mycoproteins, although incidence rates remain low [142–144]. As briefly commented on in previous sections of this chapter, regulatory bodies have established guidelines to ensure that commercially available fungal-based foods meet strict safety standards, including rigorous screenings for potential mycotoxins during fermentation [60, 145, 146].

Future research should continue investigating the long-term health effects of fungal-based diets, with a focus on microbiome modulation, metabolic health, and immune system interactions. The integration of precision fermentation techniques can further enhance the functional properties of fungal-based foods, allowing the development of personalized nutrition solutions tailored to individual dietary needs.

6 Conclusion and Future Prospects

Fungal-based foods have evolved from traditional fermentations into advanced food products that serve as viable alternatives to animal meat. Their ability to provide high-quality protein, bioactive compounds, and functional ingredients makes them essential in the future of sustainable food production. Future research efforts are expected to improve their biosafety and refine functional, organoleptic, and nutritional properties in parallel to optimized fermentation efficiency, thanks to modern biotechnological advancements such as metabolic engineering, precision fermentation, 3D scaffolding, and AI-assisted "smart" bioreactors [8, 53, 60, 147, 148].

One of the most compelling advantages of fungal-based foods is their potential to address global food security challenges. As protein demand increases, mycoprotein and self-structured mycelium-based foods offer efficient and resource-friendly solutions with a significantly lower environmental footprint than conventional livestock production [5, 8, 14]. Research has shown that fungal fermentation can reduce agricultural waste by utilizing by-products as growth substrates, enhancing sustainability [73, 149, 150]. However, several challenges remain, including consumer perception, regulatory approvals, and the need for transparent labeling to improve market acceptance [60, 78, 94, 120].

Beyond alternative proteins, the potential applications of fungal-based foods extend into several novel areas. The production of fungal-derived fats, muscle-like textures, new pigments, and flavor compounds can revolutionize plant-based dairy and alternative meat formulations, offering enhanced organoleptic properties without reliance on artificial additives. In addition, fungal food applications in clinical nutrition and personalized diets are emerging as an area of interest, with research exploring the potential for fungi-derived bioactives to support metabolic health, immune function, and cognitive performance [34, 123, 136]. As the world seeks solutions to feed 10 billion people by 2050 [151], fungal-based foods offer a promising toolset in which traditional knowledge and modern biotechnological innovation converge, offering scalable pathways to meet the sustainable dietary needs of future populations.

Acknowledgments

The authors express their gratitude to the companies that kindly provided illustrative material used in this manuscript.

Conflicts of Interest The authors declare that they have no conflict of interest. However, A.G. is employed by Planted Foods AG, a

Swiss company involved in microbial food research and commercialization. This employment has not influenced the objectivity or content of this manuscript.

Use of AI Tools The authors utilized the Writefull tool (v2.2.0, 2024) within Overleaf to facilitate grammatical and orthographic revisions of the manuscript. Despite the AI tool offering recommendations, the authors maintained complete authority over the final phrasing and substance.

Funding

This study was not supported by any dedicated grant from public, commercial, or nonprofit funding organizations.

References

1. Huang HT (2000) Science and civilisation in China, vol. 6, biology and biological technology, part 5. Fermentations and food sciences. Cambridge University Press, Cambridge
2. Shurtleff W, Aoyagi A (2020) History of tempeh and tempeh products (1815–2020): extensively annotated bibliography and sourcebook. Soyinfo Center, Lafayette
3. Tamang JP, Cotter PD, Endo A et al (2020) Fermented foods in a global age: east meets west. Compr Rev Food Sci Food Saf 19:184–217. https://doi.org/10.1111/1541-4337.12520
4. Derbyshire EJ (2020) Is there scope for a novel mycelium category of proteins alongside animals and plants? Foods 9. https://doi.org/10.3390/FOODS9091151
5. Derbyshire EJ, Delange J (2021) Fungal protein—what is it and what is the health evidence? A systematic review focusing on mycoprotein. Front Sustain Food Syst 5. https://doi.org/10.3389/FSUFS.2021.581682
6. Jones M, Gandia A, John S et al (2020) Leather-like material biofabrication using fungi. Nat Sustain:1–8. https://doi.org/10.1038/s41893-020-00606-1
7. Gandia A, van den Brandhof JG, Appels FVW et al (2021) Flexible fungal materials: shaping the future. Trends Biotechnol 39:1321–1331. https://doi.org/10.1016/j.tibtech.2021.03.002
8. Strong PJ, Self R, Allikian K et al (2022) Filamentous fungi for future functional food and feed. Curr Opin Biotechnol 76. https://doi.org/10.1016/J.COPBIO.2022.102729
9. Nussbaum N, von Wyl T, Gandia A et al (2023) Impact of malt concentration in solid substrate on mycelial growth and network connectivity in *Ganoderma* species. Sci Rep 13:1–10. https://doi.org/10.1038/s41598-023-48203-4
10. Mizuno T (1995) Bioactive biomolecules of mushrooms: food function and medicinal effect of mushroom fungi. Food Rev Intl 11:5–21. https://doi.org/10.1080/87559129509541017
11. Rathore H, Prasad S, Kapri M et al (2019) Medicinal importance of mushroom mycelium: mechanisms and applications. J Funct Foods 56:182–193. https://doi.org/10.1016/J.JFF.2019.03.016
12. Rousta N, Aslan M, Akbas MY et al (2024) Effects of fungal based bioactive compounds on human health: review paper. Crit Rev Food Sci Nutr 64:7004–7027. https://doi.org/10.1080/10408398.2023.2178379
13. Filho PFS, Andersson D, Ferreira JA et al (2019) Mycoprotein: environmental impact and health aspects. World J Microbiol Biotechnol 35:147. https://doi.org/10.1007/S11274-019-2723-9
14. Barzee TJ, Cao L, Pan Z et al (2021) Fungi for future foods. J Future Foods 1:25–37. https://doi.org/10.1016/J.JFUTFO.2021.09.002
15. Kitamura Y, Kusumoto KI, Oguma T et al (2016) Ethnic fermented foods and alcoholic

beverages of Japan. In: Ethnic fermented foods and alcoholic beverages of Asia, pp 193–236. https://doi.org/10.1007/978-81-322-2800-4_9

16. Tamang JP, Anupma A, Shangpliang HNJ (2022) Ethno-microbiology of tempe, an Indonesian fungal-fermented soybean food and koji, a Japanese fungal starter culture. Curr Opin Food Sci 48:100912. https://doi.org/10.1016/J.COFS.2022.100912
17. Fox PF, Guinee TP, Cogan TM et al (2016) Fundamentals of cheese science, 2nd edn, pp 1–799. https://doi.org/10.1007/978-1-4899-7681-9
18. Ahnan-Winarno AD, Cordeiro L, Winarno FG et al (2021) Tempeh: a semicentennial review on its health benefits, fermentation, safety, processing, sustainability, and affordability. Compr Rev Food Sci Food Saf 20: 1717–1767. https://doi.org/10.1111/1541-4337.12710
19. Nout MJ, Kiers JL (2005) Tempe fermentation, innovation and functionality: update into the third millenium. J Appl Microbiol 98:789–805. https://doi.org/10.1111/J.1365-2672.2004.02471.X
20. Han BZ, Kuijpers AF, Thanh NV et al (2004) Mucoraceous moulds involved in the commercial fermentation of sufu pehtze. Anton Leeuw Int J Gen Mol Microbiol 85:253–257. https://doi.org/10.1023/B:ANTO.0000020157.72415.B9
21. Huang X, Yu S, Han B et al (2018) Bacterial community succession and metabolite changes during sufu fermentation. LWT 97: 537–545. https://doi.org/10.1016/J.LWT.2018.07.041
22. Guan RF, Liu ZF, Zhang JJ et al (2013) Investigation of biogenic amines in sufu (furu): a Chinese traditional fermented soybean food product. Food Control 31:345–352. https://doi.org/10.1016/J.FOODCONT.2012.10.033
23. Zhao L, Liu Y, Xu Q et al (2023) Microbial community succession and its correlation with quality characteristics during gray sufu fermentation. Foods 12:2767. https://doi.org/10.3390/FOODS12142767/S1
24. Nout MJR, Aidoo KE (2011) Asian fungal fermented food. In: Industrial applications, pp 29–58. https://doi.org/10.1007/978-3-642-11458-8_2
25. Firoh AM, Naibaho J, Sugiyono et al (2024) Physico-chemical and sensory characteristics of red oncom, a traditional fermented food from Indonesia, based on the variability of ingredients and processing steps. Appl Food Res 4:100571. https://doi.org/10.1016/J.AFRES.2024.100571
26. Wijaya CH, Nuraida L, Nuramalia DR et al (2024) Oncom: a nutritive functional fermented food made from food process solid residue. Appl Sci 14:10702. https://doi.org/10.3390/APP142210702
27. Cantabrana I, Perise R, Hernández I (2015) Uses of *Rhizopus oryzae* in the kitchen. Int J Gastron Food Sci 2:103–111. https://doi.org/10.1016/J.IJGFS.2015.01.001
28. Londoño-Hernández L, Ramírez-Toro C, Ruiz HA et al (2017) *Rhizopus oryzae*—ancient microbial resource with importance in modern food industry. Int J Food Microbiol 257:110–127. https://doi.org/10.1016/J.IJFOODMICRO.2017.06.012
29. Vital RJ, Bassinello PZ, Cruz QA et al (2018) Production, quality, and acceptance of tempeh and white bean tempeh burgers. Foods 7: 136. https://doi.org/10.3390/FOODS7090136
30. Gmoser R, Fristedt R, Larsson K et al (2020) From stale bread and brewers spent grain to a new food source using edible filamentous fungi. Bioengineered 11:582–598. https://doi.org/10.1080/21655979.2020.1768694
31. Assa YA, Mayulu N, Surya R et al (2023) Soy and algae combination using Tempe fermentation method: a proposed opinion for the development of functional food. Nutrición Clínica y Dietética Hospitalaria 43:30–35. https://doi.org/10.12873/433ASSA
32. Narahara H (1994) Koji-mold and its product koji. J Brew Soc Jpn 89:954–964. https://doi.org/10.6013/JBREWSOCJAPAN1988.89.954
33. Higuchi Y, Kitamoto K (2021) Traditional and latest researches on *Aspergillus oryzae* and related koji molds. J Fungi 7:1075. https://doi.org/10.3390/JOF7121075
34. Tzachor A (2022) Novel foods for human and planetary health. Nat Food 3(4):247–248. https://doi.org/10.1038/s43016-022-00492-0
35. Lv XC, Weng X, Zhang W et al (2012) Microbial diversity of traditional fermentation starters for hong qu glutinous rice wine as determined by PCR-mediated dgge. Food Control 28:426–434. https://doi.org/10.1016/J.FOODCONT.2012.05.025
36. Zheng XW, Yan Z, Han BZ et al (2012) Complex microbiota of a Chinese "fen" liquor fermentation starter (fen-daqu), revealed by culture-dependent and culture-independent methods. Food Microbiol 31:

293–300. https://doi.org/10.1016/J.FM.2012.03.008

37. Jin G, Zhu Y, Xu Y (2017) Mystery behind Chinese liquor fermentation. Trends Food Sci Technol 63:18–28. https://doi.org/10.1016/J.TIFS.2017.02.016
38. Machida M, Yamada O, Gomi K (2008) Genomics of *Aspergillus oryzae*: learning from the history of koji mold and exploration of its future. DNA Res 15:173. https://doi.org/10.1093/DNARES/DSN020
39. Yanagihara N, Mayumi M, Yoshikawa J et al (2024) Flavor assessment of a lactic fermented vinegar described in Japanese books from the Edo period (1603–1867). Heliyon 10: e32344. https://doi.org/10.1016/J.HELIYON.2024.E32344
40. Okazaki N (2009) Traditional alcoholic beverage and its koji in Japan, China and Southeast Asia. J Brew Soc Jpn 104:951–957. https://doi.org/10.6013/JBREWSOCJAPAN.104.951
41. Kusumoto KI, Yamagata Y, Tazawa R et al (2021) Japanese traditional miso and koji making. J Fungi 7:579. https://doi.org/10.3390/JOF7070579
42. Yoshizawa K (1999) Sake: production and flavor. Food Rev Intl 15:83–107. https://doi.org/10.1080/87559129909541178
43. Gibbons JG, Salichos L, Slot JC et al (2012) The evolutionary imprint of domestication on genome variation and function of the filamentous fungus *Aspergillus oryzae*. Curr Biol 22: 1403–1409. https://doi.org/10.1016/J.CUB.2012.05.033
44. Ichikawa E, Hirata S, Hata Y et al (2020) Effect of koji starter on metabolites in Japanese alcoholic beverage sake made from the sake rice koshitanrei. Biosci Biotechnol Biochem 84:1714–1723. https://doi.org/10.1080/09168451.2020.1763154
45. Devanthi PVP, Pratama F, Pramanda IT et al (2024) Exploring the potential of *Aspergillus oryzae* for sustainable mycoprotein production using okara and soy whey as cost-effective substrates. J Fungi 10:555. https://doi.org/10.3390/JOF10080555/S1
46. Nakano H, Setoguchi S, Kawano K et al (2023) Effects of amazake produced with different *Aspergillus* on gut barrier and microbiota. Foods (12):2568. https://doi.org/10.3390/FOODS12132568
47. López-Díaz TM, Santos JA, García-López ML et al (2001) Surface mycoflora of a Spanish fermented meat sausage and toxigenicity of *Penicillium* isolates. Int J Food Microbiol 68:69–74. https://doi.org/10.1016/S0168-1605(01)00472-X
48. Toldrá F, Hui YH (2014) Dry-fermented sausages and ripened meats: an overview. In: Handbook of fermented meat and poultry, 2nd edn, pp 1–6. https://doi.org/10.1002/9781118522653.CH1
49. Lücke FK (1994) Fermented meat products. Food Res Int 27:299–307. https://doi.org/10.1016/0963-9969(94)90098-1
50. Chávez R, Fierro F, García-Rico RO et al (2011) Mold-fermented foods: *Penicillium* spp. as ripening agents in the elaboration of cheese and meat products. Mycofactories:73–98. https://doi.org/10.2174/978160805223311101010073
51. McSweeney PL, Ottogalli G, Fox PF (2017) Diversity and classification of cheese varieties: an overview. In: Cheese: chemistry, physics and microbiology, vol 1, 4th edn, pp 781–808. https://doi.org/10.1016/B978-0-12-417012-4.00031-4
52. Bomkamp C, Skaalure SC, Fernando GF et al (2022) Scaffolding biomaterials for 3D cultivated meat: prospects and challenges. Adv Sci 9. https://doi.org/10.1002/ADVS.202102908
53. Gantenbein S, Colucci E, Käch J et al (2022) Three-dimensional printing of mycelium hydrogels into living complex materials. Nat Mater 22:128–134. https://doi.org/10.1038/s41563-022-01429-5
54. Kim WJ, Kim Y, Ovissipour R et al (2024) Plant-based biomaterials as scaffolds for cellular agriculture. Fut Foods 10:100468. https://doi.org/10.1016/J.FUFO.2024.100468
55. Gordon EB, Choi I, Amanipour A et al (2025) Biomaterials in cellular agriculture and plant-based foods for the future. Nat Rev Mater:1–19. https://doi.org/10.1038/s41578-025-00800-7
56. Rühs PA, Storz F, Gómez YAL et al (2018) 3D bacterial cellulose biofilms formed by foam templating. npj Biofilms Microbiomes 4:1–6. https://doi.org/10.1038/S41522-018-0064-3
57. Lin N, Taghizadehmakoei A, Polovina L et al (2024) 3D bioprinting of food grade hydrogel infused with living *Pleurotus ostreatus* mycelium in non-sterile conditions. ACS Appl Bio Mater 7:2982–2992. https://doi.org/10.1021/acsabm.4c00048
58. Wiebe M (1995) A new use for the mycoprotein organism. Bio/technology 13:1434
59. Finnigan T (2011) Mycoprotein: origins, production and properties. In: Handbook of

food proteins, pp 335–352. https://doi.org/10.1533/9780857093639.335

60. Holt RR, Munafo JP, Salmen J et al (2024) Mycelium: a nutrient-dense food to help address world hunger, promote health, and support a regenerative food system. J Agric Food Chem 72:2697–2707
61. Tong S, Chen W, Hong R et al (2024) Efficient mycoprotein production with low co2 emissions through metabolic engineering and fermentation optimization of *Fusarium venenatum*. J Agric Food Chem 72:604–612. https://doi.org/10.1021/ACS.JAFC.3C08509
62. Hong R, Tong S, Chai M et al (2024) Enhancing mycoprotein yield: metabolic modulation of chitin synthase in *Fusarium venenatum*. J Agric Food Chem 72. https://doi.org/10.1021/ACS.JAFC.4C07105
63. Satari B, Karimi K, Taherzadeh MJ et al (2016) Co-production of fungal biomass derived constituents and ethanol from citrus wastes free sugars without auxiliary nutrients in airlift bioreactor. Int J Mol Sci 17:302. https://doi.org/10.3390/IJMS17030302
64. Svensson SE, Bucuricova L, Ferreira JA et al (2021) Valorization of bread waste to a fiber- and protein-rich fungal biomass. Fermentation 7:91. https://doi.org/10.3390/FERMENTATION7020091
65. Hashempour-Baltork F, Hosseini SM, Assarehzadegan MA et al (2020) Safety assays and nutritional values of mycoprotein produced by *Fusarium venenatum* ir372c from date waste as substrate. J Sci Food Agric 100: 4433–4441. https://doi.org/10.1002/JSFA.10483
66. Banks M, Taylor M, Guo M (2024) High throughput parameter estimation and uncertainty analysis applied to the production of mycoprotein from synthetic lignocellulosic hydrolysates. Arxiv Preprint
67. Sar T, Larsson K, Fristedt R et al (2022) Demo-scale production of protein-rich fungal biomass from potato protein liquor for use as innovative food and feed products. Food Biosci 47:101637. https://doi.org/10.1016/J.FBIO.2022.101637
68. Filho PF, Zamani A, Taherzadeh MJ (2017) Production of edible fungi from potato protein liquor (PPL) in airlift bioreactor. Fermentation (3):12. https://doi.org/10.3390/FERMENTATION3010012
69. Ahlborn J, Stephan A, Meckel T et al (2019) Upcycling of food industry side streams by basidiomycetes for production of a vegan protein source. Int J Recycl Org Waste Agric 8: 447–455. https://doi.org/10.1007/S40093-019-00317-4
70. Silva A, Rodrigues F, Prieto A et al (2023) Development of meat substitutes from filamentous fungi cultivated on residual water of tempeh factories. Molecules 28. https://doi.org/10.3390/molecules28030997
71. Upcraft T, Johnson R, Finnigan T et al (2020) Protein from renewable resources: mycoprotein production from agricultural residues. Comput Aided Chem Eng 48:985–990. https://doi.org/10.1016/B978-0-12-823377-1.50165-8
72. Risner D, McDonald KA, Jones C et al (2023) A techno-economic model of mycoprotein production: achieving price parity with beef protein. Front Sustain Food Syst 7. https://doi.org/10.3389/FSUFS.2023.1204307
73. Furlan O, de Oliveira NS, de Paula RC et al (2024) Pilot scale production of high-content mycoprotein using *Rhizopus microsporus* var. *oligosporus* by submerged fermentation and agro-industrial by-products. Bioresour Technol 413:131515. https://doi.org/10.1016/J.BIORTECH.2024.131515
74. Todhunter ME, Jubair S, Verma R et al (2024) Artificial intelligence and machine learning applications for cultured meat. Front Artif Intell 7:1424012. https://doi.org/10.3389/FRAI.2024.1424012
75. Cheng Y, Bi X, Xu Y et al (2023) Artificial intelligence technologies in bioprocess: opportunities and challenges. Bioresour Technol 369:128451. https://doi.org/10.1016/J.BIORTECH.2022.128451
76. Fronzek T, Lino FSDO, Zhao X (2023) Development of technofunctional ingredients and alternative proteins from biomass fermentation. Patent app. DE102022113445A1, 30 Nov 2023
77. Pattillo AH (2021) Enhanced aerobic fermentation methods for producing edible fungal mycelium blended meats and meat analogue compositions. Patent app. US11058137B2, 13 July 2021
78. Amara AA, El-Baky NA (2023) Fungi as a source of edible proteins and animal feed. J Fungi 9:73. https://doi.org/10.3390/JOF9010073
79. Molitorisova A, Monaco A (2023) Innovating food law with mycelium: EU regulations. SSRN Electron J. https://doi.org/10.2139/SSRN.4530242
80. Majumder R, Miatur S, Saha A et al (2024) Mycoprotein: production and nutritional aspects: a review. Sustain Food Technol 2:

81–91. https://doi.org/10.1039/D3FB00169E

81. Whiteley J, Huggins T (2023) Methods for forming directional mycelium fibers. Patent app. US20230084699A1, 16 Mar 2023
82. Miri T, Barigou M, Fryer PJ et al (2005) Flow induced fibre alignment in mycoprotein paste. Food Res Int 38:1151–1160. https://doi.org/10.1016/J.FOODRES.2005.04.005
83. Brandes D, Marison I, Akintoye M (2024) Fungal mycelium-based milk or cream substitute. Patent app. EP4298912A1, 3 Jan 2024
84. Akintoye M, Brandes D, Marison I (2024) Dairy substitute comprising mycoprotein. Patent app. WO2024251778A1, 12 Dec 2024
85. Casarramona JL, Karlsson AKS, Teixeira PG, et al (2023) Fungi-based fat tissue. Patent app. WO2023144148A1, 3 Aug 2023
86. Teixeira PG, Zhu S, Casarramona JL, et al (2024) A method of preparing a mycelium liquid extract and a dairy-free food product or food ingredient. Patent app. WO2024155225A1, 25 July 2024
87. Kozubal MA, Macur RE, Avniel YC (2020) Filamentous fungal biomats, methods of their production and methods of their use. Patent app. US10533155B2, 14 Jan 2020
88. Furey B, Slingerland K, Bauter MR et al (2022) Safety evaluation of fy protein™ (nutritional fungi protein), a macroingredient for human consumption. Food Chem Toxicol 166. https://doi.org/10.1016/j.fct.2022.113005
89. Kaplan-Bie JH, Greetham LE, Bonesteel IT, et al (2020) Increased homogeneity of mycological biopolymer grown into void space. Patent app. US20200146224A1, 14 May 2020
90. Winiski JM, Kaplan-Bie JH, McIntyre GR, et al (2022) Edible mycelia and methods of making the same. Patent app. US20220333055A1, 20 Oct 2022
91. Palermo A, Schultz R, Kasper M (2023) Composite materials comprising a polymer scaffold and methods of making and using. Patent app. WO2023244847A2, 21 Dec 2023
92. Palermo A, Schultz R (2024) Food product comprising a co-culture of bacteria and fungi. Patent app. US11986001B2, 21 May 2024
93. Spampinato E (2024) Aqua cultured foods pivots to cellulose-based platform. https://www.seafoodsource.com/aqua-cultured-foods-pivots-to-cellulose-basedplatform-nz-government-invests-nearly-usd-4-million-in-cultivated-seafoodprogram. Accessed: 31 May 2025
94. Mazac R, Meinilä J, Korkalo L et al (2022) Incorporation of novel foods in European diets can reduce global warming potential, water use and land use by over 80%. Nat Food 3:286–293. https://doi.org/10.1038/S43016-022-00489-9
95. Graham AE, Ledesma-Amaro R (2023) The microbial food revolution. Nat Commun 14:1–10. https://doi.org/10.1038/s41467-023-37891-1
96. Rekdal VM, van der Luijt CR, Chen Y et al (2024) Edible mycelium bioengineered for enhanced nutritional value and sensory appeal using a modular synthetic biology toolkit. Nat Commun 15:1–18. https://doi.org/10.1038/s41467-024-46314-8
97. Chai KF, Ng KR, Samarasiri M et al (2022) Precision fermentation to advance fungal food fermentations. Curr Opin Food Sci 47:100881. https://doi.org/10.1016/J.COFS.2022.100881
98. Li C, Zhou J, Du G et al (2020) Developing *Aspergillus niger* as a cell factory for food enzyme production. Biotechnol Adv 44:107630. https://doi.org/10.1016/J.BIOTECHADV.2020.107630
99. Mojsov KD (2016) *Aspergillus* enzymes for food industries. In: New and future developments in microbial biotechnology and bioengineering: *Aspergillus* system properties and applications, pp 215–222. https://doi.org/10.1016/B978-0-444-63505-1.00033-6
100. Zhang XY, Li B, Huang BC et al (2022) Production, biosynthesis, and commercial applications of fatty acids from oleaginous fungi. Front Nutr 9:873657. https://doi.org/10.3389/FNUT.2022.873657
101. Stöppelmann F, Chan LF, Hildebrand G et al (2024) Molecular decoding a meat-like aroma generated from *Laetiporus sulphureus*-mediated fermentation of onion (allium cepa l.). Food Res Int 192:114757. https://doi.org/10.1016/J.FOODRES.2024.114757
102. Takahashi JA, Barbosa BV, de Almeida Martins B et al (2020) Use of the versatility of fungal metabolism to meet modern demands for healthy aging, functional foods, and sustainability. J Fungi 6:223. https://doi.org/10.3390/JOF6040223
103. MycoTechnology, Inc. (2025) Cleariq™ natural flavor modulation. https://www.mycoiq.com/ingredients/clear-iq/. Accessed: 1 June 2025
104. Casey C (2025) Mycotechnology's honey truffle sweetener receives gras approval.

https://www.fooddive.com/news/mycotechnology-mycelium-honey-trufflesweetener-ingredient-citrus-caramel-flavor-taste-GRAS-clearHT/745059/. Accessed: 1 June 2025

105. Watson E (2025) Mycotechnology discontinued alt protein biz last year to focus on flavor modulation, sweetening, says CEO. https://agfundernews.com/mycotechnology-discontinued-alt-protein-bizlast-year-to-focus-on-flavor-modulation-sweetening-says-ceo. Accessed: 1 June 2025
106. Nair RB, Lennartsson PR, Taherzadeh MJ (2016) Mycelial pellet formation by edible ascomycete filamentous fungi, neurospora intermedia. AMB Express 6:31. https://doi.org/10.1186/s13568-016-0203-2
107. Veiter L, Rajamanickam V, Herwig C (2018) The filamentous fungal pellet—relationship between morphology and productivity. Appl Microbiol Biotechnol 102:2997–3006. https://doi.org/10.1007/s00253-018-8818-7
108. Whittaker JA, Johnson RI, Finnigan TJ et al (2020) The biotechnology of quorn mycoprotein: past, present and future challenges. In: Grand challenges in biology and biotechnology, pp 59–79. https://doi.org/10.1007/978-3-030-29541-7_3
109. Vegconomist (2018) Quorn opens the world's biggest alternative meat production factory. https://vegconomist.com/company-news/quorn-opens-the-worldsbiggest-alternative-meat-production-factory/. Accessed: 29 May 2025
110. European Innovation Council and SMEs Executive Agency (2021) EIC funded biotechnology company Mycorena raised €7.5 million in pre-series a funding. https://eic.ec.europa.eu/success-stories/eic-funded-biotechnologycompany-mycorena-raised-eu75-million-pre-series-funding. Accessed: 30 May 2025
111. Marston J (2022) Mycorena closes $27m in Europe's largest alt-protein Series A. https://agfundernews.com/mycorena-closes-27m-in-europes-largest-altprotein-series-a. Accessed: 30 May 2025
112. Tech.eu (2024) Naplasol acquires Mycorena to boost the mycelium and alt-protein market. https://tech.eu/2024/08/29/naplasol-acquires-mycorena-toenhance-mycoprotein-portfolio/. Accessed: 31 May 2025
113. FoodNavigator (2022) 'This is no longer moonshot technology': Swiss venture raises $8m to 'de-bottleneck' food fermentation industry. https://www.foodnavigator.com/Article/2022/04/20/this-is-no-longermoonshot-technology-swiss-venture-raises-8m-to-de-bottleneck-foodfermentation-industry/. Accessed: 30 May 2025
114. Royal Cosun (2024) Royal cosun invests in fermentation scale-up planetary. https://www.cosun.com/news/royal-cosun-invests-in-fermentation-scaleup-planetary/. Accessed: 31 May 2025
115. The Protein Brewery (2020) The protein brewery raises €22 million Series A investment. https://www.theproteinbrewery.nl/press-release-03-11-2020. Accessed: 30 May 2025
116. Future Market Insights (2024) Fungal protein market. https://www.futuremarketinsights.com/reports/fungal-protein-market. Accessed: 27 May 2025
117. Sganzerla WG, Todorov SD, da Silva AP (2022) Research trends in the study of edible mushrooms: nutritional properties and health benefits. Int J Med Mushrooms 24:1–15. https://doi.org/10.1615/INTJMEDMUSHROOMS.2022043738
118. The Business Research Company (2025) Mushroom global market report 2025. https://www.researchandmarkets.com/reports/5939329/mushroom-marketreport. Accessed: 31 May 2025
119. Granato G, Wassmann B (2024) To imitate or not to imitate? How consumers perceive animal origin products and plant-based alternatives imitating minimally processed vs ultra-processed food. J Clean Prod 472:143447. https://doi.org/10.1016/J.JCLEPRO.2024.143447
120. Fischer AR, Hilboesen O (2025) Consumer acceptance of mycelium as protein source. Food Qual Prefer 122:105304. https://doi.org/10.1016/J.FOODQUAL.2024.105304
121. Coelho MO, Monteyne AJ, Dunlop MV et al (2020) Mycoprotein as a possible alternative source of dietary protein to support muscle and metabolic health. Nutr Rev 78:486–497. https://doi.org/10.1093/NUTRIT/NUZ077
122. Muthuramalingam K, Singh V, Choi C et al (2020) Dietary intervention using (1,3)/(1,6)—glucan, a fungus-derived soluble prebiotic ameliorates high-fat diet induced metabolic distress and alters beneficially the gut microbiota in mice model. Eur J Nutr 59: 2617–2629. https://doi.org/10.1007/S00394-019-02110-5
123. Yu L, Gao Y, Ye Z et al (2023) Interaction of beta-glucans with gut microbiota: dietary

origins, structures, degradation, metabolism, and beneficial function. Crit Rev Food Sci Nutr. https://doi.org/10.1080/10408398.2023.2217727

124. Xiong RG, Zhou DD, Wu SX et al (2022) Health benefits and side effects of short-chain fatty acids. Foods 11:2863. https://doi.org/10.3390/FOODS11182863
125. Chen KI, Erh MH, Su NW et al (2012) Soyfoods and soybean products: from traditional use to modern applications. Appl Microbiol Biotechnol 96:9–22. https://doi.org/10.1007/S00253-012-4330-7
126. Nurmilah S, Frediansyah A, Cahyana Y et al (2024) Biotransformation and health potential of isoflavones by microorganisms in indonesian traditional fermented soy products: a review. J Agric Food Res 18:101365. https://doi.org/10.1016/J.JAFR.2024.101365
127. Turnbull WH, Leeds AR, Edwards GD (1990) Effect of mycoprotein on blood lipids. Am J Clin Nutr 52:646–650. https://doi.org/10.1093/ajcn/52.4.646
128. Coelho MO, Monteyne AJ, Dirks ML et al (2021) Daily mycoprotein consumption for 1 week does not affect insulin sensitivity or glycaemic control but modulates the plasma lipidome in healthy adults: a randomised controlled trial. Br J Nutr 125:147–160. https://doi.org/10.1017/S0007114520002524
129. Turnbull WH, Ward T (1995) Mycoprotein reduces glycemia and insulinemia when taken with an oral-glucose-tolerance test. Am J Clin Nutr 61:135–140. https://doi.org/10.1093/AJCN/61.1.135
130. Zhang JH, Tatsumi E, Ding CH et al (2006) Angiotensin I-converting enzyme inhibitory peptides in douchi, a Chinese traditional fermented soybean product. Food Chem 98:551–557. https://doi.org/10.1016/J.FOODCHEM.2005.06.024
131. Varnosfaderani SM, Razavi SH, Fadda AM (2019) Germination and fermentation of soybeans: two healthy steps to release angiotensin converting enzyme inhibitory activity compounds. Appl Food Biotechnol 6:201–215. https://doi.org/10.22037/AFB.V6I4.25475
132. Song JH, Kim JK, Jang HD (2014) Ferulic acid released by treatment with *Aspergillus oryzae* contributes to the cellular antioxidant capacity of wheat germ extract. Food Sci Biotechnol 23:1327–1333. https://doi.org/10.1007/S10068-014-0182-5
133. Ahmad A, Ramasamy K, Jaafar SM et al (2014) Total isoflavones from soybean and tempeh reversed scopolamine-induced amnesia, improved cholinergic activities and reduced neuroinflammation in brain. Food Chem Toxicol 65:120–128. https://doi.org/10.1016/J.FCT.2013.12.025
134. Yu J, Bi X, Yu B et al (2016) Isoflavones: anti-inflammatory benefit and possible caveats. Nutrients 8:361. https://doi.org/10.3390/NU8060361
135. Yusof HM, Ali NM, Yeap SK et al (2019) Anti-inflammatory, analgesic and acute toxicity effects of fermented soybean. BMC Complement Altern Med 19:373. https://doi.org/10.1186/S12906-019-2791-2
136. Teoh SQ, Chin NL, Chong CW et al (2024) A review on health benefits and processing of tempeh with outlines on its functional microbes. Future Foods 9:100330. https://doi.org/10.1016/J.FUFO.2024.100330
137. Nachimuthu S, Kandasamy R, Ponnusamy R et al (2019) L-ergothioneine: a potential bioactive compound from edible mushrooms. Med Mushrooms 16:391–407. https://doi.org/10.1007/978-981-13-6382-5
138. Podkowa A, Kryczyk-Poprawa A, Opoka W et al (2021) Culinary–medicinal mushrooms: a review of organic compounds and bioelements with antioxidant activity. Eur Food Res Technol 247:513–533. https://doi.org/10.1007/S00217-020-03646-1
139. Tsoupras A, Davi KG (2024) Bioactive metabolites from fungi with antiinflammatory and antithrombotic properties: current status and future perspectives for drug development. In: Fungi bioactive metabolites: integration of pharmaceutical applications, pp 427–494. https://doi.org/10.1007/978-981-99-5696-8_14
140. Zhang J, Tyler HL, Haron MH et al (2019) Macrophage activation by edible mushrooms is due to the collaborative interaction of toll-like receptor agonists and dectin-1b activating beta glucans derived from colonizing microorganisms. Food Funct 10:8208–8217. https://doi.org/10.1039/C9FO01707K
141. van Steenwijk HP, Bast A, de Boer A (2021) Immunomodulating effects of fungal beta-glucans: from traditional use to medicine. Nutrients 13:1333. https://doi.org/10.3390/NU13041333
142. Tee RD, Gordon DJ, Welch JA et al (1993) Investigation of possible adverse allergic reactions to mycoprotein ('Quorn'). Clin Exp Allergy 23:257–260. https://doi.org/10.1111/J.1365-2222.1993.TB00319.X
143. Shand AG (2002) Investigation into adverse reactions to mycoprotein. PhD thesis.

Edinburgh Medical School thesis and dissertation collection

144. Finnigan TJ, Wall BT, Wilde PJ et al (2019) Mycoprotein: the future of nutritious nonmeat protein, a symposium review. Curr Dev Nutr 3:nzz021. https://doi.org/10.1093/CDN/NZZ021

145. Dolatabadi S, Scherlach K, Figge M et al (2016) Food preparation with mucoralean fungi: a potential biosafety issue? Fungal Biol 120:393–401. https://doi.org/10.1016/J.FUNBIO.2015.12.001

146. Frisvad JC, Møller LL, Larsen TO et al (2018) Safety of the fungal workhorses of industrial biotechnology: update on the mycotoxin and secondary metabolite potential of *Aspergillus niger, Aspergillus oryzae*, and *Trichoderma reesei*. Appl Microbiol Biotechnol 102:9481–9515. https://doi.org/10.1007/S00253-018-9354-1

147. Florea A, Sipos A, Stoisor MC (2022) Applying AI tools for modeling, predicting and managing the white wine fermentation process. Fermentation 8:137. https://doi.org/10.3390/FERMENTATION8040137

148. Amore A, Philip S (2023) Artificial intelligence in food biotechnology: trends and perspectives. Front Ind Microbiol 1:1255505. https://doi.org/10.3389/FINMI.2023.1255505

149. Dhiman S, Kaur P, Narang J et al (2024) Fungal bioprocessing for circular bioeconomy: exploring lignocellulosic waste valorization. Mycology. https://doi.org/10.1080/21501203.2024.2316824

150. Rekdal VM, Villalobos-Escobedo JM, Rodriguez-Valeron N et al (2024) Neurospora intermedia from a traditional fermented food enables waste-to-food conversion. Nat Microbiol 9. https://doi.org/10.1038/S41564-024-01799-3

151. Springmann M, Clark M, Mason-D'Croz D et al (2018) Options for keeping the food system within environmental limits. Nature 562:519–525. https://doi.org/10.1038/S41586-018-0594-0

Chapter 19

Filamentous Fungi for the Valorization of Plant-Based Foods and By-products: Protocols and Applications

Mussa Makran, Sara Muñoz-Pina, Matteo Vitali, Camila Mendoza, Antonio Cilla, Ana Heredia, Jorge García-Hernández, Amparo Gamero, Ana Andrés, and Mónica Gandía

Abstract

The increasing demand for sustainable food systems has driven interest in the valorization of plant-based foods and their by-products. Filamentous fungi offer a promising biotechnological tool for transforming low-value biomass into nutritionally enriched or functionally enhanced products while reducing food waste and contributing to the development of a circular food economy.

This chapter presents a comprehensive protocol for the fungal valorization of plant-based substrates, detailing the materials, fermentation systems, and cultivation methods. Two primary substrate categories are explored: plant-based foods and their derived industrial by-products. The methodology covers the preparation of substrates, the development of fungal inoculum, and the use of various solid-state fermentation systems (using Petri dishes, containers, or fermentation bags). Specific steps such as biomass quantification, drying, and crushing of the samples are also outlined to ensure reproducibility. Emphasis is placed on the selection of fungal strains and the optimization of growth conditions to maximize valorization efficiency. The protocols and applications described support the development of circular food production models by converting plant-based foods and waste into valuable products, with potential implications as a functional ingredient in the food industry.

Key words Fermentation, Filamentous fungi, *Pleurotus*, *Aspergillus*, *Fusarium*, Plant-based foods, Mycoprotein, By-products, Valorization

1 Introduction

The global shift towards sustainable food systems has increased scientific interest in plant-based diets and the valorization of food by-products. Plant-based foods are increasingly recognized for their health benefits and low environmental impact [1]. However, despite these recognized advantages, there remains untapped potential in plant-based foods, particularly in improving the

Authors Mussa Makran and Sara Muñoz-Pina have equally contributed to this chapter.

Pablo Navarro Simarro (ed.), *Fungi*, Methods and Protocols in Food Science, https://doi.org/10.1007/978-1-0716-4989-3_19,
© The Author(s), under exclusive license to Springer Science+Business Media, LLC, part of Springer Nature 2026

bioavailability of nutrients [2]. Furthermore, the production and processing of plant-based foods generate a significant number of by-products, such as pomace, seeds, and peels, which are often discarded [3]. In this context, advances in food biotechnology offer innovative solutions to current food challenges, allowing not only the improvement of nutritional and functional properties of plant-based foods but also the conversion of by-products into high-value products. Through these strategies, food waste would be reduced, thus contributing to the development of a circular food economy and generating food products of both nutritional and health interest.

Filamentous fungi, with their remarkable metabolic versatility, present promising solutions to modern food and environmental challenges [4]. Species from genera such as *Rhizopus*, *Aspergillus*, and *Fusarium* have long been employed in the production of fermented foods. Beyond their traditional uses, their ability to transform a wide range of substrates into valuable compounds makes them an effective tool to optimize the properties of plant-based foods and valorize plant-based food by-products. In addition, other filamentous fungi, such as *Pleurotus ostreatus*, though not traditionally used in fermentation, are increasingly being investigated for their bioconversion potential and mycoprotein production on plant-based substrates. In this context, fermentation with filamentous fungi has proven to be effective in improving the nutritional value of plant-based foods; reducing the content of antinutritional factors (e.g., phytic acid and tannins), which could improve the bioavailability of essential minerals; and increasing protein digestibility [5]. Nowadays, fungi have attracted increasing attention for their ability to convert plant-based waste into protein-rich biomass, known as mycoprotein. Mycoprotein is a sustainable, high-quality protein source with immense potential for food innovation. This ingredient not only benefits human health by modulating biological processes and offering positive effects on cardiometabolic health but also promotes more sustainable agricultural practices by improving crop resilience to environmental stress and pathogens [6, 7]. Mycoprotein production, therefore, addresses the global challenge of dietary protein shortage while incorporating the principles of a circular bioeconomy, where plant-based by-products are transformed into valuable food ingredients.

In line with the scientific interest described above, this chapter offers an in-depth exploration of the various protocols, methodologies, and innovative applications of filamentous fungi as a sustainable and efficient alternative for the valorization of plant-based foods and the use of their by-products to obtain mycoprotein.

2 Materials

2.1 Materials for Valorization of Plant-Based Foods

1. *Pleurotus ostreatus* (CECT 20311).
2. Sterile gauze and pipette.
3. Laminar flow cabinet.
4. Bidistilled water.
5. 70% (v/v) ethanol.
6. Nutrient broth.
7. Malt broth and malt agar media.
8. Erlenmeyer flasks (150 mL).
9. Containers with integrated filters (types of filters are detailed in Table 1).
10. Fungus fermentation bags.
11. 72% (v/v) sulfuric acid.

Table 1
Overview of filter types employed in solid-state fermentation processes utilizing fungi as the microbial agent

Filter type	Pore size	Air exchange	Contamination risk	Type of filter	Characteristics	Typical use
T1	~0.2 μm	Low	Very low	Microporous membrane adhered to the bag or container, resistant to autoclave temperature	Ideal for sterile incubation phases	Initial incubation, sensitive strains
T2	~0.5 μm	Moderate	Low		Used when more ventilation is needed or when the risk of contamination is lower	General-purpose fermentation
T3	~0.5 μm	High (larger filter area)	Medium			Fast-growing fungi, higher oxygen demand
A	~0.2 μm	Very low	Very low	Microporous paper or synthetic membranes, autoclave resistant	Poor gas exchange. It can limit fungal growth if used for a long time	Master cultures, liquid inoculum, high sterility needs

(continued)

Table 1 (continued)

Filter type	Pore size	Air exchange	Contamination risk	Type of filter	Characteristics	Typical use
B	~5.0 μm	Very high	High	Microporous paper or special nonwoven fabric, autoclave resistant	Ideal for fungi with high oxygen demand but not recommended for substrates with high microbial load	Fruiting stage, species requiring high ventilation

Data collected from the commercial companies: Unicorn Exclusive S.L., Mushroom Solutions S.L., and ACC Mushroom Cultivation Consulting S.L.U.

12. 0.5 M sodium hydroxide: dissolve 2 g of sodium hydroxide in 100 mL of distilled water.
13. Acetylacetone reagent: mix 1 mL of acetylacetone with 50 mL of 0.5 M sodium carbonate, previously prepared by dissolving 5.3 g of anhydrous sodium carbonate in distilled water and making up to 100 mL.
14. Ehrlich's reagent: dissolve 2.67 g of 4-dimethylaminobenzaldehyde in 100 mL of a 1:1 mixture of ethanol (reagent grade) and concentrated hydrochloric acid (HCl).
15. N-Acetyl-D-glucosamine calibration line: dissolve different concentrations (from 0 to 0.5 mg/mL) in purified water.

2.2 Materials for Valorization of By-products

1. Plant-based by-products from the production of vegetable beverages, typically consisting of the solid residues (pressing cakes) remaining after filtration.
2. Distilled sterile water.
3. Fungal strains (*Aspergillus oryzae* var. *oryzae* CBS 819.72 and *Fusarium venenatum* CBS 458.93) available from fungal culture collections.
4. Culture media: Potato Dextrose Agar (PDA), Oat Agar (OA), Malt Extract Agar (MEA).

   The composition of these media is detailed in Table 2.
5. Petri dishes (round and square) or containers for solid-state fermentation and fungal cultivation.
6. Pipettes, sterile pipette tips, and tubes.
7. Sterile inoculation loops and Drigalsky handles.
8. Erlenmeyer flasks (100–250 mL).

Table 2
Composition of culture media used for fungal growth

Medium	Ingredients (g/L)
MEA	Agar (15), maltose (12.75), dextrin (2.75), glycerol (2.35), peptone (0.78)
PDA	Dextrose (20), agar (15), potato extract (4)
OA	Oat flour (60), agar (12.5)

MEA malt extract agar, *PDA* potato dextrose agar, *OA* oat agar

9. D-Glucose (for preparing 4% w/v solution for substrate supplementation).
10. Yeast extract (for preparing 2% w/v solution for substrate supplementation).
11. Miracloth, filter-sterilized for fungal spore filtration.
12. Cell counting chamber (Neubauer or similar).
13. 80% (v/v) glycerol solution for spore preservation.
14. Incubator (for maintaining 25 °C).
15. Food dehydrator or drying oven for drying samples.
16. Food processor or similar device for crushing substrates.
17. Electric grinder for processing dried samples.
18. Sterile containers for storing processed samples.
19. Laminar flow cabinet.

3 Methods

3.1 Fermentation Inoculum for Valorization of Plant-Based Food

Solid-state fermentation (SSF) typically begins with the use of a lyophilized fungal strain, *Pleurotus ostreatus* in this case, which is first reactivated in a liquid medium or on agar plates. Once the culture is successfully revived, fermentation can proceed either by directly employing the inoculum or by preparing a solid or liquid starter, depending on the process requirements and scale of production.

1. Resuspension of *Pleurotus ostreatus* (CECT 20311) lyophile to ensure the viability of the microorganism: the lyophile vial is opened in a laminar flow cabinet under sterile conditions. The exterior of the vial is disinfected (with 70% (v/v) ethanol), and the glass is carefully broken by heating it and using sterile gauze. To resuspend the lyophile, 1 mL of nutrient broth is added to the vial using a sterile pipette, and the lyophilized contents are dissolved by pipetting gently to avoid damaging the cells.

Fig. 1 *Pleurotus ostreatus* (CECT 20311) strain cultivated on (**a**) malt agar plates, (**b**) liquid fermentation medium inoculated with *Pleurotus ostreatus* strain

2. *Pleurotus ostreatus* initial culture: *Pleurotus ostreatus* from the resuspended lyophile is plated directly onto malt agar plates and incubated at 28 °C for 14 days (Fig. 1).
3. Malt agar medium: dissolve 20 g of malt extract, 20 g of glucose, 15 g of agar, and 1 g of mycopeptone in 1 L of bidistilled water. Sterilize by autoclaving at 121 °C for 20 min.
4. Malt liquid medium: dissolve 20 g of malt extract, 20 g of glucose, and 1 g of mycopeptone in 1 L of bidistilled water. Adjust the pH to 5.3 (*see* **Note 1**). Divide the content into 150 mL Erlenmeyer flask with baffles and GL45 membrane cap. Sterilize by autoclaving at 121 °C for 20 min.
5. Liquid starter: from the malt agar plates with the *Pleurotus* mycelium, a small amount is taken using the loop and suspended in 150 mL of the liquid culture medium (refer to **step 3** within this section) (*see* **Note 2**). The liquid starters are incubated in a chamber at 25 °C with constant horizontal shaking (*see* **Note 3**) for 8 days in the Erlenmeyer flasks (Fig. 1).

3.2 Containers and Systems

The selection of the appropriate SSF system is a critical factor influencing process efficiency and microbial performance. Both the filter type and the level of oxygen transfer play key roles in maintaining optimal growth conditions and preventing contamination. Table 1 provides an overview of commonly available filter types in the market. Different fermentation vessels and filter configurations may be required depending on the quantity of plant-based substrate to be processed.

1. Containers with integrated filters: ideal for small-scale fermentations (50–100 g of plant-based substrate), where the container is filled to about one-third of its volume. In these cases, a T1 filter is preferred due to its high sterility and controlled airflow. An example is depicted in Fig. 2.
2. Fungus fermentation bags: bags equipped with integrated filters, positioned either along the surface or on the top, can accommodate up to 2 kg of substrate, making them suitable

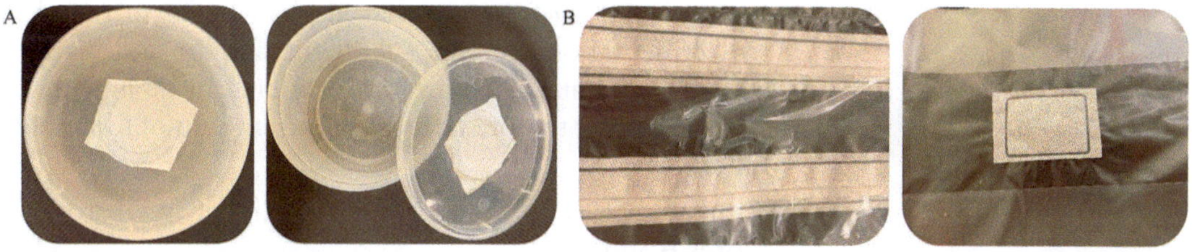

Fig. 2 Representative containers and bags used for solid-state fermentation (SSF) of plant-based substrates. (**a**) Rigid container fitted with a T1 microporous filter, suitable for small-scale applications. (**b**) Flexible cultivation bag with an integrated strip filter or a gusseted bag equipped with an A-type filter

for medium- to large-scale solid-state fermentations (Fig. 2). These systems typically employ T1, T2, or A-type filters, selected based on the specific oxygenation requirements of the microorganism. Strip-filter bags are designed to be fully filled without compromising gas exchange. In contrast, gusseted bags are generally more effective when filled to approximately one-third of their volume to ensure adequate aeration and prevent substrate compaction.

3.3 Plant-Based Preparation

Various plant-based substrates can be utilized in SSF with *Pleurotus ostreatus* as the fungal strain. Among this wide variety, legumes were chosen as a model in this work due to their high nutritional value and relevance in sustainable food systems (*see* **Note 2**).

1. Hydration: dried substrates are soaked in water (22 ± 2 °C) at a 1:6 (w/v) ratio to allow proper rehydration (*see* **Note 4**).
2. Particle size adjustment: substrates with large particle size (e.g., fava beans or common beans) are ground using a food processor until reaching an average particle size of 0.5 ± 0.1 cm.
3. Thermal pretreatment: the substrate is blanched for 2 min in boiling water, followed by immediate cooling in cold water.
4. Sterilization: the processed substrate is distributed into containers equipped with a T1 microporous filter (0.2 μm pore size) and autoclaved for 30 min at 123 °C. The final moisture content of the legumes should be adjusted to 45–55% (*see* **Note 4**).

3.4 Preparation of Starter for Solid-State Fermentation (SSF)

The fermentation of plant-based substrates with *Pleurotus ostreatus* can be initiated using different types of inoculum: (i) liquid culture, (ii) malt agar plates, or (iii) previously fermented substrate used as a solid starter.

(i) Liquid culture as inoculum:

1. Aseptically withdraw a 5 mL aliquot of *P. ostreatus*-colonized liquid culture (liquid starter, Subheading 2.1) inside a laminar flow hood.

2. Inoculate 50 g of the autoclaved plant-based substrate in the fermentation container with the aliquot and securely close the container. Multiple containers can be inoculated simultaneously to obtain the required amount of fermented biomass.
3. Incubate at 25 °C for 7 days in a digital incubator (*see* **Notes 5** and **6**).
4. After incubation, freeze the samples until further analysis or use.

(ii) Malt agar plates as inoculum:

1. Cut a colonized malt extract agar Petri dish (*Pleurotus ostreatus* initial culture, Subheading 2.1) into eight triangular sections inside a laminar flow hood.
2. Place one triangular section onto 50 g of autoclaved substrate in the sterile fermentation container. Multiple containers can be inoculated simultaneously to obtain the required amount of fermented biomass.
3. Incubate at 25 °C for 14 days (Fig. 3).
4. Freeze the samples until needed.

Fig. 3 Representative images of *Pleurotus ostreatus* mycelial growth on plant-based substrates after 7–10 days of inoculation. (**a**) Mycelial development in a 50 g container with autoclaved beans. (**b**) Mycelial colonization on lentil-based substrate in a 2 kg strip-filter bag and a gusseted-type cultivation bag

(iii) Fermented substrate as solid starter:

1. Inside a laminar flow hood, open the container with previously fermented substrate and aseptically collect 1/8 of its content.
2. Inoculate 50 g of fresh autoclaved substrate with the collected portion of fermented substrate.
3. Incubate at 25 °C for 14 days. Multiple containers can be inoculated simultaneously to obtain the required amount of fermented biomass.
4. Freeze the samples until further use.

3.5 Fermentation in Bag Containers

As previously described, SSF can also be performed using autoclavable polypropylene bags equipped with microporous filters, allowing the processing of larger quantities of plant-based substrates. In this approach, the 50 g of fermented legumes produced in containers with T1 filters serves as the solid starter culture. The procedure is as follows:

1. Soak, blanch, wash, and cool 1 kg of dried legumes as described in Subheading 3.3.
2. Divide the hydrated legumes into autoclavable polypropylene bags (~500 g per bag), placing them horizontally to maintain a maximum substrate thickness of 3 cm per bag (*see* **Note 7**; Fig. 3).
3. Place the bags on separate trays in the autoclave to prevent smashing. Autoclave at 123 °C for 90 min, then allow cooling (*see* **Note 8**).
4. Under sterile conditions in a laminar flow hood, open the sterilized bags individually and transfer their contents into a fermentation bag (strip or gusseted) (*see* **Note 9**).
5. Add 15% (w/w) of solid fermented legume starter for every 500 g of sterilized substrate.
6. Mix the contents gently by hand to ensure uniform distribution of the inoculum.
7. Seal the bag using a heat or impulse sealer and incubate at 25 ° C for 7 days in a fermentation chamber (*see* **Note 10**).
8. Following incubation, freeze the samples until further use.

The process used to obtain fungal biomass through SSF of plant-based substrates is summarized in the flowchart presented in Fig. 4.

3.6 Biomass Determination

In solid-state fungal fermentation, the amount of mycelium cannot be directly measured due to its strong connection with the solid matrix. However, estimating fungal growth is essential for both process optimization and determining the appropriate

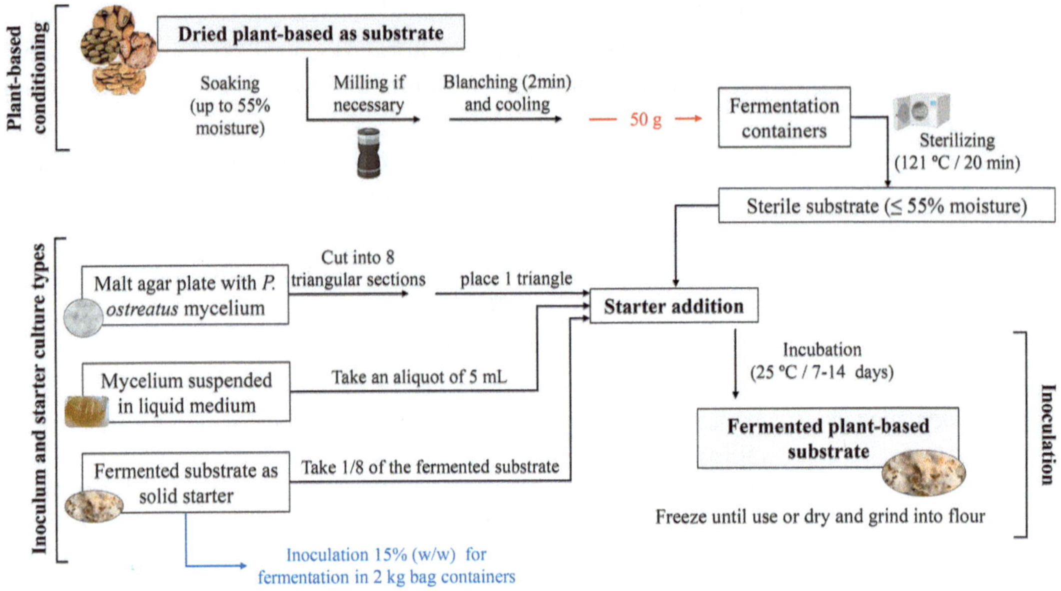

Fig. 4 Flowchart of the process to produce fungal biomass via solid-state fermentation using plant-based substrates

fermentation time. Among the various existing methods, the indirect measurement of cell wall components such as chitin and glucosamine is considered a reliable indicator of fungal biomass development.

This method involves acid hydrolysis at high temperature, in which cellular chitin is first broken down into N-acetylglucosamine (GlcNAc), followed by deacetylation to produce D-glucosamine (GlcN). This glucosamine contains a primary amine group, which reacts with Ehrlich's reagent, allowing its detection via spectrophotometry. To perform the assay, follow the steps described by Aidoo et al. [8] and Tomaselli Scotti et al. [9]:

1. Weigh 100 mg of fermented dry sample and mix with 2.4 mL of 72% sulfuric acid. Incubate at 25 °C for 24 h.
2. Dilute the mixture with 55 mL of distilled water and autoclave at 121 °C for 2 h to complete the hydrolysis.
3. Adjust the pH of the hydrolysate to 7 using 10 M and 0.5 M sodium hydroxide solutions.
4. Mix 1 mL of the neutralized hydrolysate with 1 mL of acetylacetone reagent in a sealed glass tube. Heat the mixture in a boiling water bath for 20 min, then cool to room temperature.
5. Add 6 mL of ethanol, followed by 1 mL of Ehrlich's reagent.
6. Incubate the reaction mixture at 65 °C for 10 min, then cool to room temperature. Measure the absorbance at 530 nm using a spectrophotometer (*see* **Note 11**).

7. Express results as mg of glucosamine per gram of dry sample, based on the calibration curve with standard glucosamine solutions (from 0 to 0.5 mg/mL of glucosamine).

3.7 By-products Preparation

The primary raw materials used as substrates in this study consisted of by-products from the food industry, specifically solid residues obtained during the production of plant-based beverages from various raw materials. Conditioning of the by-products was carried out as follows:

1. Obtain by-products from the production of plant-based beverages. These are typically the solid residues remaining after the process of washing and rehydrating the raw materials, grinding with water, and pressing/filtering to separate the liquid from the solid.
2. Immediately after collection, store the by-product cakes at approximately −18 °C to prevent microbial growth or degradation until further processing.
3. Prior to use, thaw the frozen by-products in a refrigeration unit at approximately 4 °C for 24 h.
4. After thawing, divide the by-products into distinct batches for different pre-treatment methods:
   (i) Direct use method: package the first batch directly in individual aluminum foil packages, each containing approximately 30 g of substrate. Sterilize these packages in an autoclave at 121 °C for 15 min.
   (ii) Dehydration and milling method: subject the second batch to a dehydration process by placing it in a drying oven at 55 °C for 24 h. Grind the dehydrated residue into smaller particles using a suitable food processor or mill. Package the ground substrate and sterilize in an autoclave at 121 °C for 15 min.
5. Keep both sterilized batches of substrate sealed and store at room temperature until the time of inoculation with the fungal strains.

3.8 Growth of Fungal Strains for Valorization of By-products

Among the most widely used filamentous fungi for food-grade mycoprotein production are members of the family Ascomycota, specifically *Fusarium venenatum* CBS 458.93 and *Aspergillus oryzae* var. *oryzae* CBS 819.72. The recovery of lyophilized spores and preparation of fungal cultures can be carried out following these steps:

1. Heat the narrow tip of the ampule containing the lyophilized fungi using a flame source, keeping it to the flame for approximately 10 s. Immediately place a few drops of sterile water on the heated tip to crack the glass. Once the ampule is open, remove the cotton plug using a sterile clamp.

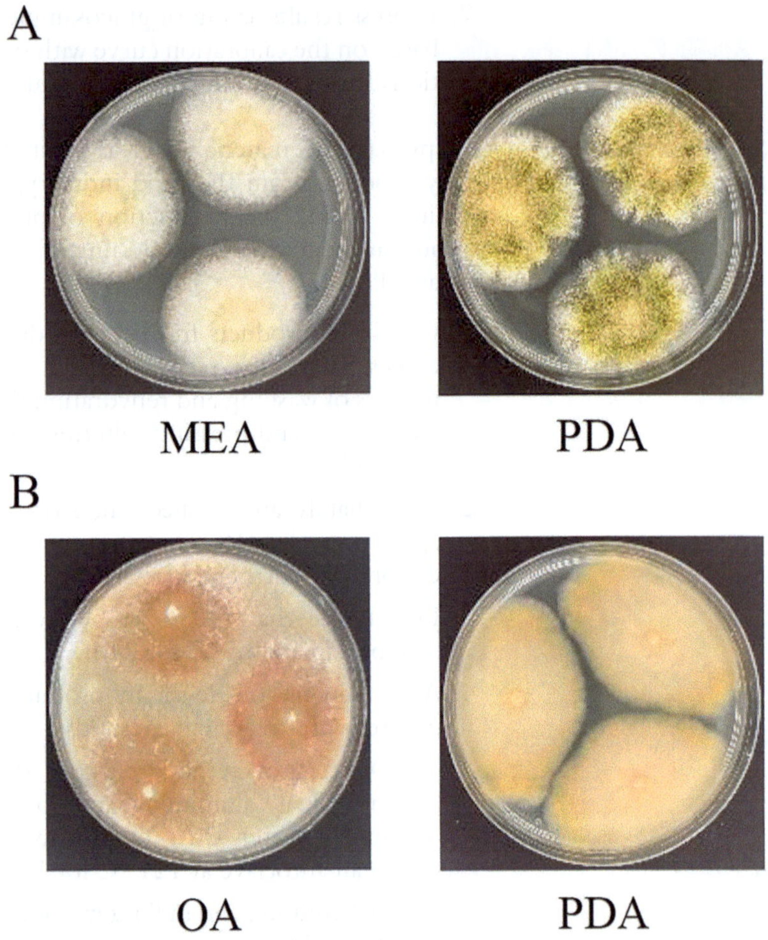

Fig. 5 Representative images showing the growth of different fungal strains 6 days after inoculation. (**a**) Growth of *Aspergillus oryzae* var. *oryzae* CBS 819.72 strain on Malt Extract Agar (MEA) and Potato Dextrose Agar (PDA) plates. (**b**) Growth of *Fusarium venenatum* CBS 458.93 strain on Oat Agar (OA) and PDA plates

2. Add approximately 0.5 mL of distilled sterile water to the ampule using a pipette. Carefully aspirate and expel the suspension to resuspend the lyophilized material. After resuspending the lyophilized material in sterile water, inoculate a small amount (approximately 5 μL) of the suspension at three points on solid medium agar plates, to obtain isolated colony growth (Fig. 5), or distribute a larger amount (approximately 20 μL) over the entire surface of the plate to obtain growth covering the entire plate. For *F. venenatum*, inoculate on PDA and OA plates. For *A. oryzae*, inoculate on PDA and MEA plates following standard mycological procedures, as recommended by the fungal culture collection (in this case, according to Westerdijk Fungal Biodiversity Institute, Utrecht, The Netherlands).

3. The following steps refer to routine methods that have been described in other assays (*see* preparation of fungal spores in [10]). Incubate the inoculated plates in the dark at 25 °C for approximately 5–7 days to allow sufficient growth of the fungal strains until spores appear. Maintain and propagate the strains routinely by preparing spore suspensions from the inoculated plates. Each fungus has an optimum growth temperature.
4. Prepare the spore suspension by adding distilled sterile water (approximately 8 mL) to the surface of cultured agar plate and scraping the fungal mycelium with a sterile loop to release the spores. Other types of solutions can be used (*see* **Note 12**). Collect the solution using a pipette and filter it through a sterile Miracloth filter placed in a sterile tube or pipette to remove large debris. In some cases, the filtration must be repeated.
5. Determine the spore concentration using a cell counting Neubauer chamber and dilute the spores to the required concentration.
6. Preserve 100 μL aliquots of the spore suspension in 80% v/v glycerol solution and store at −80 °C for subsequent use.
7. For new propagation of the strains, inoculate another agar plates in duplicate with 50 μL of the aliquots of glycerol stock of fungal spores recovered and thawed, spreading it with the help of a sterile Drigalsky handle.
8. Incubate the inoculated agar plates in the dark at 25 °C for 5–7 days, until the propagation process is repeated.
9. To prepare the fungal inoculum for growth on the by-products, use the spore suspensions recovered from the culture plates, with the initial concentration previously calculated. Dilute this initial spore suspension with distilled sterile water, transferring the appropriate volume to a new sterile container to obtain a final spore concentration of 5×10^5 spores/mL. Use these prepared spore suspensions immediately to inoculate the different substrates to ensure viability and avoid any possible contamination.

3.9 Solid-State Fermentation (SSF) with By-products

Fungal cultivation will be carried out under SSF conditions. Uninoculated control samples will be included to monitor potential microbial contamination and provide baseline values for comparative analysis. The process was conducted as follows:

1. SSF will be carried out in sterile square Petri dishes or containers, with the substrate evenly distributed across the surface. Three experimental conditions could be tested to evaluate the effect of substrate particle size (*see* **Note 13**) and moisture content on fungal growth:
   (i) Sterile, unprocessed substrate: 30 g per dish.
   (ii) Sterile, dried and ground substrate: 20 g per dish, adjusted to 75% moisture content by adding 15 mL of distilled sterile water.

(iii) Sterile, dried and ground substrate with nutrient supplementation: 20 g per dish, moistened with 15 mL of a sterile solution containing 4% (w/v) glucose and 2% (w/v) yeast extract (*see* **Note 14**).

2. Each substrate formulation will be inoculated with 15 mL of a spore suspension of *Aspergillus oryzae* or *Fusarium venenatum* at a concentration of 5×10^5 spores/mL (*see* Subheading 3.8, Growth of fungal strains, **step 9**, for fungal inoculum preparation).
3. Inoculated plates or containers will be incubated at 25 °C for 7–10 days under static conditions.

3.10 Drying and Crushing of Samples

Following the incubation period for SSF, fungal biomass samples, including their respective control samples (i.e., uninoculated substrates) (Fig. 6), will undergo post-processing:

1. Samples will be dried in a dehydrator at 50 °C for 24 h using open Petri dishes to facilitate moisture removal.
2. Once dried, the samples will be ground using an electric grinder until a fine powder is obtained.
3. The resulting powdered material will be transferred to sterile containers and stored at room temperature under fresh and dry conditions until subsequent analytical procedures are performed.

Control samples will be subjected to the same drying, milling, and storage protocols as the inoculated samples to ensure consistency and allow for reliable comparative analysis. Both samples and controls can be used for further analysis (*see* **Note 15**).

Figure 7 presents a flowchart illustrating the process used to obtain fungal biomass via SSF from the plant-based by-products.

4 Notes

1. The pH is adjusted to encourage the mycelium in the liquid medium to grow as small, spherical aggregates. The pH affects growth by influencing the diameter of these aggregates: when the pH is increased, the mycelium grows faster and forms fewer, but larger, spheres.
2. If constant horizontal agitation is not maintained, the mycelium will grow on the surface of the liquid, forming a secondary layer, like its growth on a Petri dish.
3. Once a liquid-phase culture is established, it can be used to inoculate fresh liquid medium. To do this, a small volume of the colonized liquid (approximately 5 mL) is transferred into 150 mL of fresh culture medium and incubated with agitation at 25 °C.

Fig. 6 Fungal growth in different by-products 10 days after inoculation. (**a**) Growth in carob by-product. (**b**) Growth in rice by-product. (**c**) Growth in tiger nut by-product. The samples represent, from left to right, uninoculated control, *A. oryzae* var. *oryzae* CBS 819.72 growth, and *F. venenatum* CBS 458.93 growth, respectively

4. The exact soaking time and whether soaking is necessary depend on the specific substrate. Preliminary tests should be conducted to ensure that, after autoclaving, the final moisture content is between 45% and 55%, which is optimal for proper mushroom growth. Although *Pleurotus ostreatus* can grow at lower moisture levels, excessive humidity can hinder its development and significantly increase the risk of contamination.

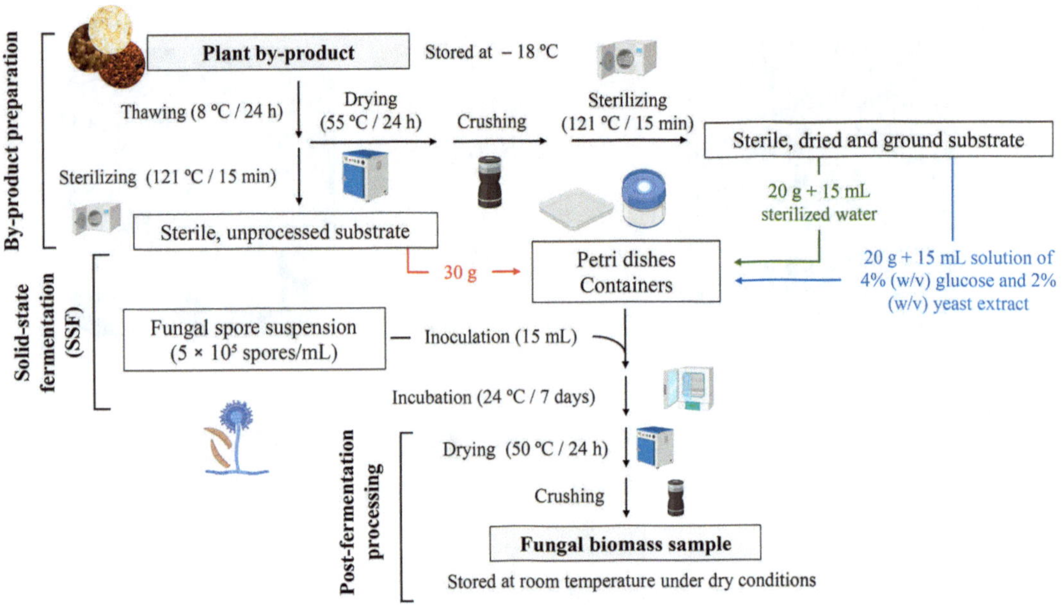

Fig. 7 Flowchart of the process to produce fungal biomass via solid-state fermentation using plant-derived by-products as substrates

For example, lentils typically require 40 min of soaking, while beans may need between 8 and 13 h and fava beans up to 22 h.

5. The incubation time depends on the substrate and the amount of starter added.
6. *P. ostreatus* can grow in temperatures between 25 °C and 28 °C.
7. The legumes are distributed into bags with a maximum thickness of 3 cm to ensure effective autoclaving, allowing heat to penetrate the entire substrate while also preventing the legumes from being smashed.
8. Cool down the substrate as quickly as possible to minimize the risk of growth from any remaining contaminant spores.
9. No residual water should be present. If found, remove it prior to fermentation.
10. The incubation time depends on the substrate and the amount of starter added.
11. It is important to highlight that a blank sample (unfermented substrate) should also be processed under the same conditions and its result value subtracted from that of the fermented substrates to obtain a more accurate estimation of fungal-derived glucosamine.
12. To collect the spores from the solid medium plates on which the fungus has been grown, a solution of Tween 20 (0.02% v/v), a nonionic surfactant, can be added, which acts as a dispersing agent, facilitating the recovery of the spores.

13. To examine the effects of by-product particle size, the by-product can be crushed to decrease the particle size, as suggested in the methods section. This reduction can significantly improve the surface-to-volume ratio of the substrate, thus facilitating greater accessibility of nutrients and oxygen to the fungi, increasing growth, fermentation rates, and higher yields of fermentation products [11, 12]. However, the relationship between particle size and fermentation efficiency is not always linear. Too small particle sizes can reduce the efficiency of the fermentation process by increasing compaction [13]. Therefore, it should be noted that the optimal particle size for SSF depends on the specific fungal strains and substrates used, as well as the desired fermentation products [12].
14. To improve protein level in the case of by-products with low nutritional content, these can be supplemented with carbon and nitrogen sources (as described in the methods section), as each fungal species may have specific preferences in terms of substrate and growth conditions, underlining the importance of tailoring fermentation processes to each fungal-substrate combination to optimize mycoprotein production.
15. The fungal biomass obtained can be analyzed both to determine the amount of fungal growth (*see* Subheading 3.6) and to measure its protein content, using routine methods (e.g., Kjeldahl method).

Acknowledgments

This study forms part of the AGROALNEXT program (**AGROALNEXT/2022/030** and **AGROALNEXT/2022/047**) with the support of the MCIN and funding from the European Union Next Generation EU (PRTR-C17.I1) and the Generalitat Valenciana (Spain). Matteo Vitali has a research staff contract in the latter aforementioned project (CPI-22-735). We acknowledge the technical assistance of Juan Francisco García-Moreno.

References

1. Viroli G, Kalmpourtzidou A, Cena H (2023) Exploring benefits and barriers of plant-based diets: health, environmental impact, food accessibility and acceptability. Nutrients 15: 4723. https://doi.org/10.3390/nu15224723
2. Huang J, Zhang M, Mujumdar AS, Semenov G, Luo Z (2024) Technological advances in protein extraction, structure improvement and assembly, digestibility and bioavailability of plant-based foods. Crit Rev Food Sci Nutr 64:11556–11574. https://doi.org/10.1080/10408398.2023.2240892
3. Pattnaik M, Pandey P, Martin GJ, Mishra HN, Ashokkumar M (2021) Innovative technologies for extraction and microencapsulation of bioactives from plant-based food waste and their applications in functional food development. Foods 10(2):279. https://doi.org/10.3390/foods10020279

4. Wikandari R, Hasniah N, Taherzadeh MJ (2022) The role of filamentous fungi in advancing the development of a sustainable circular bioeconomy. Bioresour Technol 345: 126531. https://doi.org/10.1016/j.biortech.2021.126531
5. Samtiya M, Aluko RE, Puniya A, Dhewa T (2021) Enhancing micronutrients bioavailability through fermentation of plant-based foods: a concise review. Fermentation 7(2):63. https://doi.org/10.3390/fermentation7020063
6. Majumder R, Miatur S, Saha A, Hossain S (2024) Mycoprotein: production and nutritional aspects: a review. Sustain Food Technol 2(1):81–91. https://doi.org/10.1039/D3FB00169E
7. Tarafdar JC (2024) Mycoprotein nanoparticles in agriculture. Fungal Biol 128(8):2493–2502. https://doi.org/10.1016/j.funbio.2024.07.003
8. Aidoo KE, Hendry R, Wood BJB (1981) Estimation of fungal growth in a solid state fermentation system. Eur J Appl Microbiol Biotechnol 12(1):6–9. https://doi.org/10.1007/BF00508111
9. Tomaselli Scotti C, Vergoignan C, Feron G, Durand A (2001) Glucosamine measurement as indirect method for biomass estimation of *Cunninghamella elegans* grown in solid state cultivation conditions. Biochem Eng J 7:1–5. https://doi.org/10.1016/S1369-703X(00)00090-5
10. Vazquez-Vilar M, Gandía M, García-Carpintero V, Marqués E, Sarrion-Perdigones-A, Yenush L et al (2020) Multigene engineering by GoldenBraid cloning: from plants to filamentous fungi and beyond. Curr Prot Mol Biol 130:e116. https://doi.org/10.1002/cpmb.116
11. Deswal D, Khasa YP, Kuhad RC (2011) Optimization of cellulase production by a brown rot fungus *Fomitopsis* sp. RCK2010 under solid state fermentation. Bioresour Technol 102: 6065–6072. https://doi.org/10.1016/j.biortech.2011.03.032
12. Singhania RR, Saini R, Adsul M, Saini JK, Mathur A, Tuli D (2015) An integrative process for bio-ethanol production employing SSF produced cellulase without extraction. Biochem Eng J 102:45–48. https://doi.org/10.1016/j.bej.2015.01.002
13. Kumar S, Sharma HK, Sarkar BC (2011) Effect of substrate and fermentation conditions on pectinase and cellulase production by *Aspergillus niger* NCIM 548 in submerged (SmF) and solid-state fermentation (SSF). Food Sci Biotechnol 20:1289–1298. https://doi.org/10.1007/s10068-011-0178-3

Chapter 20

Rhizopus spp. and the Evolution of Tempeh from Tradition to Modern Uses

Laura Alejandra Fernandez Castaneda, Jaqueline Auer, Maud Langton, and Galia Zamaratskaia

Abstract

Fungi play a crucial role in various industrial applications, particularly in food production and pharmaceuticals. They produce enzymes such as amylases used in baking, proteases used in cheese-making, and penicillin for pharmaceutical purposes. The ability of *Rhizopus* species to convert substrates into valuable compounds stems from their distinctive carbohydrate-active enzyme profile. Fungal enzymes are key to developing sustainable food systems, offering healthier, minimally processed alternatives. Particularly, the fungus *Rhizopus oligosporus* drives the fermentation of traditional foods like tempeh, a staple in local diets of Indonesia. This fungus is capable of metabolizing a wide range of carbon and nitrogen sources and forming a dense, white cotton-like matrix that binds beans together. Tempeh is traditionally made using soybeans and has gained global attention due to its culinary versatility, nutritional benefits, and rich umami-like flavors, as well as its chewier texture that increases its palatability compared to boiled beans. This makes it a highly attractive option as a plant-based protein. Tempeh production process involves the use of legumes or cereals and simple steps such as cleansing, soaking, hull removal, boiling, acidifying, tightly packing, and fermentation for around 2 days with a food-grade inoculum. Tempeh fermentation increases the bioavailability of proteins, fiber, vitamins, and minerals and produces isoflavones and bioactive peptides with health benefits. Therefore, studies have proven that consuming tempeh provides antidiabetic effects, reduces cholesterol, enhances cognition, offers antiaging benefits, improves gut health, and reduces cardiovascular disease risk. Traditional fermentation techniques align with the clean-label movement while promoting health benefits and environmental sustainability. This chapter presents scientific evidence on the health and environmental benefits of tempeh-like products, details the production process, and explores culinary applications, including tips for fermentation and fungi enthusiasts.

Key words Fungi, Enzymes, Tempeh, Fermentation, *Rhizopus oligosporus*

1 Introduction

1.1 Fungal Fermentation as a Sustainable Path to Plant-Based Nutrition

Diets based on plant-based foods have the potential to significantly reduce environmental impacts [1] and promote human health, in contrast to diets reliant on animal products. Numerous studies, including research by the World Health Organization (WHO) and the American Institute for Cancer Research, have linked high

Pablo Navarro Simarro (ed.), *Fungi*, Methods and Protocols in Food Science, https://doi.org/10.1007/978-1-0716-4989-3_20,
© The Author(s), under exclusive license to Springer Science+Business Media, LLC, part of Springer Nature 2026

consumption of red and processed meat with an increased risk of cardiovascular disease, type 2 diabetes, certain types of cancer, and a higher mortality rate [2]. In addition to these health concerns, animal products generate almost twice as much greenhouse gas emissions as plant-based alternatives. They are also major contributors to biodiversity loss, deforestation, and water pollution [3]. Despite the increasing interest in plant-based diets, the market for plant-based foods remains smaller than the meat and dairy markets, but it has seen a surge in demand, particularly in the past decade [4].

The plant-based food market experienced rapid growth in the 2010s, particularly in 2018 and 2019, driven by shifting consumer preferences, greater availability of plant-based options, and increasing awareness about sustainability [4]. However, by 2024, growth in this sector began to slow due to inflationary pressures and higher prices compare to animal-based alternatives. Moreover, consumer hesitancy stems from issues related to taste, texture, and perception of being overly processed and less nutritional [5]. In contrast, foods like tofu, tempeh, and natto retain and enhance the nutritional value of plant-based ingredients, offering healthier alternatives based on ancestral fermentation techniques [5].

1.1.1 Back to Traditional Plant-Based Food

Supporting the transition toward planetary health diets, fungal fermentation is crucial as it enables the production of high-quality proteins from agricultural products and by-products. Microbial fermentation has been a traditional practice for centuries across various cultures [6]. For instance, in Europe, fungi like *Penicillium* are used in the ripening of cheeses such as Brie and Roquefort, developing complex flavors and textures through fermentation. In Africa, cereals and porridges are often spontaneously fermented by bacteria and fungi such as yeasts, providing essential nutrients while improving digestibility as seen in fermented millet porridge. In Asia, fungal fermentation is used in the production of foods like soy sauce, miso, and tempeh (Fig. 1 [7, 8]). Fungi such as *Aspergillus oryzae* and *Rhizopus oligosporus* are used to enhance flavor and texture, increase digestibility, reduce bloating associated with bean consumption, and overall provide better nutritional value [9].

1.1.2 Tempeh

Tempeh, also spelled tempe, traditionally made from soybeans, has been a staple in Asian diets for centuries. It is gaining worldwide popularity due to its health benefits, umami flavor, and chewy texture. The fermentation process improves the digestibility of soybeans and increases the bioavailability of minerals and vitamins, such as calcium, iron, and B vitamins. It was suggested that tempeh offers antidiabetic effects, cholesterol reduction, cognitive enhancement, and anticancer properties, which are attributed to the presence of bioactive peptides and abundant isoflavones, such as genistein and daidzein, as well as its high dietary fiber content

Fig. 1 (**a**) Fresh *Tempeh* from soybean. (**b**) Frozen tempeh made from locally produce faba bean at Swedish University of Agricultural Sciences (Photo by Niklas Storm 2025 [8])

compared to unfermented soybeans. Evidence for these benefits has been demonstrated in clinical trials, in vitro studies, and animal studies. Comparisons between beans and tempeh fermentation show significantly higher antitumor activities associated with consuming tempeh [10–13]. Furthermore, its sustainable production, requiring less land and water than animal-based protein sources, positions tempeh as a relevant option in the search for low-environmental-impact protein alternatives [14, 15].

1.1.3 Nutritional Aspects

Tempeh is a nutrient-dense fermented food, rich in high-quality protein (~20%), moderate fat (~10%), low carbohydrates (~8%), and dietary fiber (~2%). It provides calcium (~111 mg), iron (~2.7 mg), magnesium (~81 mg), and folate (~24 mcg) per 100 g of fresh product. Fermentation enhances nutrient bioavailability [16], reduces anti-nutritional factors such as phytic acid [17] and tannins naturally present in plant materials, improves protein digestibility [16], and increases essential branched-chain amino acids. It also elevates B vitamins such as thiamine, riboflavin, niacin, biotin, and folate and may produce vitamin B12, addressing a common deficiency concern among plant-based eaters. Additionally, it supports gut health by reducing flatulence-inducing oligosaccharides and improving digestion [18]. Table 1 provides a detailed comparison of the nutritional content of beef products and plant-based alternatives. Tempeh has higher protein content that non-fermented plant-based alternatives, such as tofu, as well as higher levels of fiber and micronutrients [19].

Table 1
Nutritional content per 100 g of food: comparative amounts of meat and soybean meat alternatives

Nutrient (unit)	Beef entrecôte raw	Hamburger beef fried frozen product	Beef mincemeat raw 15% fat	Soybean boiled	Tofu bean curd	Tempeh (soybean)
Energy (kcal)	142	260	182	128	70	207
Energy (kJ)	594	1 087	762	536	291	865
Fat (g)	6.1	20.3	11.3	5.7	4.2	10.8
Sum of saturated fatty acids (g)	3	9.2	5	0.8	0.6	2.4
Protein (g)	21.7	14	20.1	11	8	19
Carbohydrates (g)	0	5.4	0	5.8	1.2	8.2
Fiber (g)	0	0.8	0	5	0.4	1.3
Water (g)	71.5	57.2	67.8	71	86.8	59.6
Ash (g)	1	2.3	0.9	1.5	0.8	1.6
Iron (mg)	2.1	1.6	2	2.7	0.9	2.7
Magnesium (mg)	22	19	20	88	130	81
Calcium (mg)	5	11	8	73	86	111

Adapted from Livsmedelsverket—Swedish Food Agency, Food Composition Database, version 2024–12

Replacing soybeans with alternative legumes or cereals preserves the nutritional benefits derived from fungal and microbial fermentation, reduces environmental impact by utilizing regionally available crops, and promotes agricultural sustainability across diverse regions [20].

1.1.4 Health Aspects

The health benefits from tempeh have been widely studied, highlighting its role in bone health, metabolism, and gut microbiota, among others. Below is a summary of some clinical, animal, and in vitro studies supporting its nutritional and functional properties (Table 2).

1.1.5 Tempeh Production

Tempeh is produced by fermenting soaked, dehulled, and boiled soybeans or other plant-based ingredients using a starter culture containing *Rhizopus* species (Fig. 2 with modifications [20]). Traditional methods may incorporate additional molds, yeasts, and bacteria, whereas industrial production relies on a pure *R. oligosporus* culture to ensure consistency and meet hygienic standards [15].

Table 2
Health-promoting properties of tempeh: a scientific overview

Study type	References
Clinical trials (human studies)	
Calcium bioavailability and bone health: a study on 20 postmenopausal Malaysian women found calcium from tempeh to be as bioavailable as milk	[21]
Isoflavone absorption, bioavailability, and anti-aging effects: soybean tempeh isoflavones improved estrogen levels and quality of life in women	[22–24]
Cognitive function and aging: observational and intervention studies in elderly Indonesians and individuals with mild cognitive impairment linked tempeh consumption to improved memory global cognitive function	[25, 26]
Gut microbiota and immunity: consumption of 100 g of steamed tempeh daily for 16 days increased *Akkermansia muciniphila* and improved gut barrier function	[27]
Muscle recovery, weight gain, and metabolic health: tempeh extract drink reduced muscle damage and improved recovery	[28, 29]
Appetite regulation and insulin response: a study in overweight women showed that tempeh reduced ghrelin levels and increased insulin response	[30]
Animal studies	
Iron absorption and bone health: tempeh fermentation increased iron bioavailability and supported bone density in rats	[25]
Cholesterol and heart health: tempeh polyphenols and free fatty acids lowered cholesterol by inhibiting HMG-CoA reductase	[26, 27]
Anticancer and anti-aging: tempeh isoflavones boosted immunity, suppressed tumors, and provided anti-aging benefits	[28, 29]
Blood pressure regulation: mung bean tempeh and GABA reduced hypertension in rats	[30]
Gut and brain health: tempeh-fed zebrafish showed improved gut bacteria and increased immunoglobulin A (IgA)	[31]
In vitro studies	
Antioxidant and neuroprotective effects: reduction of oxidative stress and improving microglial cell function	[32, 33]
Antitumor and anticancer effects: in vitro studies demonstrated that tempeh isoflavones inhibited breast, cervical, and ovarian cancer cell proliferation	[34]
Gut microbiota and probiotic effects: Tempeh fermentation promoted beneficial bacteria like *Bacillus subtilis* and *Bacillus stearothermophilus* while inhibiting harmful microorganisms	[35, 36]

The production process consists of two main stages (Fig. 5). First, pre-treatment involves rinsing, soaking, hull removal, boiling or steaming, draining, and cooling the soybeans. Figures 2 and 5 illustrate the standard tempeh-like production flowchart, adaptable

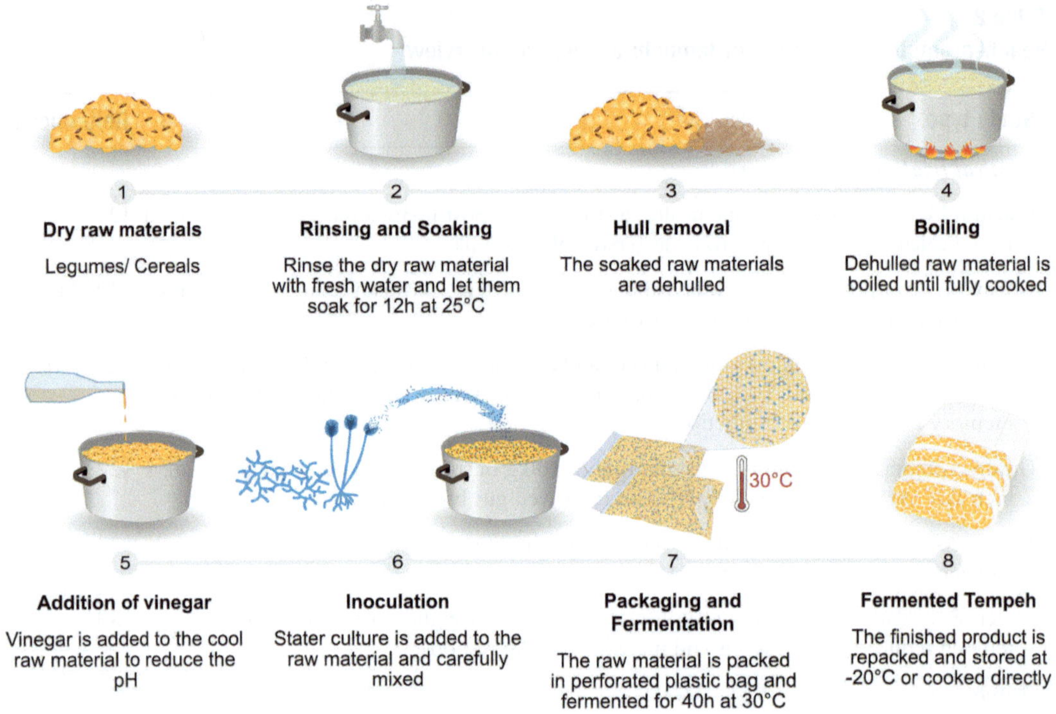

Fig. 2 Step-by-step process of making *tempeh*-like product using legumes and/or cereals. (Illustrated by Jaqueline Auer 2025)

to locally available legumes or cereals for greater environmental and economic sustainability.

Table 3 summarizes a variety of raw materials that have undergone tempeh-like fermentation, emphasizing their improved nutritional profiles and decreased levels of anti-nutrients. For instance, faba beans fermented with *Rhizopus oligosporus* exhibited higher amino acid content and lower polyphenol levels. In the case of African yam bean, fermentation resulted in increased protein content and reduced levels of harmful cyanogenic glycosides. These results underscore the effectiveness of tempeh-like fermentation in enhancing the nutritional value of different substrates.

Some of the nutritional changes due to tempeh fermentation could be explained by the changes in the microstructure. The changes in the cell wall of the plant material are due to the production of enzymes by the fungi. This modifies the plant-based cell wall and increases the bioaccessibility of nutrients such as proteins [50]. Figure 3 shows tempeh made from faba beans and how the cell walls, which are composed of fibers, are disturbed and opened, releasing protein into the medium.

Table 3
Nutritional improvements and anti-nutritional reductions in raw materials via tempeh-like fermentation

Substrates	Strains and references	Nutritional impact and others
Faba bean finely ground	*R. oligosporus* (ATCC 22959, ATCC 42222, ATCC 48012) [37] *R. oligosporus and Aspergillus oryzae* [38]	Increase of amino acids by tenfold and γ-aminobutyric acid Decreased values of total polyphenols, tannins, and antiradical activity
Whole-grain barley	*R. oligosporus* (ATCC 64063) and *L. plantarum SR3.60* [39]	Development of complex volatile compounds
Soybeans	*L. plantarum* and *R. oligosporus* [40]	Modulation of serum glucose and lipid levels in rats
Soybean (control), chickpea, lentils (red and green), and beans (white, black, and broad)	*R. oligosporus* [41]	Chickpea tempeh was more popular for taste preferences in sensory evaluation compared to traditional soybean tempeh
African yam bean (black, marble, and white)	*R. oligosporus* (DSMZ 1964) [42]	Increase of protein and starch content Decreased cyanogenic glycoside content, phytic acid, and tannins
Soybean with chickpeas and red kidney beans	*R. oligosporus* [43]	Sensory evaluation shows that soybean and red kidney bean were the most preferred with an acceptance of 7.54 over 10 Decreased fermentable oligosaccharides and polyols; no changes in trypsin inhibitors
Hemp-pressed cake	*R. oligosporus strain* ATCC 48012 [44]	Increase in antioxidant activity and total phenolic content; essential amino acids
Jack bean	*R. oligosporus RAPRIMA® mold* [45]	Increase in protein content by 67% and enhanced protein digestibility in vitro by 49%
Tenebrio molitor (mealworm), pupae, and mix with soybean	*R. oligosporus* [46]	Mealworm-based tempeh had greater amounts of bioavailable iron compared to raw material
Black beans	*R. oligosporus* [47]	Tempeh increases protein content and phenolic compound bioaccessibility and antioxidant activity
Koro kratok	*R. oligosporus* [48]	Tempeh enhanced the release of angiotensin-converting enzyme (ACE) inhibitory peptides
Barley and oat tempeh	*R. oligosporus* [49]	

(continued)

Table 3 (continued)

Substrates	Strains and references	Nutritional impact and others
		Barley tempeh showed slower starch hydrolysis compared to oat tempeh
Yellow peas, gray peas, and faba beans	*L. plantarum* soaking and *R. oligosporus*	Increase of protein content and in vitro protein digestion compared to non-fermented legumes
Faba bean and whole-grain oat mix	*L. plantarum* soaking and *R. oligosporus*	Improvement of sensory profile and increase on overall acceptability

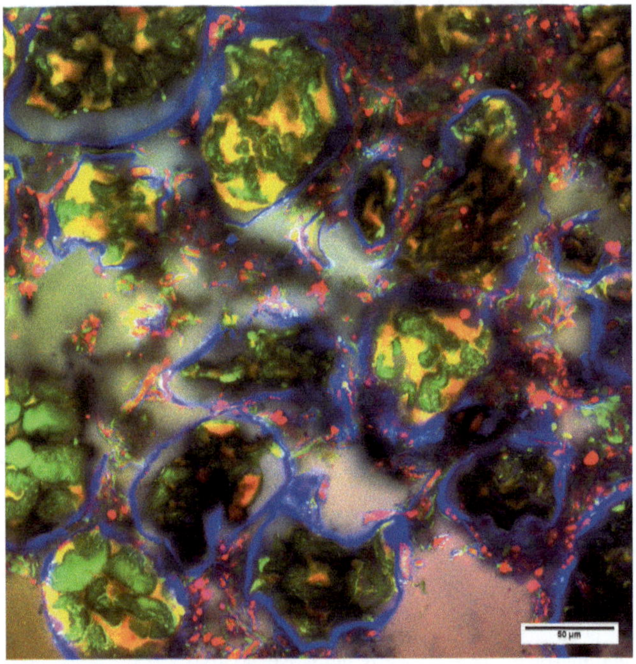

Fig. 3 Confocal laser scanning microscopy (CLSM) image of faba bean tempeh food, showing cell walls in blue, protein in red, and starch in green. The micrograph was taken using a 40× objective lens.)

1.2 Rhizopus *Species*

Rhizopus species, thermotolerant molds within the *Mucoromycota* phylum and a sister group to *Dikarya*, are ubiquitous on organic substrates, particularly decaying fruits and plant debris. Their rapid sporulation enables them to thrive in transient ecological niches. In food fermentation, such as tempeh production, *Rhizopus oligosporus* is widely used as a starter culture. During solid-state fermentation, the fungus releases enzymes that break down carbohydrates and proteins in the substrate, providing energy and building blocks for growth [51, 52].

The enzymatic activity in fermentation not only enhances digestibility but also influences the texture, taste, and nutritional quality of the product. Aside from soybeans, other legumes, cereals, and by-products can also serve as substrates. Additionally, *R. oligosporus* plays a role in industrial enzyme production and waste treatment. However, its use in food production requires strict quality control to prevent contamination [51].

1.2.1 Secretome of Rhizopus *spp.*

Genomics and proteomic technologies help us understand how fungi grow, adapt, and interact with their environment. These tools reveal the genetic instructions and proteins that fungi use to break down materials around them. For example, fungi like *R. oligosporus* produce enzymes that act like tools, breaking complex materials into nutrients they can use and building new edible structures. Fungi are remarkable for their ability to selectively digest materials, guided by their unique genomic blueprint. This explains the science behind traditional foods like tempeh. *Rhizopus* species, especially *R. oligosporus*, use powerful enzymes to adapt to their environment and transform substrates into nutrient-rich, digestible products. Here are some of the latest key discoveries summarized from Vellozo-Echevarría et al. (2024) [52].

Some of the most abundant carbohydrate active enzymes in *Rhizopus species*:

1. The most abundant enzymes produced are carbohydrate esterases (CE4) and glycoside hydrolases (GH18). CE4 enzymes play a key role in chitin deacetylation, converting chitin into chitosan, which is essential for fungal cell wall remodelling, flexibility, and rapid growth. GH18 chitinases degrade chitin and modify fungal cell walls, enhancing structural adaptability and promoting ecological competitiveness.
2. Glycoside hydrolases (GH28) and carbohydrate esterases (CE8) specialize in breaking down pectin, a key component of plant cell walls, releasing nutrients and showcasing strong pectinolytic capabilities. Glycoside hydrolases (GH16) target hemicellulose, breaking down glucans in plants like cereals, while polysaccharide lyases (PL14) likely act as glucuronan lyases, modifying cell walls to support fungal growth and ecological adaptation.

1.2.2 Flavor

Rhizopus oligosporus, the primary fungus used in tempeh production, plays a key role in developing its distinctive aroma and flavor. During fermentation, enzymatic activity releases free amino acids, fatty acids, and nitrogen compounds, contributing to a nutty, mushroom-like, and slightly beany flavor. Lipid breakdown by fungal enzymes generates aldehydes, such as hexanal, and mushroom-scented compounds like 1-octen-3-ol [39, 53].

Frying tempeh enhances these aromas, producing roasted, popcorn-like, and fatty notes through the formation of compounds like 2-acetyl-1-pyrroline, pyrazines, and dimethyl trisulfide. These reactions, driven by fermentation and heat, transform soybeans into a flavorful, aromatic food with a rich sensory profile, highlighting the critical role of microbial and enzymatic processes [53].

1.3 Culinary Uses

Indonesia is the largest producer and consumer of tempeh, with an average consumption of 43 g per capita per day [54]. Tempeh is enjoyed as a snack, side dish, or protein-rich addition to vegetable-based meals. Popular preparations include tempeh goreng (crispy fried tempeh), tempeh orek (sweet and savory stir-fry), and tempeh bacem (slow-cooked with coconut sugar and spices). Tempeh can be stored long term in the freezer and is quick to prepare, requiring only about 5 min for cooking, making it a convenient and versatile ingredient (Fig. 4).

1.3.1 Quick and Healthy Tempeh Bowl (2 Servings, 20-min Preparation)

Ingredients

200 g of *tempeh*, 1/2 avocado, 1/2 red bell pepper, sliced, chopped broccoli, 1 tbsp peanut seeds, peas, edamame, a cup of rice, salt, chili flakes, marinade for tempeh (in advance), 1 tbsp olive oil, 1 tbsp soy sauce, juice of 1 lemon, and 1 minced clove of garlic.

Instructions

1. Marinate tempeh in olive oil, lemon juice, and soy sauce for 10 min or over the night.
2. Cook tempeh in hot frying pan for 2–3 min per side, until golden color.
3. Plate tempeh with the other ingredients.
4. Sprinkle chili flakes.

1.4 Future Perspectives Industrial Scale

Research and producers' efforts have been dedicated on optimizing solid-state fermentation with mixed cultures to improve plant-based substrates like protein concentrates, isolates, and nutrient-rich by-products with poor functional properties such as undesirable texture and mouthfeel. This process enhances techno-functional properties and addresses nutritional gaps in plant-based foods by increasing essential amino acids, improving protein digestibility, and reducing anti-nutritional factors [18]. These advancements aim to make plant-based foods more appealing, nutrient-dense, and comparable to animal-based alternatives.

The use of *Lactiplantibacillus plantarum* (formerly *Lactobacillus plantarum*) in fermentation has demonstrated significant

Fig. 4 *Tempeh*: full plant-based dish inspiration. (**a**) Roasted natural tempeh on salad. (**b**) Fried tempeh with rice and steamed vegetables. (**c**) Marinated sweet chili tempeh with vegetables and rice. (**d**) Grilled tempeh in a burger.

potential to enhance the nutritional and sensory properties of plant-based substrates. In legumes, *L. plantarum* effectively reduces off-flavor compounds such as hexanal and 2-furfural while generating desirable aroma compounds such as acetic acid, diacetyl, and ethyl acetate, which contribute sour and buttery notes [55, 56].

Lactobacillus plantarum could be used as a mixed microorganism in combination with fungal fermentation for tempeh production. Studies indicate that *L. plantarum* does not inhibit the growth of *Rhizopus microsporus*, regardless of whether it is applied during the soaking and/or inoculation process [20, 57]. Its presence plays a critical role in improving texture by increasing chewiness and contributing to flavor development in tempeh. Together, these microbes enhance umami and savory profiles, reduce undesirable compounds, and form a cohesive, cake-like texture, showcasing their synergistic potential in fermented foods [20].

Recent research on barley and rice fermented with fungi, such as *Aspergillus oryzae* and *Rhizopus oligosporus*, has shown improved protein quality by increasing lysine content by up to 33% and enhancing the protein digestibility-corrected amino acid score and indispensable amino acid index, despite reduced protein digestibility in barley [58]. The process also produced significant fungal biomass, highlighting its potential for improving the nutritional value of plant-based diets [59]. Similarly, fermentation improved pea protein isolate, with *Zygosaccharomyces rouxii* increasing protein solubility for beverages. Strains such as *Aspergillus oryzae* and *Saccharomyces cerevisiae* enhanced foam capacity, while reduced gelation capacity offers potential for industrial applications. Exploring shorter fermentation times and pre-wetting effects could further optimize these processes [60].

2 Materials

2.1 Tempeh Production Method

2.1.1 Plant Materials

1. Legumes: Soybean, chickpeas, black beans, faba beans, lentils yellow peas, lupins, gray peas, peanuts, cowpeas, and any other legumes (*see* **Note 1**).
2. Cereals: Whole-grain oats, barley (*see* **Note 2**).

2.1.2 Solutions

1. Tap water for soaking and rinsing.
2. Sodium bicarbonate solution: 5 g in 1.2 L of tap water (optional for soaking).
3. Acetic acid or vinegar solution: 5 g in 1.2 L of tap water (optional for soaking).
4. Lactic acid bacteria (LAB) inoculum (e.g., *L. plantarum*; optional for soaking).
5. Vinegar (2 tsp. per 1000 g of boiled legumes/cereals for pH reduction).

2.1.3 Starter Culture

1. *Rhizopus oligosporus* tempeh starter (1 tsp. (around 5 g) per 1000 g of boiled legumes/cereals).

2.1.4 Packaging Materials

1. Food-grade plastic bags with perforations or banana leaves (optional; traditional method).

2.1.5 Specialized Equipment

1. Cooking pot for boiling (10 L capacity).
2. Perforation tool for packaging (e.g., clean itches teeth).
3. Incubator capable of maintaining 30 °C.
4. Clean surface for draining and cooling legumes/cereals.

2.2 *Rhizopus oligosporus: Spore Production*

2.2.1 Media and Solutions

1. Malt extract agar (MEA).
2. Sterile 0.9% NaCl solution.

2.2.2 Sterile Equipment and Supplies

1. Sterile inoculating loop.
2. Sterile Petri dishes.
3. Sterile tubes for collecting spore suspension.
4. Centrifuge tubes.

2.2.3 Equipment

1. Centrifuge.
2. Bürker counting chamber or hemocytometer.
3. Fermentation/incubation room or incubator with temperature control (37 °C).
4. Sterilization equipment (autoclave).
5. UV lamp for workspace sanitation.

2.2.4 Additional Tools

1. Laboratory bottle with a loose cap for media preparation.
2. Laminar flow hood for sterile work.
3. Disposable plastic loops for spore harvesting.

3 Methods

3.1 Tempeh Production Method

This guide outlines the steps for producing tempeh, (Fig. 4) drawing on methods from Ahnan-Winarno et al. (2021) and Fernandez Castaneda et al. (2024) [15, 20], with modifications to optimize fermentation, ensure food safety, and maintain an efficient production process.

3.1.1 Pre-treatment

Rinsing Legumes must be thoroughly washed to remove soil, bacteria (e.g., *Bacillus* spp.), and processing contaminants. Use abundant water and vigorous movement to eliminate dirt and microbial load.

Soaking Soaking softens legumes, reduces cooking time, and decreases soluble anti-nutritional compounds such as oligosaccharides and phytic acid. Soaking initiates spontaneous fermentation, which will degrade indigestible compounds.

(i) Use at least twice the volume of water for soaking.

(ii) Cover the container or pot used for fermentation to avoid additional environmental microbes.

(iii) Soak for 8–16 h at room temperature.

(iv) Optional soaking treatments for improved texture and nutrient absorption. In this case, 500 g of raw material is specified, as it could be legumes, cereals, or a combination of both.

(v) Sodium bicarbonate: Dissolve 5 g or 1 tsp. on 1.2 L of tap water (reduces boiling time by 50% on legumes [20]).

(vi) Acidic soak: 5 g or 1 tsp. of acetic acid or vinegar on 1.2 L of tap water (improve tempeh flavor [61]).

(vii) Lactic acid bacteria (LAB) inoculation: Enhances flavor and digestibility as *L. plantarum* [62]. The amount of LAB will depend on the producer's recommendation (recommended for semi-industrial or industrial production).

(viii) Discard soaking water before boiling to remove excess antinutrients.

3.1.2 Hull Removal and Boiling

Hull Removal or Dehulling The hulls of legumes, such as soybeans and yellow peas, are rich in insoluble fiber and cellulose, which are indigestible and hinder fermentation by creating a barrier that inhibits fungal growth. Removing the hull ensures better nutrient access for *Rhizopus* fungi and improves texture.

Boiling or Steaming Boiling enhances digestibility, removes bitterness, and eliminates bacteria that may interfere with fermentation; another heat treatment option is steaming. Below is a list describing the approximate boiling times required for various legumes and cereals (Table Table 4). After soaking and boiling, legumes typically double in weight; for example, 500 g of dried legumes become 1000 g when boiled. For cereals, the weight change depends on the type; for instance, whole-grain oats retain approximately the same weight when dried and boiled.

(i) Pour approximately 1.5 L of tap water into a cooking pot and bring it to a vigorous boil.

(ii) Add the legumes or cereals and begin timing the cooking process.

(iii) Periodically check the texture of the legumes or cereals to prevent overcooking (*see* **Note 3**).

(iv) After boiling, discard the hot water and rinse the legumes or cereals thoroughly with cold water.

(v) Discard the cold water and allow the legumes or cereals to cool down.

Table 4
Boiling times for various legumes and cereals with and without sodium bicarbonate

	Soaking times	
Legume type	In water	In sodium bicarbonate
Soybean	≥60 min	40–45 min
Black beans	>60 min	40 min
Faba beans	16 min	8 min
Yellow peas	>30 min	15 min
Lupins	>70 min	40–50 min
Gray peas	≥17	10 min
Whole-grain oats	10 min	10 min
Barley	10 min	10 min
Peanuts	>30 min	30 min

Some data based on Fernandez et al. (2024) [20]

3.1.3 Draining and pH Reduction

Draining It is crucial to remove excess water, as the initial moisture content directly affects the success of solid-state fermentation. Excessive moisture can promote the growth of spoilage microorganisms, while the right balance ensures optimal fungal growth and fermentation quality.

(i) Spread legumes/cereals evenly on a clean surface and pat dry.

(ii) Ensure legumes/cereals are dry but still warm (~37 °C) to facilitate fermentation.

pH Reduction Adding a small amount of an acidifying agent, such as low-pH vinegar, before introducing the starter culture helps lower the pH of the legumes or cereals. This creates an environment that inhibits the growth of pathogenic and spoilage bacteria, facilitating the efficient colonization and growth of *Rhizopus oligosporus* during fermentation.

(i) Add 2 tsp. of vinegar to 1000 g of legumes/cereals before inoculation to lower the pH.

(ii) Mix gently to avoid breaking the legumes/cereals.

3.1.4 Inoculation and Packaging

Inoculation Traditionally in Indonesia, a piece of finished tempeh or reused wrapping material containing *Rhizopus* spores is used to initiate a new fermentation [15]. Today, powdered starter cultures

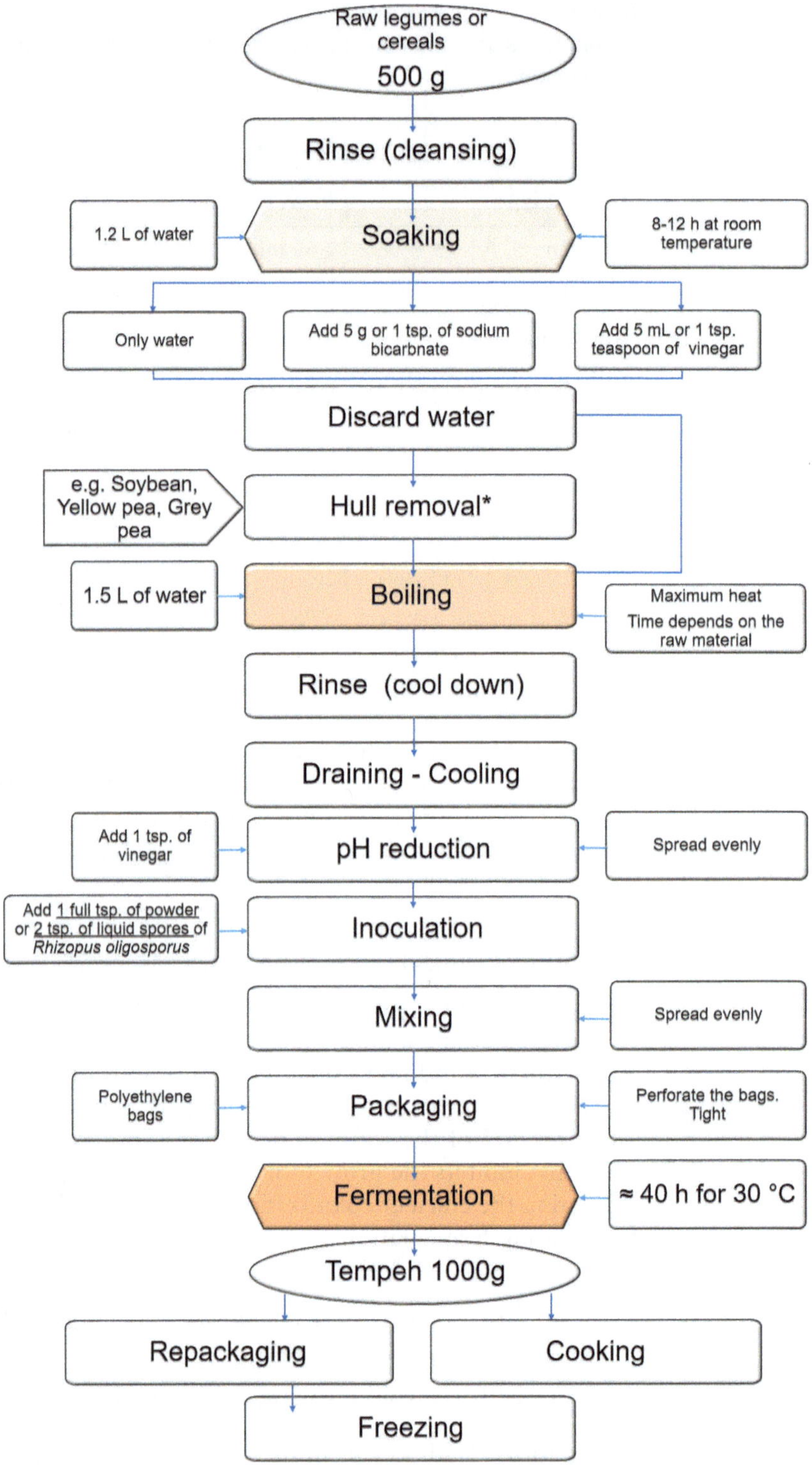

Fig. 5 Standardized workflow for Tempeh production. *Optional steps may vary, depending on the raw material; for instance, soybean requires it. Soaking, boiling, and fermentation are critical steps responsible for the nutritional enhancement of the raw materials.

of dried *Rhizopus oligosporus* are readily available online, simplifying the process. For laboratory-scale production, spore cultivation and harvesting to create a liquid inoculum are detailed below.

(i) Use 1 tsp. of tempeh starter per 1000 g of boiled legumes/ cereals.

(ii) Evenly sprinkle the starter culture over the legumes and mix gently to ensure uniform distribution.

Packaging Packaging is essential to maintain proper aeration and support fungal growth.

(i) Use food-grade plastic bags or traditional materials like banana leaves (*see* **Note 4**).

(ii) Tempeh is typically rectangular, measuring 15–20 cm long, 8–12 cm wide, and 1.5–3 cm thick, allowing even fermentation and proper mycelium growth.

(iii) Perforate the plastic bags with small holes evenly to allow oxygen exchange, essential for fungal respiration.

3.1.5 Fermentation (Incubation Phase)

Fermentation biotransforms the inoculated legumes/cereals into tempeh by promoting fungal growth under controlled conditions.

(i) Incubate the packaged legumes/cereals at 30 °C for 40–48 h, depending on the legume/cereal type.

(ii) Ensure proper ventilation during incubation to prevent excessive bacterial growth (*see* Fig. 6).

(iii) Tempeh is ready when it forms a firm cake with a white mycelium coating, indicating successful fermentation. Small patches of black mycelium may appear, which is normal and expected as the spores mature.

3.1.6 Storing and Cooking

Storage Proper storage maintains tempeh's freshness and extends its usability.

(i) Refrigerate in a sealed container for 3–4 days.

(ii) For long-term storage, freeze tempeh and thaw for 15 min before use.

Cooking Cooking enhances the flavor and texture of tempeh, making it suitable for various dishes.

(i) Slice or cube tempeh before cooking.

(ii) Pan-fry, stir-fry, roast, or deep-fry using oil for about 2 min per side until golden brown and crispy.

Fig. 6 Conventional *tempeh* fermentation setup: Tempeh could be fermented in rectangular blocks on wooden racks, ensuring proper airflow for mycelium growth. (Photo by Laura F. Castaneda from Attempe Sleman company, Yogyakarta, Indonesia)

(iii) Optional: Marinate tempeh before cooking to add flavor and depth to the dish (see some easy recipes below in Subheading 3).

3.2 Rhizopus oligosporus*: Spore Production Microbial Inoculum (Lab-Scale Production Method)*

This guide details the process of cultivating *Rhizopus oligosporus* spores on malt extract agar (MEA) media for SSF applications, ensuring high-quality, viable spores under controlled conditions (Fig. 7) based on Fernandez et al. (2024) with modification [20].

3.2.1 Preparation of MEA Slopes

1. Prepare 250 mL of MEA on a laboratory bottle with a cap.
2. Sterilize the bottle containing MEA dilution in an autoclave (remember to have lose cap).
3. After autoclaving, allow the medium to cool slightly. Before it solidifies, pour it into sterilized Petri dishes under a previously sanitized laminar flow hood.

3.2.2 Inoculation

1. Obtain *Rhizopus oligosporus* strains growing on MEA plates (provided or cultured beforehand). The commercial powder as starter culture for tempeh can be used. Then, you will grow your own spores from it.

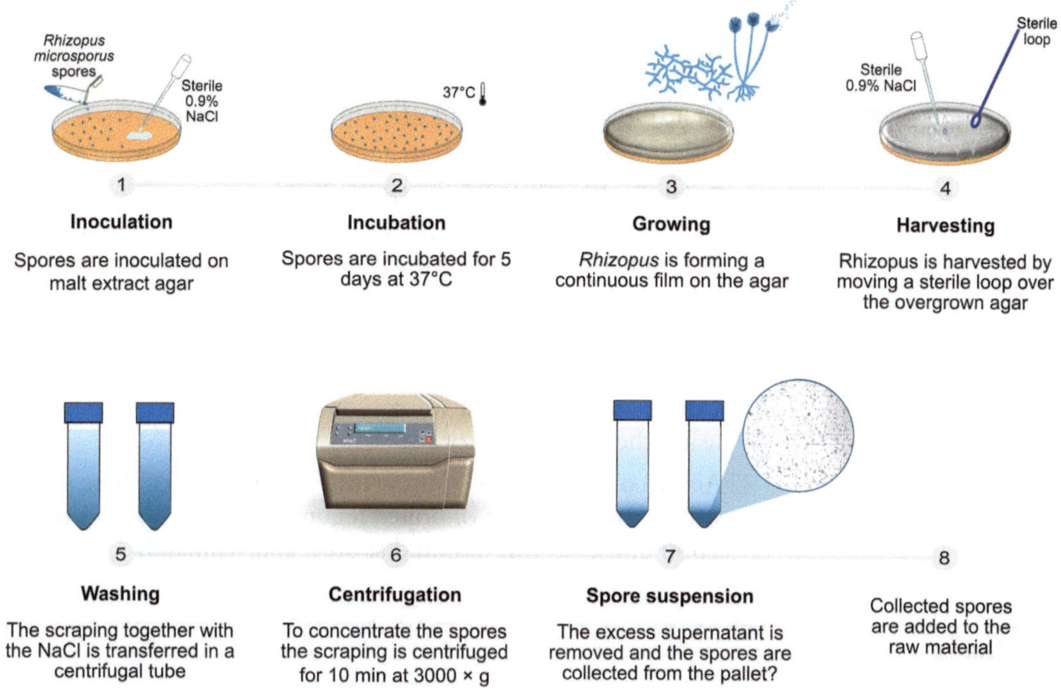

Fig. 7 Overview of *Rhizopus oligosporus* spore cultivation and harvesting process for tempeh-like solid-state fermentation.

2. Using a sterile inoculating tool, transfer spores from the MEA plates onto the surface of the new prepared MEA Petri dishes.
3. Ensure even distribution of spores across the slope to promote uniform fungal growth.
4. Add some drops of the sterile 0.9% NaCl solution.

3.2.3 Incubation

1. Incubate the inoculated MEA Petri dishes at 37 °C.
2. Keep the lid loose to allow sufficient oxygen exchange. **Important**: Tight lid will inhibit sporulation due to oxygen limitation.
3. Monitor fungal growth daily. After 4–5 days, the mycelium will develop a middle-to-dark gray color, indicating mature sporangia and spore readiness.

3.2.4 Harvesting Spores

1. Harvest spores when the fungal growth is middle-to-dark gray.
2. Add approximately 50 mL of 0.9% NaCl solution into the Petri dish to wet the mycelium.
3. Using a sterile disposable plastic loop, gently rub the surface of the mycelium to release spores into the solution. The liquid will turn gray as spores are suspended.

3.2.5 Washing Spores

1. Repeat **step 4** at least three times more and collect all in the same centrifuge tube.

3.2.6 Centrifugation

1. Centrifuge at 3000×*g* for 10 min to pellet the spores.
2. Discard most of the supernatant (liquid part), leaving around 5–10 mL to resuspend the spore pellet.
3. Resuspend the pellet in the remaining supernatant by shaking. For higher purity, repeat the process one more time.

3.2.7 Quantifying Spores

1. Use counting spores using a Bürker chamber or hemocytometer.
2. Count spores under a microscope and calculate total spores using the dilution factor.
3. Aim to obtain 5×10^6 spores/mL. Then, 1 mL of these spores will be used in 100 g of boiled legumes/cereals. Therefore, for 1000 g, use 10 mL of liquid spores, approximately 2 tsp.

4 Notes

1. Legume: Any other legume can be used, making this method highly versatile and adaptable to various legume types.
2. Cereals: Other cereals suitable for use include wheat, sorghum (milo), rye, corn, triticale, etc. Cereals are potential substrates for tempeh fermentation; however, their complex cell wall structures, particularly the presence of cellulose and hemicellulose, inhibit the breakdown by fungi during fermentation. Pre-treatment methods like soaking, steaming, or crushing help soften these walls, making it easier for the fungi to colonize and ferment the cereals efficiently. These physical modifications increase nutrient availability, improving the fermentation process and the final tempeh product [63].
3. Boiling time: Prevention of overcooking is essential to ensure proper mycelium growth. Ideally, the bean should maintain its shape on the outside while remaining soft on the inside. A slightly al dente texture may also be preferable.
4. Packaging: Biodegradable alternatives such as waru, jati, banana, guava, and cassava leaves provide adequate permeability, support mycelium growth, and may enhance the sensory attributes of the final product. Unlike plastic, these natural materials reduce waste and mitigate the risk of microplastic migration into the food product. However, their availability varies by region, highlighting the importance of locally sourced, sustainable packaging solutions.

Acknowledgments

The authors acknowledge financial support from the project HealthFerm, which is co-funded by the European Union under the Horizon Europe grant agreement No. 101060247, and from the Swiss State Secretariat for Education, Research and Innovation (SERI) under contract No. 22.00210. The views and opinions expressed however are those of the author(s) only and do not necessarily reflect those of the European Union or the European Research Executive Agency (REA). Neither the European Union nor REA can be held responsible for them. Moreover, this study was financially supported by Trees and Crops for the Future (TC4F), a Strategic Research Area at SLU, supported by the Swedish Government.

References

1. Crippa M, Solazzo E, Guizzardi D, Monforti-Ferrario F, Tubiello FN, Leip A (2021) Food systems are responsible for a third of global anthropogenic GHG emissions. Nat Food 2: 198–209. https://doi.org/10.1038/s43016-021-00225-9
2. Bouvard V, Loomis D, Guyton KZ, Grosse Y, El Ghissassi F, Benbrahim-Tallaa L, Guha N, Mattock H, Straif K (2015) Carcinogenicity of consumption of red and processed meat. Lancet Oncol 16:1599–1600. https://doi.org/10.1016/S1470-2045(15)00444-1
3. Xu X, Sharma P, Shu S, Lin T-S, Ciais P, Tubiello FN, Smith P, Campbell N, Jain AK (2021) Global greenhouse gas emissions from animal-based foods are twice those of plant-based foods. Nat Food 2:724–732. https://doi.org/10.1038/s43016-021-00358-x
4. Andreani G, Sogari G, Marti A, Froldi F, Dagevos H, Martini D (2023) Plant-based meat alternatives: technological, nutritional, environmental, market, and social challenges and opportunities. Nutrients 15:452. https://doi.org/10.3390/nu15020452
5. He J, Evans NM, Liu H, Shao S (2020) A review of research on plant-based meat alternatives: driving forces, history, manufacturing, and consumer attitudes. Compr Rev Food Sci Food Saf 19:2639–2656. https://doi.org/10.1111/1541-4337.12610
6. Cantabrana I, Perise R, Hernández I (2015) Uses of *Rhizopus oryzae* in the kitchen. Int J Gastron Food Sci 2:103–111. https://doi.org/10.1016/j.ijgfs.2015.01.001
7. Eka A (2019) Indonesia plans to export more tempeh. News Desk (The Jakarta Post). https://www.thejakartapost.com/news/2019/02/01/indonesia-plans-to-export-of-more-tempeh.html
8. Niklas Storm (2024) Favaböna i förvandling—blir framtidens mat i SLU:s labb. Swedish University of Agricultural Sciences. https://www.slu.se/ew-nyheter/2024/11/tempeh/. Accessed 22 Jan 2025
9. Tamang JP, Cotter PD, Endo A, Han NS, Kort R, Liu SQ, Mayo B, Westerik N, Hutkins R (2020) Fermented foods in a global age: east meets west. Compr Rev Food Sci Food Saf 19: 184–217. https://doi.org/10.1111/1541-4337.12520
10. Kiriakidis S, Högemeier O, Starcke S, Dombrowski F, Hahne JC, Pepper M, Jha HC, Wernert N (2005) Novel tempeh (fermented soyabean) isoflavones inhibit *in vivo* angiogenesis in the chicken chorioallantoic membrane assay. Br J Nutr 93:317–323. https://doi.org/10.1079/BJN20041330
11. Jayachandran M, Xu B (2019) An insight into the health benefits of fermented soy products. Food Chem 271:362–371. https://doi.org/10.1016/j.foodchem.2018.07.158
12. Nurkolis F, Qhabibi FR, Yusuf VM, Bulain S, Praditya GN, Lailossa DG, Al Mahira MFN, Prima EN et al (2022) Anticancer properties of soy-based Tempe: a proposed opinion for future meal. Front Oncol 12. https://doi.org/10.3389/fonc.2022.1054399
13. Kumar MR, Azizi NF, Yeap SK, Abdullah JO, Khalid M, Omar AR, Osman Mohd A, Leow ATC et al (2022) Clinical and preclinical studies of fermented foods and their effects on Alzheimer's disease. Antioxidants 11:883. https://doi.org/10.3390/antiox11050883

14. Springmann M (2024) A multicriteria analysis of meat and milk alternatives from nutritional, health, environmental, and cost perspectives. Proc Natl Acad Sci 121. https://doi.org/10.1073/pnas.2319010121
15. Ahnan-Winarno AD, Cordeiro L, Winarno FG, Gibbons J, Xiao H (2021) Tempeh: a semicentennial review on its health benefits, fermentation, safety, processing, sustainability, and affordability. Compr Rev Food Sci Food Saf 20:1717–1767. https://doi.org/10.1111/1541-4337.12710
16. Zhang Y, Wei R, Azi F, Jiao L, Wang H, He T, Liu X, Wang R, Lu B (2022) Solid-state fermentation with *Rhizopus oligosporus* RT-3 enhanced the nutritional properties of soybeans. Front Nutr 9:1–14. https://doi.org/10.3389/fnut.2022.972860
17. Kumar V, Sinha AK, Makkar HPS, Becker K (2010) Dietary roles of phytate and phytase in human nutrition: a review. Food Chem 120:945–959. https://doi.org/10.1016/j.foodchem.2009.11.052
18. Rizzo G (2024) Soy-based tempeh as a functional food: evidence for human health and future perspective. Front Biosci-Elite 16:3. https://doi.org/10.31083/j.fbe1601003
19. Nutritional Composition of Foods version (2024) Livsmedelsverket: Swedish Food Agency. https://soknaringsinnehall.livsmedelsverket.se/. Accessed 23 Jan 2025
20. Fernandez Castaneda LA, Auer J, Leong SLL, Newson WR, Passoth V, Langton M, Zamaratskaia G (2024) Optimizing soaking and boiling time in the development of tempeh-like products from faba bean (*Vicia faba* L.). Fermentation 10. https://doi.org/10.3390/fermentation10080407
21. Haron H, Shahar S, O'Brien KO, Ismail A, Kamaruddin N, Rahman SA (2010) Absorption of calcium from milk and tempeh consumed by postmenopausal Malay women using the dual stable isotope technique. Int J Food Sci Nutr 61:125–137. https://doi.org/10.3109/09637480903348080
22. Xu X, Wang H-J, Murphy PA, Hendrich S (2000) Neither background diet nor type of soy food affects short-term isoflavone bioavailability in women. J Nutr 130:798–801. https://doi.org/10.1093/jn/130.4.798
23. Cassidy A, Brown JE, Hawdon A, Faughnan MS, King LJ, Millward J, Zimmer-Nechemias-L, Wolfe B, Setchell KD (2006) Factors affecting the bioavailability of soy isoflavones in humans after ingestion of physiologically relevant levels from different soy foods. J Nutr 136:45–51. https://doi.org/10.1093/jn/136.1.45
24. Khosravi A, Razavi SH (2021) Therapeutic effects of polyphenols in fermented soybean and black soybean products. J Funct Foods 81:104467. https://doi.org/10.1016/j.jff.2021.104467
25. Handajani YS, Turana Y, Yogiara Y, Widjaja NT, Sani TP, Christianto GAM, Suwanto A (2020) Tempeh consumption and cognitive improvement in mild cognitive impairment. Dement Geriatr Cogn Disord 49:497–502. https://doi.org/10.1159/000510563
26. Hogervorst E, Mursjid F, Priandini D, Setyawan H, Ismael RI, Bandelow S, Rahardjo TB (2011) Borobudur revisited: soy consumption may be associated with better recall in younger, but not in older, rural Indonesian elderly. Brain Res 1379:206–212. https://doi.org/10.1016/j.brainres.2010.10.083
27. Stephanie S, Ratih NK, Soka S, Suwanto A (2017) Effect of tempeh supplementation on the profiles of human intestinal immune system and gut microbiota. Microbiol Indones 11:11–17. https://doi.org/10.5454/mi.11.1.2
28. Jauhari M, Sulaeman A, Riyadi H, Ekayant I (2013) Effect of administering tempeh drink on muscle damage recoveries after resistance exercise in student athletes. Pak J Nutr 12:924–928. https://doi.org/10.3923/pjn.2013.924.928
29. Permatasari O, Murwani R, Rahfiludin MZ (2018) Tempe nuggets provision improves energy adequacy and protein intake in underweight Underfive children. Curr Res Nutr Food Sci J 6:89–96. https://doi.org/10.12944/CRNFSJ.6.1.09
30. Noer ER, Dewi L, Kuo C-H (2021) Fermented soybean enhances post-meal response in appetite-regulating hormones among Indonesian girls with obesity. Obes Res Clin Pract 15:339–344. https://doi.org/10.1016/j.orcp.2021.06.005
31. Chen Y-C, Tao N-L, Hu S-Y, Tsai H-Y, Liao S-C, Tsai W-L, Hu C-Y (2021) Effect of tempeh on gut microbiota and anti-stress activity in zebrafish. Int J Mol Sci 22:12660. https://doi.org/10.3390/ijms222312660
32. Chang C, Hsu C, Chou S, Chen Y, Huang F, Chung Y (2009) Effect of fermentation time on the antioxidant activities of tempeh prepared from fermented soybean using Rhizopus oligosporus. Int J Food Sci Technol 44:799–806. https://doi.org/10.1111/j.1365-2621.2009.01907.x
33. Xiao Y, Fan J, Chen Y, Rui X, Zhang Q, Dong M (2016) Enhanced total phenolic and isoflavone aglycone content, antioxidant activity and DNA damage protection of soybeans processed by solid state fermentation with *Rhizopus*

oligosporus RT-3. RSC Adv 6:29741–29756. https://doi.org/10.1039/C6RA00074F

34. Elkhalifa AEO, Banu H, Khan MI, Ashraf SA (2023) Integrated network pharmacology, molecular docking, molecular simulation, and *in vitro* validation revealed the bioactive components in soy-fermented food products and the underlying mechanistic pathways in lung cancer. Nutrients 15:3949. https://doi.org/10.3390/nu15183949
35. Kiers JL, Nout MJR, Rombouts FM, Nabuurs MJA, van der Meulen J (2002) Inhibition of adhesion of enterotoxigenic *Escherichia coli* K88 by soya bean tempe. Lett Appl Microbiol 35:311–315. https://doi.org/10.1046/j.1472-765X.2002.01182.x
36. Kuligowski M, Jasińska-Kuligowska I, Nowak J (2013) Evaluation of bean and soy tempeh influence on intestinal bacteria and estimation of antibacterial properties of bean tempeh. Pol J Microbiol 62:189–194
37. Polanowska K, Grygier A, Kuligowski M, Rudzińska M, Nowak J (2020) Effect of tempe fermentation by three different strains of *Rhizopus oligosporus* on nutritional characteristics of faba beans. LWT 122:109024. https://doi.org/10.1016/j.lwt.2020.109024
38. Gautheron O, Nyhan L, Torreiro MG, Tlais AZA, Cappello C, Gobbetti M, Hammer AK, Zannini E et al (2024) Exploring the impact of solid-state fermentation on fava bean flour: a comparative study of *Aspergillus oryzae* and *Rhizopus oligosporus*. Foods 13:2922. https://doi.org/10.3390/foods13182922
39. Mei Feng X, Ostenfeld Larsen T, Schnürer J (2007) Production of volatile compounds by Rhizopus oligosporus during soybean and barley tempeh fermentation. Int J Food Microbiol 113:133–141. https://doi.org/10.1016/j.ijfoodmicro.2006.06.025
40. Huang Y-C, Wu B-H, Chu Y-L, Chang W-C, Wu M-C (2018) Effects of tempeh fermentation with *Lactobacillus plantarum* and *Rhizopus oligosporus* on streptozotocin-induced type II diabetes mellitus in rats. Nutrients 10:1143. https://doi.org/10.3390/nu10091143
41. Erkan SB, Gürler HN, Bilgin DG, Germec M, Turhan I (2020) Production and characterization of tempehs from different sources of legume by *Rhizopus oligosporus*. LWT Food Sci Technol 119:108880. https://doi.org/10.1016/j.lwt.2019.108880
42. Azeke MA, Fretzdorff B, Buening-Pfaue H, Betsche T (2007) Nutritional value of African yambean (*Sphenostylis stenocarpa*, L): improvement by solid substrate fermentation using the tempeh fungus Rhizopus oligosporus. J Sci Food Agric 87:297–304. https://doi.org/10.1002/jsfa.2721
43. Tan ZJ, Abu Bakar MF, Lim SY, Sutimin H (2024) Nutritional composition and sensory evaluation of tempeh from different combinations of beans. Food Res 8:138–146. https://doi.org/10.26656/fr.2017.8(2).088
44. Aktaş H, Raczyk M, Aydemir LY, Polanowska K (2025) Chemical and bioactive changes during fermentation and *in vitro* digestion of hemp pressed cake using *Rhizopus oligosporus*. LWT 215:117264. https://doi.org/10.1016/j.lwt.2024.117264
45. Yarlina VP, Djali M, Andoyo R, Nurmilah S, Lani MN (2024) Metagenomic insights into enhancing protein content and digestibility in Jack bean (*Canavalia ensiformis*) tempeh: unraveling microbial dynamics during fermentation. Appl Food Res 4:100588. https://doi.org/10.1016/j.afres.2024.100588
46. Wilson JW, Thompson TW, Wei Y, Chaparro JM, Stull VJ, Nair MN, Weir TL (2024) Comparison of the *in vitro* iron bioavailability of tempeh made with *Tenebrio molitor* to beef and plant-based meat alternatives. Nutrients 16:2756. https://doi.org/10.3390/nu16162756
47. Wang K, Gao Y, Zhao J, Wu Y, Sun J, Niu G, Zuo F, Zheng X (2022) Effects of *in vitro* digestion on protein degradation, phenolic compound release, and bioactivity of black bean tempeh. Front Nutr 9. https://doi.org/10.3389/fnut.2022.1017765
48. Pertiwi MGP, Marsono Y, Indrati R (2020) *In vitro* gastrointestinal simulation of Tempe prepared from koro kratok (*Phaseolus lunatus* L.) as an angiotensin-converting enzyme inhibitor. J Food Sci Technol 57:1847–1855. https://doi.org/10.1007/s13197-019-04219-1
49. Alminger ML, Eklund-Jonsson C, Kidman S, Langton M (2012) Starch microstructure and starch hydrolysis in barley and oat Tempe during *in vitro* digestion. Food Digest 3:53–62. https://doi.org/10.1007/s13228-012-0027-8
50. Di Stefano E, Tsopmo A, Oliviero T, Fogliano V, Udenigwe CC (2019) Bioprocessing of common pulses changed seed microstructures, and improved dipeptidyl peptidase-IV and α-glucosidase inhibitory activities. Sci Rep 9:15308. https://doi.org/10.1038/s41598-019-51547-5
51. Jennessen J, Schnürer J, Olsson J, Samson RA, Dijksterhuis J (2008) Morphological characteristics of sporangiospores of the Tempe fungus *Rhizopus oligosporus* differentiate it from

other taxa of the *R. microsporus* group. Mycol Res 112:547–563. https://doi.org/10.1016/j.mycres.2007.11.006

52. Vellozo-Echevarría T, Barrett K, Vuillemin M, Meyer AS (2024) Mini-review: the distinct carbohydrate active enzyme secretome of *Rhizopus* spp. represents fitness for mycelium remodeling and solid-state plant food fermentation. ACS Omega 9:34185–34195. https://doi.org/10.1021/acsomega.4c04378
53. Jeleń H, Majcher M, Ginja A, Kuligowski M (2013) Determination of compounds responsible for tempeh aroma. Food Chem 141:459–465. https://doi.org/10.1016/j.foodchem.2013.03.047
54. Astuti T, Sitompul M, Faadhilanisyah A (2023) Tempeh consumption patterns in Indonesian family and contribution nutritional adequacy. AcTion: Aceh Nutr J 8:518. https://doi.org/10.30867/action.v8i4.1006
55. Murphy RM, Stanczyk JC, Huang F, Loewen ME, Yang TC, Loewen MC (2023) Reduction of phenolics in faba bean meal using recombinantly produced and purified *Bacillus ligniniphilus* catechol 2, 3-dioxygenase. Bioresour Bioprocess 8. https://doi.org/10.1186/s40643-023-00633-8
56. Adeyemo SM, Onilude AA (2013) Enzymatic reduction of anti-nutritional factors in fermenting soybeans by *Lactobacillus plantarum* isolates from fermenting cereals. Niger Food J 31: 84–90. https://doi.org/10.1016/s0189-7241(15)30080-1
57. Feng XM, Eriksson A, Schnurer J (2005) Growth of lactic acid bacteria and *Rhizopus oligosporus* during barley tempeh fermentation. 104:249–256. https://doi.org/10.1016/j.ijfoodmicro.2005.03.005
58. Zwinkels J, Wolkers-Rooijackers J, Smid EJ (2023) Solid-state fungal fermentation transforms low-quality plant-based foods into products with improved protein quality. LWT 184: 114979. https://doi.org/10.1016/j.lwt.2023.114979
59. Liu C, Wei J, Shi M, Huang X, Wang Z, Liu Q, Lang T, Zhu Z (2024) Metabolomic analysis reveals the positive effects of *Rhizopus oryzae* fermentation on the nutritional and functional constituents of adlay millet seeds. Sci Rep 14: 17435. https://doi.org/10.1038/s41598-024-68478-5
60. Shi D, Liu E, Jafarian Z, Stone AK, Kryachko Y, Zhou L, Kimmel J, House JD et al (2025) Impacts of solid-state fermentation on functional properties of pea protein isolate. J Food Meas Charact. https://doi.org/10.1007/s11694-024-03086-5
61. Dahlan HA, Nambu Y, Putri SP, Fukusaki E (2022) Effects of soaking tempe in vinegar on metabolome and sensory profiles. Meta 12. https://doi.org/10.3390/metabo12010030
62. Magdalena S, Hogaputri JE, Yulandi A, Yogiara Y (2022) The addition of lactic acid bacteria in the soybean soaking process of tempeh. Food Res 6:27–33. https://doi.org/10.26656/fr.2017.6(3).304
63. Cai S, Gao F, Zhang X, Wang O, Wu W, Zhu S, Zhang D, Zhou F, Ji B (2014) Evaluation of γ-aminobutyric acid, phytate and antioxidant activity of tempeh-like fermented oats (*Avena sativa* L.) prepared with different filamentous fungi. J Food Sci Technol 51:2544–2551. https://doi.org/10.1007/s13197-012-0748-2

Chapter 21

Production of Neurosporaxanthin-Rich Extract from *Fusarium fujikuroi* with Potential Uses as Food or Feed Additive

M. Carmen Limón, Ismael Borrego-Serrano, Dámaso Hornero-Méndez, and Javier Avalos

Abstract

Neurosporaxanthin is a carotenoid with a carboxylic moiety, produced by some fungi such as those from the genera *Neurospora* and *Fusarium*, which possesses strong antioxidant activity and is a source of vitamin A. The fungus *Fusarium fujikuroi* has been the subject of detailed research on the genes involved in neurosporaxanthin biosynthesis and its regulatory mechanisms, which resulted in the discovery of the key repressor gene *carS*. The ease of identifying carotenoid-overproducing mutants affected in this gene in some *Fusarium* species has been complemented by the recent development of culture media for these mutants, which allow the accumulation of this carotenoid in large quantities, as well as appropriate HPLC separation conditions for its analysis and purification. This chapter describes the methodology available to produce and purify extracts rich in neurosporaxanthin, from using a simple mutagenesis technique to obtain suitable overproducing strains to generating the final extract, including its further purification, thus facilitating the potential industrial application of these neurosporaxanthin-rich extracts as food additives, nutraceuticals, or cosmetics.

Key words Xanthophyll, UV mutagenesis, Submerged culture, Bioreactor, Carotenoid extraction, HPLC, Neurosporaxanthin, Provitamin A

1 Introduction

Carotenoids are fat-soluble pigments of terpenoid origin that participate as accessory pigments in photosynthesis in all plants and green algae. In addition, they are synthesized by many non-photosynthetic microorganisms, including bacteria and fungi, where they play very diverse functions [1]. Some carotenoids are used as food additives because of their beneficial health effects, mainly due to their antioxidant properties and, in some cases, also because of their role in vision, either as precursors of vitamin A or as protective pigments in the macula [2]. In the case of vitamin A (retinol), it is important not only for its role in vision but also as a

Pablo Navarro Simarro (ed.), *Fungi*, Methods and Protocols in Food Science, https://doi.org/10.1007/978-1-0716-4989-3_21,
© The Author(s), under exclusive license to Springer Science+Business Media, LLC, part of Springer Nature 2026

precursor of retinoic acid, an important regulatory signal in chordates [3]. The most widely used carotenoid in the food industry is β-carotene, a yellow pigment present in the diet due to its presence in many foods—naturally in many vegetable products—and at the same time the most efficient source of vitamin A [4] through its oxidative cleavage by specific carotenoid oxygenase enzymes [5]. Lycopene, a precursor of β-carotene, contributes to the red color of tomato fruits and tomato-derived food products, as well as other foods [6]. Other carotenoids, such as the orange-colored astaxanthin, find applications mainly as an animal feed additive [7]. Astaxanthin, which is responsible for the pigmentation of salmonids and crustaceans, is widely used in aquaculture and, because of its properties, its consumption has numerous beneficial effects on human health [8].

Although these carotenoids can be obtained by chemical synthesis, there is a growing tendency to use biotechnological sources for their production [9]. For this purpose, algal biomass and fungal biomass are used—algae are grown in open-air cultures, while fungi are grown in bioreactors. In addition to β-carotene, lycopene, and astaxanthin, there are other carotenoids of microbial origin whose properties give them an enormous biotechnological potential yet to be exploited [10]. A noteworthy example is neurosporaxanthin, a carboxylic carotenoid produced by some fungi, including those of the genera *Neurospora* and *Fusarium* [9]. In the latter, the species *Fusarium fujikuroi* has been used as a model for the study of its biosynthetic pathway and its regulatory mechanisms [11]. Recent studies have shown that neurosporaxanthin has a high antioxidant capacity [12] and is efficiently assimilated after ingestion in mice, where it has also been shown to be metabolized by the enzyme BCO1 to produce vitamin A [13]. Studies on the regulation of neurosporaxanthin synthesis in this fungus allowed the identification of a regulatory gene that maintains the pathway at a low level of activity and whose loss of function by mutation gives rise to strongly pigmented strains [14]. This is due to the accumulation of abnormally high amounts of neurosporaxanthin and carotenoid precursors, which allows their easy identification in mutagenic screening, without having to resort to genetic engineering techniques or the generation of transgenic strains. The availability of these mutants has allowed the development of optimized culture conditions with the highest yields of production of this carotenoid described to date [12]. This chapter describes the methods for obtaining this type of mutant, its cultivation under optimized conditions for efficient neurosporaxanthin production, and its extraction procedure. The result is a carotenoid-rich extract with potential applications as a food additive, nutraceutical, and cosmetic, subject to compliance with current regulatory standards.

2 Materials

2.1 Cultivation of *Fusarium fujikuroi*

Strains used in this chapter:

1. *Fusarium fujikuroi* IMI 58289 wild-type strain (previously, *Gibberella fujikuroi*), obtained from the Imperial Mycological Institute, Kew, UK.
2. *F. fujikuroi* strain SG39. An overproducer of carotenoids with a mutation in the gene *carS* [14].

Sterilize all media and materials in an autoclave at 120 °C for 20 min.

1. DG medium [15]: 30 g of glucose, 3 g of $NaNO_3$, 1 g of KH_2PO_4, 0.5 g of $MgSO_4{\cdot}7H_2O$, 0.5 g of KCl, 16 g of agar, and 2 mL of microelement solution in 1 L of distilled water. Autoclave. This is the standard minimal medium for *F. fujikuroi*.
2. Microelement solution for DG medium: dissolve 0.5 mg of HBO_4, 5 mg of $CuSO_4$, 10 mg of $FeCl_3$, 1 mg of $MnCl_2$, 1 mg of $NaMoO_4$, and 100 mg of $ZnSO_4$ in 100 mL of distilled water [15].
3. DG-Asn medium: a DG medium with 3 g of asparagine instead of $NaNO_3$. Used for growing *F. fujikuroi* from frozen conidia stocks.
4. Optimized medium [12] for neurosporaxanthin production (hereafter referred to as "optimized medium"): 80 g of sucrose, 0.5 g of NH_4NO_3, 2.4 g of KH_2PO_4, 1 g of $MgSO_4{\cdot}7H_2O$, and 2 mL of ICI microelement solution in 1 L of distilled water.
5. ICI microelement solution: 100 mg of $FeSO_4{\cdot}7H_2O$, 15 mg of $CuSO_4{\cdot}5H_2O$, 161 mg of $ZnSO_4{\cdot}7H_2O$, 1.5 mg of $MnSO_4{\cdot}H_2O$, and 10 mg of $(NH_4)_6Mo_7O_{24}{\cdot}4H_2O$ in 100 mL of distilled water [16].
6. EG medium [15]: 1 g of glucose, 1 g of yeast extract, 1 g of NH_4NO_3, 1 g of KH_2PO_4, 0.5 g of $SO_4Mg{\cdot}7H_2O$, and 16 g of agar in 1 L of distilled water. Autoclave. EG medium was used to obtain spores (also called conidia in *Fusarium*).
7. Sterile VitraPOR® borosilicate filter crucibles, porosity 1, 15 mL volume, ROBU® (Glasfilter-Geräte GmbH, Hattert, Germany).
8. Sterile 250 mL flasks.
9. Bürker counting chamber with a Neubauer pattern.
10. For UV exposure, a dark chamber is needed, equipped with a UV lamp. In our case we used a Philips TUV 15 W/G15 T8 long-life lamp at a distance from the Petri dishes of *ca.* 60 cm.

2.2 Carotenoid Extraction and Analysis

1. Porcelain mortar and pestle.
2. Retsch MM400 oscillating ball mill (Retsch, Haan, Germany) or similar.
3. 25 mL stainless-steel grinding jars (Retsch, Haan, Germany).
4. 15 mm diameter stainless-steel balls (Retsch, Haan, Germany).
5. 50 mL centrifuge tubes (Falcon type or similar).
6. 1.5 mL microcentrifuge tubes (Eppendorf type or similar).
7. Acetone (HPLC-grade).
8. Lyophilizer Virtis SP Scientific Sentry 2.0 (SP Scientific, SP Industries, PA, USA) or equivalent.
9. HPLC system, Waters e2695 Alliance chromatograph fitted with a Waters 2998 photodiode array detector (Waters Cromatografía, S.A., Barcelona, Spain), or a chromatographic system with similar configuration.

2.3 General Equipment

The general equipment used in our case is indicated, but other equivalent equipment may be used instead.

1. Platform shaker, New Brunswick Innova (Eppendorf, Hamburg, Germany), and orbital shaker, S200D (Comecta®).
2. Vortex, Heidolph REAX 2000 (Heidolph, Schwabach, Germany).
3. Refrigerated benchtop Allegra™ X-22R centrifuge with rotor for 50 mL tubes (Beckman Coulter™, Brea, CA, USA).
4. Refrigerated microcentrifuge Eppendorf 5424R (Eppendorf) for 1.5-mL Eppendorf tubes.
5. Laminar flow hood Telstar® SA (Terrassa, Barcelona, Spain).
6. Standard fume hood, Flowlan GV, Indelab®, S.L. (Esparza de Galán, Navarra, Spain).
7. Rotavapor® (Büchi Labortechnik AG, Switzerland).
8. Stirred benchtop bioreactor, model F0 BABY (Bionet, Spain), with a vessel of 7.1 L and a working volume of 3–5 L.

3 Methods

3.1 Generation of UV-Induced *Fusarium* Carotenoid-Overproducing Strains

3.1.1 Conidia Preparation for UV Mutagenesis

1. Thaw a sample of a frozen spore stock of the starting wild-type strain, streak an aliquot onto DG-Asn medium to isolate individual colonies, and incubate for 3 days at 30 °C.
2. Obtain fresh conidia (Fig. 1a) from the selected strain. In the case of low-conidiation strains such as IMI58289, a sporulation medium must be used, but standard DG medium allows sufficient sporulation for other strains (*see* **Note 1**).

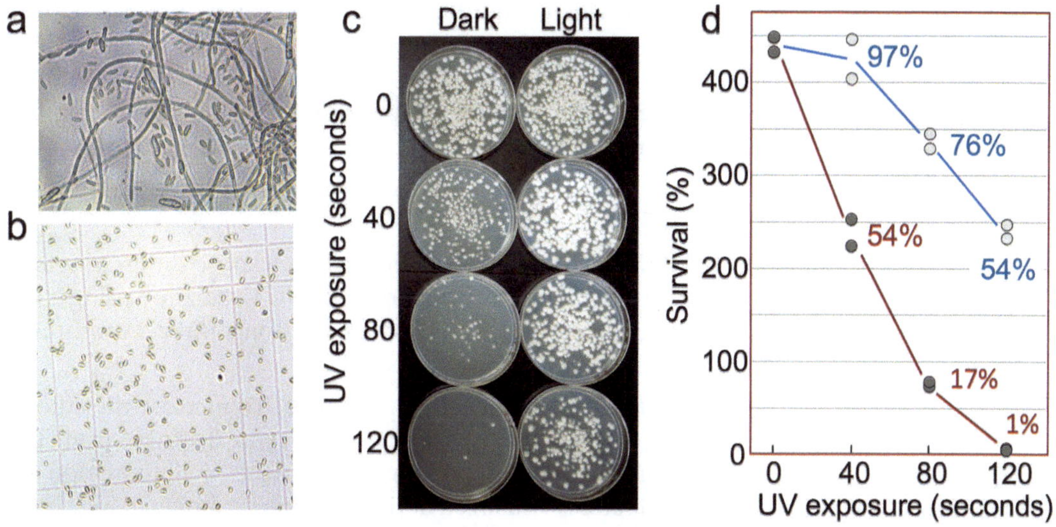

Fig. 1 Result of UV exposure on *F. fujikuroi* microconidia in a representative experiment. (**a**) Aspect under the microscope of a sample containing hyphae and microconidia. (**b**) Conidia sample in a counting chamber. The inner square is 200 × 200 μm. (**c**) Petri dishes inoculated with ca. 500 conidia, exposed for the indicated time to UV radiation, and incubated for 3 days at 30 °C in the dark or under illumination. (**d**) Effect of UV on conidia survival. Light (gray points, blue lines): plates were kept under illumination after UV exposure. Dark (black points, red lines): plates were kept in the dark. The number of colonies in two independent plates is represented, and survival averages are indicated. The whole experiment was done with the wild-type FKMC1995 strain of *F. fujikuroi*

3. Inoculate the chosen medium with *F. fujikuroi* wild type. Use sterile toothpicks to inoculate in 7 symmetrical points on Petri dishes from a 3-day-old mycelial colony (*see* **Note 2**).
4. Incubate the Petri dishes under illumination in a chamber at 22 °C for 7 days (*see* **Note 3**). In our case, we used 5 OSRAM SubstiTUBE Value LED TUBE T8 EM VAL 1200 15 W/865 tubes at 60 cm, providing *ca.* 6 W/m^2 white light, *ca.* 360 l×.
5. Add 10 mL of sterile 10% glycerol to each Petri dish (*see* **Note 4**). Rub the surface of the sporulated culture with a sterile spatula. Harvest the spores with a sterile glass pipette using a pipettor.
6. Filter the spore solution through a borosilicate (size-1 pore) filter on top of a sterile 250 mL flask. Transfer the filtered spores to a 50 mL Falcon tube.
7. Centrifuge the spore solution at 4500 rpm for 20 min at 4 °C in a benchtop centrifuge (Beckman Coulter™). Remove the supernatant and resuspend the pellet of spores in 1 mL of sterile water.
8. Calculate the conidia concentration of the sample using a Bürker counting chamber (Fig. 1b, *see* **Note 5**).

9. Using adequate dilutions in distilled water, prepare a fresh conidia suspension at a concentration of 5×10^4 conidia/mL.

3.1.2 UV Calibration Curve

1. Prepare an appropriate number of Petri dishes of DG minimal medium (*see* **Note 6**).
2. Spread 100 μL of the conidia suspension on each Petri dish (*see* **Note 7**).
3. To obtain a survival calibration curve, expose different sets of Petri dishes to increasing exposures to UV radiation using a UV chamber (*see* **Note 8**). An example using three exposures is shown in Fig. 1c.
4. Incubate the plates at 30 °C for 3 days. To increase the efficacy of UV exposure, keep the Petri dishes in darkness from this time until mutant screening. The effect of illumination on survival rates is visualized in the experiment shown in Fig. 1c, d.
5. Count the number of colonies on each plate (see representative experiment in Fig. 1b), and calculate survival rates relative to the average number of colonies on the nonexposed control (*see* **Note 9**).

3.1.3 Mutant Screening

1. To search for mutants, prepare a sufficiently large number of Petri dishes. It is recommended to screen >10,000 surviving colonies, and therefore at least 50 plates will be required (*see* **Note 10**).
2. Inoculate each plate with 100 μL of the conidia suspension.
3. Irradiate the plates with an appropriate UV dose. UV exposure time should be chosen from the calibration curve to allow an approximate survival of 50% (*see* **Note 11**).
4. Incubate the plates in a thermoregulated chamber in the dark (*see* **Note 12**).
5. Carry out mutant screening. Any deep-pigmented colony will stand out in the nonpigmented background of colonies (Fig. 2).
6. Select those with orange pigmentation and transfer them to a new Petri dish by seeding with a sterile toothpick (*see* **Note 13**).

3.2 Production of Biomass of a *Fusarium* Carotenoid-Overproducing Strain

After selecting a carotenoid-overproducing mutant—in this case, SG39—biomass was obtained by culturing in flasks or a bioreactor. In both cases, conidia were needed for the inoculum and were obtained following the steps indicated in Subheading 3.1.1.

1. Harvest conidia from three plates of EG medium after growing for 7 days as described [17].

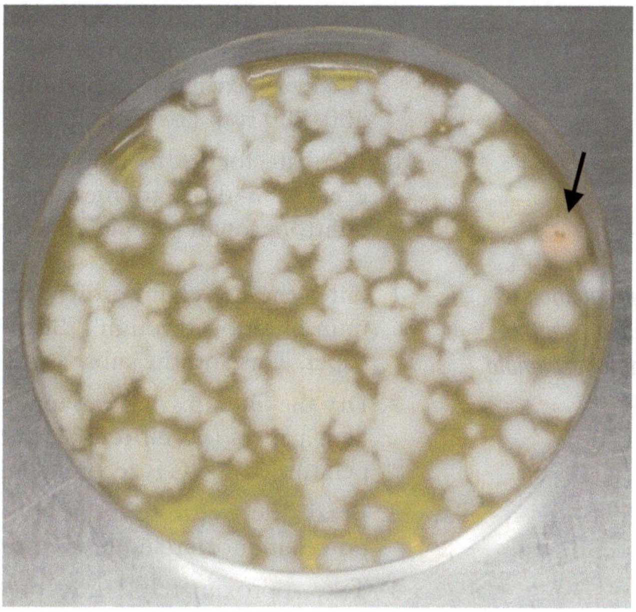

Fig. 2 Screening of carotenoid overproducer mutants after UV treatment of *F. fujikuroi* microconidia. The arrow points to an over-pigmented colony

3.2.1 Culture in Flasks

1. Count the spores using a Bürker counting chamber and prepare conidia suspension at a concentration of 10^6 conidia/mL.
2. Take 1 mL of conidia suspension to inoculate 100 mL of fresh optimized medium in a 250 mL Erlenmeyer flask.
3. Incubate the culture in darkness for 4 weeks at 30 °C in an orbital shaker at 150 rpm (*see* **Note 14**).

3.2.2 Inoculum Preparation and Culture in Bioreactor

1. Inoculate a 1 L Erlenmeyer flask with 500 mL of DG medium with 1.7×10^5 conidia/mL.
2. Incubate for 3 days at 28 °C and 200 rpm. In our case, an orbital shaker S200D (Comecta®) was used.
3. Calibrate the pH sensor with adequate buffer solutions (*see* **Note 15**).
4. Fill the bioreactor with 4.5 L of optimized medium and place the probes for pH and dissolved oxygen in the tank (*see* **Note 16**).
5. Add 0.2% antifoaming agent (*see* **Note 17**).
6. Autoclave the bioreactor at 120 °C for 20 min (*see* **Note 18**).
7. Inoculate the tank with 500 mL of the 3-day grown *Fusarium* inoculum (*see* **Note 19**).
8. Use appropriate culture conditions for your strain. Maintain the temperature of the bioreactor at 28 °C and an air flux of 1 vvm (1 L of air per L of medium per min) (*see* **Note 20**).

9. Establish the minimum dissolved oxygen concentration by controlling the agitation. In our case, we used at least 30% dissolved oxygen by agitation at a range of stirring speeds of 500–700 rpm (*see* Note **21**).
10. Incubate for up to 2 weeks. Take samples periodically, e.g., every 2–3 days, to measure carotenoid production and to control growth parameters such as biomass (*see* **Note 22**).

3.3 Production of Neurosporaxanthin-Enriched Carotenoid Extract

1. Filter the mycelium grown in the optimized medium by vacuum filtration through a Büchner funnel using standard filter paper. This and the subsequent steps are depicted in Fig. 3.
2. Dry the filtered sample as much as possible between absorbing filter papers (*see* **Note 23**); freeze at the lowest possible temperature (–80 °C is recommended); and lyophilize overnight.
3. Break the freeze-dried sample into small fragments using a porcelain mortar and pestle (*see* **Note 24**).
4. Take approximately 1 g of broken dry mycelium and place it into a 25 mL stainless-steel grinding jar (*see* **Note 25**).

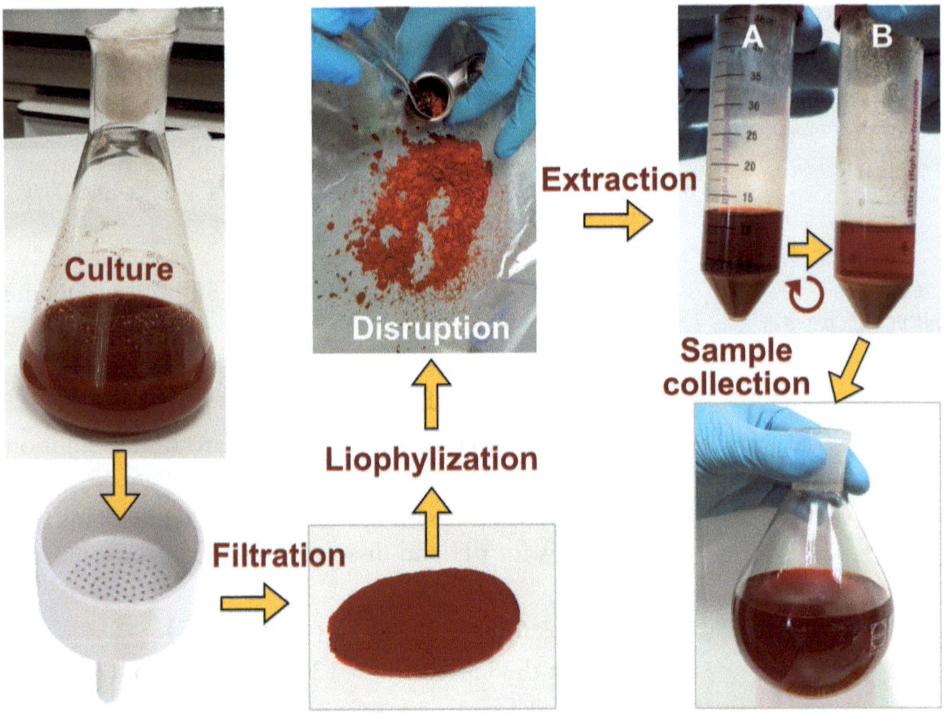

Fig. 3 Process of carotenoid extraction. The mycelium of a culture is separated from the medium by filtration in a funnel, partially dried, lyophilized, and broken to a fine powder in a cell disruptor equipment. Powder samples were then extracted with acetone and centrifuged. Finally, supernatants were collected in an evaporation flask for vacuum concentration in a rotary evaporator

5. Add two 15-mm-diameter stainless-steel balls and 8 mL of acetone (HPLC-grade).
6. Close the jar tightly to prevent leakage during grinding and place it in the oscillating ball mill (*see* **Note 26**).
7. Grind the samples for 1 min at 30 Hz. Repeat the extraction cycle three times, allowing a 10-min pause between pulses (*see* **Note 27**). During this step the carotenoid pigments will be extracted into the acetone (*see* **Note 28**).
8. Transfer the resulting slurry into a clean 50 mL polypropylene centrifuge tube by using a Pasteur pipette (*see* **Note 29**).
9. Centrifuge the extract at 4500× *g* for 5 min at 4 °C, then transfer the clear supernatant into a 250 mL rotary evaporator flask.
10. Add 15 mL of acetone to the tube containing the pellet from the previous step and extract for 2 min with the aid of a vortex. Centrifuge the mix and repeat the extraction process until no more color is extracted.
11. Transfer and collect the supernatants (extracts) into a 250 mL glass evaporation flask, and dry them using a rotary evaporator at a temperature not exceeding 30 °C.
12. Dissolve the dry extract in HPLC-grade acetone to produce a concentrated neurosporaxanthin-rich extract, and check the carotenoid composition by HPLC (Subheading 3.4).
13. Store the concentrated extract under a nitrogen atmosphere at −20 °C, or lower, until use (*see* **Note 30**).

3.4 HPLC Analysis of Neurosporaxanthin in the Carotenoid Extract

1. An aliquot of the concentrated extract was analyzed by HPLC following the protocol described by Hornero-Méndez et al. [18]. Briefly, carotenoids were separated on a C18 reverse-phase column (Mediterránea SEA18, 3 mm, 20 × 0.46 cm; Teknokroma Analítica, Barcelona, Spain) at 25 °C, using a Waters e2695 HPLC system equipped with a Waters 2998 photodiode array detector and controlled with Empower2 software (Waters Cromatografía, S.A., Barcelona, Spain) (*see* **Note 31**). The mobile phase used was a binary elution gradient (HPLC-grade acetone and deionized water) with a flow rate of 1 mL/min. The UV/Vis signal was recorded at 450 nm, a wavelength of high absorbance for most colored carotenoids.
2. The online UV/Vis spectra between 300 and 600 nm wavelength were acquired using the photodiode array detector and used for the correct assignment of the peaks. In our case, the data of the *F. fujikuroi* extract show a high proportion of neurosporaxanthin and very low amounts of precursor carotenoids (Fig. 4).

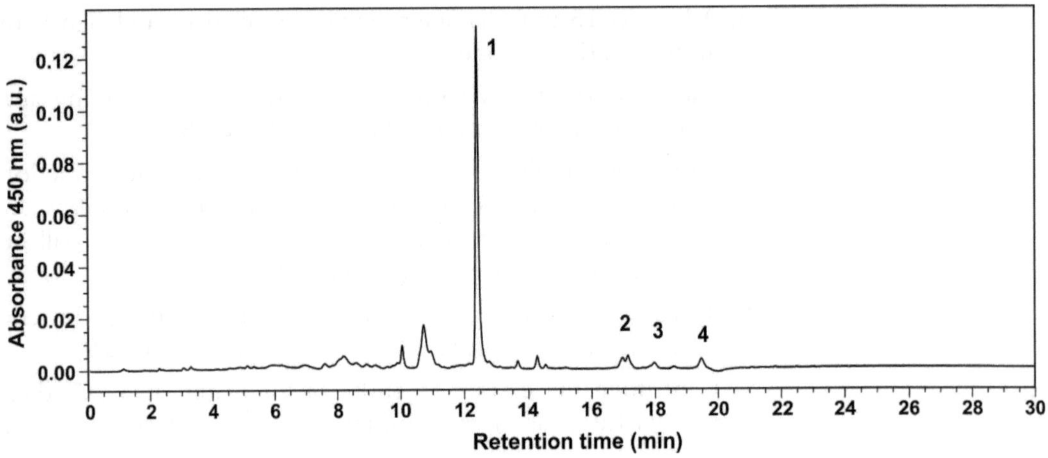

Fig. 4 Chromatogram of HPLC separation of carotenoids accumulated by the overproducer strain SG39 in the bioreactor after growing for 11 days at 28 °C in optimized medium. Peaks assignation: 1, neurosporaxanthin; 2, torulene; 3, γ-carotene; 4, β-carotene

4 Notes

1. In *F. fujikuroi*, different wild-type strains have different conidiation capacities. For example, the FKMC1995 strain produces abundant conidia under different culture conditions, while the IMI58289 strain requires specific conditions that favor conidiation. Other *Fusarium* species, e.g., *Fusarium oxysporum*, normally conidiate more abundantly.
2. Inoculation at seven points on the same plate improves conidiation efficiency and reduces the number of plates required.
3. Petri dishes should not be too close to the light source to avoid overheating of the cultures.
4. The glycerol solution is used to facilitate the suspension of hydrophobic conidia.
5. Conidia precipitate slowly in water suspensions. In this step and during subsequent conidia manipulations, shake conidia suspensions vigorously in vortex before use to ensure homogenization.
6. To check the reproducibility of the results, use at least two Petri dishes for each irradiation time and subsequent light/dark condition.
7. Homogeneous distribution of the conidia on the agar surface is highly desirable to facilitate subsequent visualization of the colonies. It is recommended to spread the 100 μL of inoculum using sterile glass beads.

8. The UV lamp must be turned on 1 min before exposure to get it warm. Plastic lids of Petri dishes are opaque to short UV radiation, so they must be temporally removed during UV exposure.
9. If the number of colonies on the control Petri dishes is lower than expected, the most likely cause is that not 100% of the conidia are germinating. In this case, the germination rate can be improved by replacing $NaNO_3$ with the same amount of organic nitrogen source. Asparagine (DG-Asn medium) works well for us.
10. The pigmentation of about 200–300 colonies per Petri dish can be analyzed individually. If necessary, and given the striking phenotype sought, larger numbers of colonies can be screened on each plate, and putative orange spots can still be detected among the weakly pigmented colonies background.
11. High survival conditions reduce the incidence of mutations in the screened colonies. This results in a lower frequency of overproducing mutants, but these will have a lower number of additional mutations, which could produce other alterations in the phenotype.
12. Minimize exposure of the Petri dishes to light during transfer from the irradiation site to the incubation chamber.
13. A puncture in the center of the colony and a subsequent puncture at the transfer site on the fresh plate are sufficient. If the colony is not completely isolated from other colonies, nonpigmented sectors may appear in the replicate, in which case further purification steps may be required.
14. Incubation in the dark is recommended. If the shaker is exposed to light, the flasks can be covered with aluminum foil and black plastic without blocking the air passage through the cotton-plugged top.
15. The equipment has sensors for pH and oxygen submerged in the medium which are previously calibrated. For example, for the pH sensor, two buffer solutions of pH 4 and pH 7 are used.
16. Once both sensors are calibrated, they are placed in the reactor before sterilizing the culture medium so that the whole set is sterile before inoculating the medium.
17. The use of anti-foaming agent will prevent the possible loss of culture in the form of foam through gas outlet.
18. Ensure that the culture medium in the bioreactor has cooled down before continuing to the next step.
19. Inoculation must be done in a sterile environment, e.g., in a laminar flow hood.

20. Set a temperature range of ±1 in the supervisory control and data acquisition (SCADA) system. In case of a rise or fall in the culture temperature, the warning alarm will be activated.
21. Intense agitation may affect cell viability. In our case, since the organism is a filamentous fungus, excessive agitation could affect hyphal growth. In case of insufficient agitation, the minimum dissolved oxygen concentration can also be controlled by adjusting the air flow range. In cases where oxygen demand is very high, both an air flow range and an agitation range could be used at the same time.
22. Record dissolved oxygen and pH values along with the culture time. The initial pH of the culture was 4.3, and it was not controlled during the cultivation.
23. It is advisable to dry the sample manually as much as possible to facilitate the subsequent lyophilization process.
24. The starting sample is a compact dry mycelial clump that needs to be partially broken to handle it properly in the subsequent breaking steps.
25. The biomass of the original sample should be higher. So, this step will have to be repeated as many times as necessary until the total of the available sample is broken.
26. The use of a PTFE gasket is necessary to ensure a tight seal, preventing leakage and loss of sample and solvent during grinding.
27. Carotenoids are sensitive to high temperatures. During the milling process, the samples can heat up considerably, so 10-min pauses without shaking are important to allow the samples to cool down. Grinding jars can be kept in ice or liquid nitrogen, before grinding, to minimize temperature rise.
28. Grinding jars should be allowed to settle for 30 s in a vertical position before opening.
29. In this step and in subsequent manipulations, exposure of the samples with carotenoids to light should be reduced as much as possible to minimize their photodegradation. If available, red or green ambient light should be used, as carotenoids are less sensitive to these wavelengths.
30. It is highly recommended to store the extract hermetically sealed in a nitrogen-enriched atmosphere to avoid as much as possible the oxidative degradation of carotenoids.
31. The HPLC equipment should be able to deliver at least binary solvent gradients, control constant column temperature, and have a UV/Vis spectrophotometric diode array detector (DAD) for the online acquisition of UV/Vis spectra of the eluted compounds.

Acknowledgments

This research was supported by Grant PID2022-140414NB-I00 funded by MICIU/AEI/10.13039/501100011033 and by ERDF/EU. I. B.-S. received an "Industrial Doctorate" grant (DIN2018-010196) for the training of doctors in companies under the State Training Subprogram within the framework of the State Plan for Scientific and Technical Research and Innovation 2017–2020. M.C.L., D. H.-M., and J.A. are members of the CaRed Excellence Network from the Ministry of Science and Innovation (RED2022-134577-T).

References

1. Rodríguez-Concepción MR, Avalos J, Bonet ML et al (2018) A global perspective on carotenoids: metabolism, biotechnology, and benefits for nutrition and health. Prog Lipid Res 70:62–93. https://doi.org/10.1016/j.plipres.2018.04.004
2. Eggersdorfer M, Wyss A (2018) Carotenoids in human nutrition and health. Arch Biochem Biophys 652:18–26. https://doi.org/10.1016/j.abb.2018.06.001
3. Petkovich M, Chambon P (2022) Retinoic acid receptors at 35 years. J Mol Endocrinol 69:T13–T24. https://doi.org/10.1530/JME-22-0097
4. Weber D, Grune T (2012) The contribution of β-carotene to vitamin A supply of humans. Mol Nutr Food Res 56:251–258. https://doi.org/10.1002/mnfr.201100230
5. Amengual J, Widjaja-Adhi MAK, Rodriguez-Santiago S et al (2013) Two carotenoid oxygenases contribute to mammalian provitamin A metabolism. J Biol Chem 288:34081–34096. https://doi.org/10.1074/jbc.M113.501049
6. Khan UM, Sevindik M, Zarrabi A et al (2021) Lycopene: food sources, biological activities, and human health benefits. Oxidative Med Cell Longev 2021:2713511. https://doi.org/10.1155/2021/2713511
7. Stachowiak B, Szulc P (2021) Astaxanthin for the food industry. Molecules 26:2666. https://doi.org/10.3390/molecules26092666
8. Ambati RR, Phang SM, Ravi S et al (2014) Astaxanthin: sources, extraction, stability, biological activities and its commercial applications – a review. Mar Drugs 12:128–152. https://doi.org/10.3390/md12010128
9. Avalos J, Nordzieke S, Parra O et al (2017a) Carotenoid production by filamentous fungi and yeasts. In: Sibirny AA (ed) Biotechnology of yeasts and filamentous fungi. Springer, pp 225–279. https://doi.org/10.1007/978-3-319-58829-2_8
10. Avalos J, Pardo-Medina J, Marente J et al (2025) Biotechnological production and applications of fungal carotenoids. In: Sibirny AA (ed) Biotechnology of yeasts and filamentous fungi, 2nd edn. Springer, Berlin. (in press)
11. Avalos J, Pardo-Medina J, Parra-Rivero O et al (2017b) Carotenoid biosynthesis in *Fusarium*. J Fungi 3:1–16. https://doi.org/10.3390/jof3030039
12. Parra-Rivero O, de Barros MP, Prado MDM et al (2020) Neurosporaxanthin overproduction by *Fusarium fujikuroi* and evaluation of its antioxidant properties. Antioxidants (Basel) 9:1–17. https://doi.org/10.3390/antiox9060528
13. Miller AP, Hornero-Méndez D, Bandara S et al (2023) Bioavailability and provitamin A activity of neurosporaxanthin in mice. Commun Biol 6:1068. https://doi.org/10.1038/s42003-023-05446-1
14. Rodríguez-Ortiz R, Limón MC, Avalos J (2013) Functional analysis of the *carS* gene of *Fusarium fujikuroi*. Mol Gen Genomics 288:157–173. https://doi.org/10.1007/s00438-013-0739-7
15. Avalos J, Casadesús J, Cerdá-Olmedo E (1985) *Gibberella fujikuroi* mutants obtained with UV radiation and N-methyl-N'-nitro-N-nitrosoguanidine. Appl Environ Microbiol 49:187–191. https://doi.org/10.1128/aem.49.1.187-191.1985
16. Geissman TA, Verbiscar AJ, Phinney BO et al (1966) Studies on the biosynthesis of gibberellins from (−)-kaurenoic acid in cultures of *Gibberella fujikuroi*. Phytochemistry 5:933–947

17. Marente J, Ortega P, Pardo-Medina J et al (2020) Modulation of activity of a carotenoid pathway through the use of the TeT-on regulatory system: application in the fungus *Fusarium fujikuroi*. In: Rodríguez-Concepción M, Welsch R (eds) Plant and food carotenoids: methods and protocols. Springer, New York, pp 343–360. https://doi.org/10.1007/978-1-4939-9952-1_26

18. Hornero-Méndez D, Limón MC, Avalos J (2018) HPLC analysis of carotenoids in neurosporaxanthin-producing fungi. Methods Mol Biol 1852:269–281. https://doi.org/10.1007/978-1-4939-8742-9_16

INDEX

Pablo Navarro Simarro (ed.), *Fungi*, Methods and Protocols in Food Science, https://doi.org/10.1007/978-1-0716-4989-3,
© The Editor(s) (if applicable) and The Author(s), under exclusive license to Springer Science+Business Media, LLC, part of Springer Nature 2026

MIX
Papier aus verantwortungsvollen Quellen
Paper from responsible sources
FSC® C105338

If you have any concerns about our products,
you can contact us on
ProductSafety@springernature.com

In case Publisher is established outside the EU,
the EU authorized representative is:
Springer Nature Customer Service Center GmbH
Europaplatz 3, 69115 Heidelberg, Germany

Printed by Libri Plureos GmbH
in Hamburg, Germany